INTELLIGENT AND ACTIVE PACKAGING FOR FRUITS AND VEGETABLES

INTELLIGENT AND ACTIVE PACKAGING FOR FRUITS AND VEGETABLES

Edited by
Charles L. Wilson

CRC Press is an imprint of the
Taylor & Francis Group, an **informa** business

CRC Press
Taylor & Francis Group
6000 Broken Sound Parkway NW, Suite 300
Boca Raton, FL 33487-2742

Printed in the United States of America on acid-free paper
10 9 8 7 6 5 4 3 2 1

International Standard Book Number-10: 0-8493-9166-0 (Hardcover)
International Standard Book Number-13: 978-0-8493-9166-8 (Hardcover)

Library of Congress Cataloging-in-Publication Data

Intelligent and active packaging for fruits and vegetables / Charles L. Wilson, editor.
p. cm.
Includes bibliographical references and index.
ISBN-13: 978-0-8493-9166-8 (alk. paper)
ISBN-10: 0-8493-9166-0 (alk. paper)
1. Fruit--Packaging. 2. Vegetables--Packaging. I. Wilson, Charles L. II. Title.

TP440.I557 2007
634--dc22 2007004463

Visit the Taylor & Francis Web site at
http://www.taylorandfrancis.com

and the CRC Press Web site at
http://www.crcpress.com

Dedication

To Miriam, my beloved wife, whose love and warmth sustains me

Contents

Preface

Packaging of fruits and vegetables is a dynamic field. It is driven largely by consumer demands for fresher, safer, more nutritious, and convenient produce. Increased importation and exportation of fresh commodities requires packaging that maintains fruits and vegetables in a fresh state for longer periods during transport. The exponential growth in the semiprocessing of fresh produce requires specialized packaging that maintains the quality and nutrition of its contents. Increased concern and awareness of fruit and vegetable contamination from foodborne pathogens and toxins places added incentives to create packaging to prevent, detect, or eliminate this problem. Innovative packaging technologies are critically needed to meet these various challenges.

A diverse group of scientists around the globe are accepting these challenges. *Frontiers of Intelligent and Active Packaging for Fruits and Vegetables* contains the thinking of world leaders in this new technology. It is hoped that this book will accelerate efforts to bring fresh, safe, and nutritious produce to the consumer.

Definitions for *intelligent* and *active* packaging are evolving as these two fields evolve. Most seem to agree that active packaging connotes a package that "responds" to a suboptimal physiological or environmental condition in the package and improves it. Intelligent packaging, on the other hand, involves a package sensing conditions in a package and communicating this information to a human or an appliance. Intelligent packages are dependent on us or an appliance to respond.

Active packaging systems have been developed that respond to suboptimal environmental or physiological conditions within the food package. These systems usually involve the scavenging or absorption of undesirable compounds such as oxygen, carbon dioxide, ethylene, excessive water, and taints. Other active packaging systems add or release compounds such as carbon dioxide, antioxidants, and preservatives into the headspace of the package. These various compounds are generally administered by sachets, labels, or films. Michael Rooney, a pioneer in the development of active food packaging, gives its history in Chapter 2. He brings us up to date on the present status of this field.

As mentioned above, intelligent packaging is dependent on a sensor and some means of communicating the information that the sensor detects. A variety of sensors have been developed based on chemical, enzymatic, immunochemical, or mechanical reactions. These sensors can be placed in or on the package. They can be used to detect and communicate such information as time/temperature conditions and history, oxygen and carbon dioxide levels, package leakage or spoilage, commodity ripeness and freshness, microbial growth, and specific foodborne human pathogen identification.

Radio frequency identification (RFID) communication systems have opened up multiple possibilities for the development of intelligent packaging. Utilizing this technology, packages can communicate information about their condition to us or appliances from a distance, and perhaps some day will be able to converse with one another. Presently, the expense of wireless communication for packaging is limiting the general application of this technology. However, costs are rapidly declining and our imaginations should be the limit as to the technology's future applications. Dr. Aaron Brody, who has been at the forefront of both active and intelligent packaging technology, addresses the history of intelligent packaging and prospects for its future in Chapter 1.

Application-specific integrated circuit (ASIC) technology is emerging as an alternative to RFID use in packaging. ASIC is viewed as a simpler and more economical technology based on custom-designed silicon chips. ASIC devices may help overcome the economic threshold posed presently by the use of RFID technology. Bill Roberts with Sealed Air Corporation discusses the use of RFID devices for temperature monitoring and the potential for ASIC devices in Chapter 11.

Modified atmosphere packaging (MAP) has become an important technology for the preservation of fruits and vegetables. In MAP the gaseous environment in the package (oxygen, carbon dioxide, and nitrogen) is modified to create an optimum mixture to extend the shelf life and quality of the commodity. This can be accomplished either by flushing the existing gasses and replacing them with a desired mixture or allowing the gases generated (carbon dioxide) and consumed (oxygen) in the package through the respiration of living systems to modify the atmosphere. Packaging plastics with varying permeabilities to oxygen and carbon dioxide can help stabilize this system. Dr. Randy Beaudry takes a fresh look at MAP and how it can serve as a basis for active and intelligent packaging for fruits and vegetables in Chapter 3. Jan Thomas Rosnes et al. explore the use of MAP packaging in combination with other technologies in Chapter 6. In Chapter 7, Morten Sivertsvik discusses how MAP technologies have been used for commodities other than fruits and vegetables such as fish and meat. Elhadi Yahia explores the special needs of developing countries for MAP technology in Chapter 14.

Plastics that constitute containers for fruits and vegetables are a key part of any MAP system. They serve as the traffic cop that controls the diffusion of oxygen, carbon dioxide, other gases, and moisture into and out of the container. Yachuan Zhang and his colleagues explain how this system works and give us models for designing plastics that will meet specific needs for various fruits and vegetables in Chapter 8. Nehemia Aharoni and associates show us how the freshness of fruits and vegetables can be extended by the use of high-water-vapor-permeable films in Chapter 5. Shlomo Navarro et al. (Chapter 10) explore how natural nontoxic insect repellent plastics can be used to protect packaging against insect attack. Kay Cooksey in Chapter 9 examines the interactions of plastic packages and food and discusses the safety implications of these interactions.

Where are the new frontiers?

Frontiers are generally opened up by problems in search of technologies and, in some cases, by technologies in search of applications. New frontiers for the use of intelligent and active packaging for fruits and vegetables are emerging from both drivers. Cris Spillet reminds us in Chapter 13 that as we develop new packaging for fruits and vegetables, we should find a consumer need and fill it. One pressing need for new technologies in the packaging of fruits and vegetables has been reemphasized recently by a nationwide outbreak of *Escherichia coli* poisoning from tainted spinach.

Intelligent packaging is needed that will detect foodborne human pathogens on produce and alert the consumer of the contamination. On the other hand, active packaging is needed that will prevent or reduce the growth of human foodborne pathogens in the package. There is extensive research under way addressing these two approaches in dealing with microbial contamination of food containers. Scientists worldwide are studying the release of volatile antimicrobial compounds into the headspace of food storage containers as a way to reduce microbial growth. Plastics for food containers are also being developed that contain antimicrobial compounds. A general problem has arisen with such technologies: food can be tainted when the volatile antimicrobial compounds in the headspace are administered in high enough concentrations — even though the compounds may also effectively reduce microbial growth and extend the freshness of the commodity.

Commercial packaging for fruits and vegetables that claims it can detect specific human foodborne pathogens in the package — or guarantee the elimination of human foodborne pathogens — is not presently in the marketplace. Although considerable progress has been made in these arenas, the bar has to be set high for the performance of such technologies. If the consumer is going to rely on the performance of technologies that detect or eliminate human pathogens to determine the safety of the produce they eat, such technologies have to perform 100% of the time. This is a challenging threshold for intelligent and active packaging researchers.

Another frontier of intelligent and active packaging is traceability. Concerns over accidental and, in the case of bioterrorism, deliberate contamination of food in packages require us to be able to trace the origins of the contamination. RFID has been incorporated into smart labels that can communicate with remote transponders. The chief benefits of smart labels are that they can both transmit information and have information written onto them. David Phillips of Axess Technologies discusses the potential of this technology along with other methods to trace, authenticate, and code packaging in Chapter 12.

Nanotechnology has the potential to revolutionize packaging. It involves the use of materials on an extremely small scale — at sizes of millionths of a millimeter. Nanoparticles generally have entirely different characteristics than their macromolecular counterparts. They can completely change the characteristic of materials to which they are added. In packaging, nanoparticles have been used to modify gas and moisture permeation characteristics of plastics, enhance their barrier properties, improve their mechanical strength, and make their surfaces antimicrobial. According to Dr. Manuel Marquez, a senior scientist at Kraft Foods, "nanotechnology is going to have broad, sweeping applications that have the potential to significantly improve the quality and safety of food."

Major packaging companies have research under way on the use of nanotechnology in packaging, and there are many applications already in use. According to a market report by Helmut Kaiser, the sales of nano-related packaging products have risen worldwide from $150 million in 2002 to $860 million in 2004. While there were less than 40 nanopackaging products on the market 3 years ago, this number is beyond 250 today. Much of the technology will require years of further development, but nano-related packaging to extend the freshness of commodities is starting to enter the marketplace. Sharper Image is presently marketing a food storage bag and container that the company claims extends the freshness of produce through antimicrobial nanoparticles of silver embedded in the plastic.

As reported in AzoNanotechnology (www.azonano.com/details. asp?ArticleID=857), scientists at Kraft, Rutgers University, and the University of Connecticut are working on nanoparticle films with embedded sensors that will detect food pathogens. They are utilizing an electronic tongue technology. Sensors can detect substances in parts per trillion and will trigger a color change in the package to alert the consumer of microbial contamination. AzoNanotechnology also reports that researchers in the Netherlands are developing active packaging that will release a preservative into the food when it begins to spoil. This release-on-command preservative packaging operates by means of a bioswitch developed through nanotechnology.

Rapid advances in packaging utilizing nanotechnology have overrun regulations to protect the consumer from possible harm from this technology. As stated in a report by the ETC Group, "unfortunately, the U.S. government has so far acted as a cheerleader — not a regulator — in addressing the nanotech revolution." The report argues that a regulatory vacuum persists today, even though hundreds of products containing engineered nanomaterials have been commercialized. Michael Cole and Lynn Bergeson (Chapter 15) address the safety and regulatory issues surrounding nanotechnology as it applies to packaging. They emphasize that attention needs to be paid to how nanoparticles differ in their migration from their macromolecular counterparts. They caution, for instance, that antimicrobial nanoparticles may be so small that they would be able to breech the blood–brain barrier in humans. Clearly, regulations to protect the consumer must evolve along with the development of innovative technologies for intelligent and active packaging. At the same time, regulations should not impede the application of food preservation technologies that are safe and would benefit the consumer. Regulatory restrictions have impeded the use of active and intelligent packaging in Europe. W.D. van Dongen and Nico de Kruijf address this in Chapter 17. Jerome Heckman questions the European approach to this problem in Chapter 16.

It is my hope that this book will help accelerate the development of intelligent and active packaging for fruits and vegetables. The authors are to be congratulated for providing an excellent base of knowledge upon which researchers, regulators, and administrators can advance this important technology.

Recognition is given to the U.S./Israeli Binational Agricultural Research and Development Fund (BARD), which supported a workshop titled "Active and Intelligent Packaging of Fruits and Vegetables" in Shepherdstown, West Virginia in 2004, co-hosted by Samir Droby from Israel and myself. Many of the authors in this book were participants at this workshop.

Charles L. Wilson
Editor

About the Editor

Charles L. Wilson, Ph.D. is currently president and CEO of Wilson Associates International LLC, a consulting group advising on the development of biologically based technologies to preserve food. He recently retired from the USDA ARS Appalachian Fruit Research Station, in Kearneysville, West Virginia, after 37 years of service with the federal government. Dr. Wilson's research and teaching career spans over 47 years, during which he has been at the forefront of research and thinking on the use of biological technologies to control plant diseases, weeds, and the postharvest preservation of food.

In the early part of his career, Dr. Wilson was a professor at the University of Arkansas (1958 to 1968), where he received the Arkansas Alumni Award for distinguished teaching and research. He subsequently served on the faculty at Ohio State University from 1970 to 1979. Dr. Wilson joined the USDA ARS Appalachian Fruit Research Station in 1980, where he initiated a research program to find biologically based alternatives to synthetic pesticides for the control of fruit diseases. This internationally recognized research program yielded a variety of innovative technologies for the control of postharvest diseases of fruits and vegetables that include the use of antagonistic microorganisms, natural plant-derived fungicides, and induced resistance.

Dr. Wilson and the associates of Wilson Associates International LLC are presently advising companies on the use of natural plant- and animal-derived antimicrobials for the control of plant diseases, weeds, insects, and foodborne human pathogens. They are also advising companies on the incorporation of natural antimicrobials into food packaging to extend the freshness of commodities placed in the package. In conjunction with investigators worldwide, Dr. Wilson has authored a series of international patents involving the use of antagonistic yeast and natural compounds to control postharvest diseases. Dr. Wilson has worked closely with industry to bring about the large-scale testing and commercialization of this technology and is presently consulting in this area.

In 1984, Dr. Wilson received the Washington Academy of Sciences distinguished service award in the biological sciences for "pioneering research in understanding and manipulating plant diseases." He was also elected a fellow of the academy. In 1988, he received the ARS-NAA Scientist of the Year Award for "innovative research on biological control of postharvest diseases of fruit." In 1994, Dr. Wilson was named a fellow of the American Phytopathology Society, and in 1996 he received the Award of Excellence from the Federal Laboratory Consortium for Technology Transfer for his role in developing the first Environmental Protection Agency (EPA)-registered antagonistic yeast biofungicide for the control of postharvest diseases of fruits and vegetables.

Dr. Wilson has authored or coauthored over 200 scientific publications, 18 patents, and 2 books on gardening. He has previously coedited books for Academic Press (*Exotic Plant Pests and North American Agriculture*) and CRC Press (*Biological Control of Postharvest Diseases: Theory and Practice* and *Microbial Food Contamination*).

Contributors

Uzi Afek
Department of Postharvest Science
of Fresh Produce
ARO, The Volcani Center
Bet Dagan, Israel

Zion Aharon
Department of Postharvest Science
of Fresh Produce
ARO, The Volcani Center
Bet Dagan, Israel

Nehemia Aharoni
Department of Postharvest Science
of Fresh Produce
ARO, The Volcani Center
Bet Dagan, Israel

Sam Angel
Food Technology International
Consultancy (FTIC) Ltd.
Bney-Brak, Israel

Randolph Beaudry
Department of Horticulture
Michigan State University
East Lansing, Michigan

Lynn L. Bergeson
Bergeson & Campbell, P.C.
Washington, D.C.

Aaron L. Brody
Packaging/Brody, Inc.
Duluth, Georgia
and
University of Georgia Department
of Food Science and Technology
Athens, Georgia

Daniel Chalupowicz
Department of Postharvest Science
of Fresh Produce
ARO, The Volcani Center
Bet Dagan, Israel

Michael F. Cole
Bergeson & Campbell, P.C.
Washington, D.C.

Kay Cooksey
Clemson University
Clemson, South Carolina

Nico de Kruijf
TNO Quality of Life
Zeist, The Netherlands

Elazar Fallik
Department of Postharvest Science
of Fresh Produce
ARO, The Volcani Center
Bet Dagan, Israel

Simcha Finkelman
Food Technology International Consultancy (FTIC) Ltd.
Bney-Brak, Israel

J.H. Han
Department of Food Science
University of Manitoba
Manitoba, Canada

Jerome Heckman
Keller and Heckman, LLP
Washington, D.C.

M.A. Horsham
Food Science Australia
North Ryde
New South Wales, Australia

Z. Liu
Agri-Food Innovation and Adaptation Knowledge Centre
Manitoba Agriculture
Food and Rural Initiatives
Manitoba, Canada

Dalia Maurer
Department of Postharvest Science of Fresh Produce
ARO, The Volcani Center
Bet Dagan, Israel

Shlomo Navarro
Food Technology International Consultancy (FTIC) Ltd.
Bney-Brak, Israel

Janeta Orenstein
Department of Postharvest Science of Fresh Produce
ARO, The Volcani Center
Bet Dagan, Israel

David Phillips
Axess Technologies Ltd.
Boston, Massachusetts

Bill Roberts
Sealed Air Corporation

Victor Rodov
Department of Postharvest Science of Fresh Produce
ARO, The Volcani Center
Bet Dagan, Israel

Michael L. Rooney
Formerly of CSIRO/Food Science Australia
North Ryde, New South Wales, Australia

Jan Thomas Rosnes
Norconserv
Stavanger, Norway

A.D. Scully
Food Science Australia
North Ryde
New South Wales, Australia

Morten Sivertsvik
Norconserv
Stavanger, Norway

Torstein Skåra
Norconserv
Stavanger, Norway

Cris Tina Spillett
Clorox Company
Pleasanton, California

William D. van Dongen
TNO Quality of Life
Zeist, The Netherlands

Elhadi M. Yahia
Facultad de Ciencias Naturales
Universidad Autónoma de Querétaro
Querétaro, México

Dov Zehavi
Food Technology International
Consultancy (FTIC) Ltd.
Bney-Brak, Israel

Yachuan Zhang
Department of Food Science
University of Manitoba
Manitoba, Canada

chapter one

A chronicle of intelligent packaging

Aaron L. Brody

Contents

1.1 In the beginning

Once upon a time in food packaging — long, long ago — pottery and glass were the benchmarks. These heavy, hard structures could be closed with pottery, glass, or cork to protect against theft, animals, and ether, the purported perpetrator of food and drink spoilage. Wine or beer resulting from protecting the grapes, grapejuice, or grain could be drunk later, and sometimes even tasted better than the original. Sometimes the resulting liquid tasted like vinegar and could be used for further food preservation. Milk curdled to cheese, a more stable form of this nutrient-laden elixir. Might the goatskin — or the calf stomach — packaging have enhanced the product contents? Were these antiquities the beginnings of active and intelligent packaging?

During the early 19th century came Nicolas Appert and his disciples. This French confectioner employed pottery to package and heat to preserve food (for the historically challenged, he is credited with inventing canning) to save Napoleon and his armies and earn his francs. What did Appert know of how to push the phlogiston out of food to preserve it?

Almost moments later in history, England's Peter Durand became the father of the cylindrical metal can. And Thomas Kensett linked processing and canning — to "drive out the putrefactants" with steam.

During the mid-19th century, Guy-Lussac's theory of oxygen affecting food storage was not far from fact, leading to overt expulsion of oxygen.

All of these empirical pioneers concluded that removal of something, whether phogiston, putrefactants, or oxygen, from food was better than leaving it in.

During the late 19th century, another Frenchman, Louis Pasteur, discovered pasteurization. He postulated and demonstrated that food is a nutrient that supports the growth of microorganisms, that microorganisms can grow on food and spoil them, and that heating can destroy (some) microorganisms and preserve the food — a lot of proofs. If the package remained closed, microorganisms could not reenter to spoil food. Was this another beginning of intelligent packaging?

Although most modern food scientists and technologists have never heard the names of MIT's Samuel Cate Prescott and William Lyman Underwood, except on Institute of Food Technologists' awards, both men really lived and were the pioneers of food technology as we know it today. During the 1890s, these two giants (in importance, not necessarily stature) determined that spoilage of canned foods was due to (Pasteur's) microorganisms. They also quantified the heat and heat penetration required to destroy those destructive creatures within hermetically closed metal cans so that the "inevitable proportion" of spoiled cans was virtually eliminated. Another active/intelligent packaging historical milestone?

C. Olin Ball and Charles Stumbo and their scientific descendants actuated quantitative calculations to optimize the thermal effects. Ball and his pragmatic colleague, William M. Martin, postulated and prototyped a bizarre alternative to canning during the 1930s: processing **outside** the package, sterilizing the package separately, and assembling in a sterile environment — **aseptic** canning. This concept was further developed during the 1950s by Sweden's Hans Rausing with his tetrahedral Tetra Pak, which required really good gas barrier plastic plus metal foil food packaging plus fusion heat sealing.

1.2 The second half of the 20th century

In the latter half of the late 20th century, few of us in food packaging thought of sealed gas/moisture barrier packages as passive. We, the fledgling food packaging technologists, addressed barrier with all manner of metal, glass, polymer, and combinations, and redefined barrier with water, water vapor, oxygen, carbon dioxide, and oil. But we were not satisfied with the results — we did not achieve infinite preservation time. So we persuaded our bosses to compromise the product distribution time and temperature and possibly even the product quality and, to a much lesser degree, safety, and thus were able to match and sometimes optimize product, distribution, and packaging.

From 1950 forward, few of us (very few and very young) food packaging technologists considered our packaging "dumb," especially when we occasionally affixed maximum temperature indicators to the outer surfaces, or when the first of the time–temperature integrators was affixed to package surfaces during the 1960s. These primitive devices suggested to inventors, entrepreneurs, and device suppliers forecasts of all packages visibly signaling temperature abuse.

In those (good old) days when television sets and computers operated with vacuum tubes, military rations required very long, 3-year distribution. Whether we recognized it or not, biochemical deterioration, largely oxidation, adversely affected the packaged food products. The conclusion was that barrier was not enough to deliver long shelf life, and so residual interior environmental oxygen had to be removed, leading to the first modern oxygen scavenger (hydrogen gas and palladium) and active packaging, the subject of chapters 2 and 3 in this text.

When television was exclusively black and white through a cathode ray tube, the controlled atmosphere surrounding the product ensured that the environment in and around food contained the correct quantities and proportions of oxygen, carbon dioxide, and water vapor to help preserve the fresh or minimally processed food. Temperature was **controlled** because controlled atmosphere and its direct descendant, **modified** atmosphere, do not function effectively without temperature control — the lower the better — above the freezing point.

While teenagers (and others) in the 1940s and 1950s listened to music from 33 1/3 and 45 rpm vinyl discs, moisture control was silica gel desiccant in porous paper sachets to keep the metal contents from rusting. Of course, the primary package had to be moisture resistant, and the contained product had to be valuable enough to warrant the added cost. Or the military ration was intended to be in its distribution channels for the usual many years. The desiccant sachet usually contained an admonition, "Do NOT eat" — sometimes, but not always, obeyed. Purge or liquid-absorbent "diaper" pads appeared in the bottom of pulp, paperboard, and oriented or expanded polystyrene trays for fresh meat, poultry, fish, and produce.

Smart, interactive, active, and intelligent packaging began to flood the publications, if not packages, during the 1970s. The implication was that barrier packaging might not be so smart, but it was passive. During the mid-1980s, Dan Farkas, then with the University of Delaware, coined the term *smart packaging* for an obscure and now long-forgotten conference. Soon thereafter, the classical seminal paper by the University of Minnesota's LaBuza and Breene was presented at, of all places, the Icelandic Conference on Nutritional Impact of Food Processing in Reykjavik. Shortly after, Ted LaBuza presented a précis at a New York City gathering, and the entire discipline of active and intelligent packaging exploded into technological consciousness.

Twenty-first century food packaging may be characterized by innovation for the future: enhanced barrier against oxygen ingress (or egress for high oxygen packages), controlled relative humidity, and controlled internal environment. In the definitional sense, active packaging senses change and modifies package properties in response, and intelligent packaging senses change and overtly signals that change.

1.3 *Intelligent packaging*

An intelligent package senses change in the internal or external package environment information to enable a user to offer benefits such as more convenience, increased safety, or higher quality (or better retention of quality).

Examples of intelligent packaging include:

- Time–temperature and other indicators that can imply/signal the user about the quality of the packaged product (reported as far back as the 1960s)
- A biosensor, in theory, which can inform the user of the growth of microorganisms or even a specific microorganism in the package
- A bar code to help communicate information for more precise reheating or cooking of the contained food in an appliance (conceived in the 1930s, but not implemented until the 1970s)
- An ethylene sensor, probably for the ripeness of fresh fruit
- Nutritional attributes of the contained food
- Gas concentrations in modified atmosphere packages (coincided with the advent of oxygen scavengers during the late 1980s)

If you can measure a variable, you can probably control it, eventually.

1.3.1 *Location indicators*

The current most widespread applications for radio frequency identification (RFID) for distribution inventory control originally were developed during the 1990s:

- Site identification
- Duration of distribution
- Quantity at the site/time
- Product identification

In 2004, Wal-Mart began phasing in the use of RFID devices on incoming inventories (on distribution packaging) to assist in tracking within its system. And the U.S. military has mandated RFID to keep track of its supplies in the field.

1.3.2 *Oxygen indicators*

Oxygen indicators are able to signal the presence or absence of oxygen, for example:

- An oxygen scavenger has performed according to specification.
- The probability of content oxidation has been decreased.
- Sufficient oxygen is present to obviate or retard respiratory or microbiological anaerobiosis, or to maintain oxymyoglobin color of fresh red meat.

1.3.3 *Microbiological growth/spoilage indicators*

Very rapid microbiological growth or presence indicators do not exist in the real commercial world, and rapid microbiological **spoilage** sensors for packagers do not exist in commerce. And certainly, rapid, i.e., instant, microbiological pathogen indicators do not exist in the commercial world.

Consumer sense of smell, visual observations, and judgment today still represent the best signals of food spoilage or hazard. Some devices can sense the presence of amines or sulfides, both stinks resulting from microbiological deterioration of proteins.

The alternative to sensors is tedious laboratory analysis, but developmental work is under way to try to create such instant devices.

1.3.4 *Temperature experience*

Time and temperature indicators (TTIs) have been commercial for five decades. Among these are mechanical recorders, electronic recorders, and physicochemical devices that sense temperature change as a function of time. Such devices are often directly correlated to product content, shelf life, quality, and microbiological experience. TTIs are now required for many military rations and some pharmaceuticals, among other things.

Following Arrhenius law relating biological activity to temperature, time–temperature history, in theory, can be correlated to the extent of food deterioration, food quality, and safety. The lower the temperature, the slower the deterioration rate, and so the better the food quality is retained. The shorter the distribution time, the better the food quality is retained. TTI can signal the endpoint based on the time–temperature history. Unlike the temperature limit indicator, TTIs are not triggered when the temperature reaches a certain limit, but the lower limit.

Maximum temperature, i.e., threshold, indicators have been in commercial use for many decades. Mostly, such devices are applied to indicate that a specified temperature has been reached in a heating process such as retorting. Some maximum temperature indicators are now applied for refrigerated and frozen foods, especially ice cream.

Current commercial TTI products include:

- LifeLines' Fresh-Check, based on polymerization reaction
- 3M MonitorMark, based on dye diffusion
- Vitsab® TTI, based on enzymatic lipase color change

LifeLines' Fresh-Check indicators are color-changing labels that respond to cumulative exposure to temperature. They are attached to packages to help consumers determine that the products are still fresh at the point of purchase and at home. The indicator center gradually and irreversibly darkens, rapidly at higher temperatures and slowly at lower temperatures. The TTI can be tailored to match a specific product's shelf life characteristics.

3M MonitorMark TTIs are activated by pulling an activation strip. Upon exposure to temperatures above the threshold, the indicator's window irreversibly turns color. 3M MonitorMark TTIs contain layers of paper and film, adhesive, and a porous wick indicator track strip, one end of which is positioned over a reservoir pad containing a blue-dyed chemical with a desired melting point. Before activation, a removable activation strip separates the indicator track from the reservoir.

1.3.5 *Direct shelf life sensing*

In 2006, developmental technologies to determine and measure shelf life or remaining shelf life included:

- Remote RFID signaler
- Chipless sensor/signaler with internal paper battery
- Smart active level (SAL) with chip
- Sense time–temperature integral
- Signal sent to reader
- Display time–temperature record or integral
- A sensor to sense the change and a transducer to transmit the information to the signaler (required)

1.4 Universal product code

The concept of the Universal Product Code (UPC) originated during the 1930s as a graduate student thesis. Not until the 1960s was the concept converted into some practical systems. During the 1970s and 1980s, the development of precision bar codes capable of identifying every Stock Keeping Unit (SKU) and accompanying laser readers made the concept both feasible and commercial. Today's UPCs include the manufacturer's and SKU's identification numbers.

The newer (2004) Portable Data File (PDF) 417 is a **two-dimensional** symbol. The symbol can carry up to 1.1 kbyte of machine-readable data in a space the same size as a standard bar code. A PDF scanner is required for PDF symbols. Ordinary bar code scanners will not work.

A bar code symbol contains an access key to the database. The full data record is in a host database. A PDF symbol contains a full data record and can contain a number of different data types.

1.5 Food spoilage sensors

Still in development, intelligent packaging for food preservation senses the age or quality level of the food and signals the exterior. Food spoilage signalers have been demonstrated in the laboratory. Usually, however, secondary measurements are used to reflect food spoilage, e.g., time–temperature integrational history.

Ripeness sensors such as the ripeSense label reportedly sense aromatics or other valuable emissions, such as ethylene, emitted from ripening fruit. The device signals ripeness by label visual cue/color change for fruit that does not change external color during ripening, such as pears, melons, and avocados.

Biosensors detect and signal quality or spoilage by sensing food deterioration end products such as amines or sulfides. Spoilage odor detectors are semicommercial. Immunodetection is microbe specific.

1.6 Inventory control

Such devices, in very limited commercial use for consumers, determine the presence, absence, or impending decrease in quantity of a packaged food in pantry/refrigerator inventory and signal the consumer that the pantry is empty or near empty, thus alerting him or her to restock. The devices may simultaneously or sequentially signal the retail grocer that the household is low on product and to remind by telephone or e-mail or to send replenishment stock. The device may also signal the distribution center that stock is low or out. An indicator on the package may signal as the empty package is discarded.

Inventory control tracks packaged food inventory in the home. Such devices can theoretically sense when an empty package is discarded and place a tentative order with the retailer/delivery service, subject to confirmation by the consumer. The information may be complemented by interaction with the retailer, e.g., special requests or change order schedule.

Intelligent packaging may provide product information such as recipes specific to a consumer's kitchen equipment or nutritional characteristics, integrating the individual consumer's nutritional needs, taking into account the age and time–temperature experience, and overtly signaling the individual consumer as to dietary value.

For the past few years, intelligent packaging has permitted automatic self-checkout with direct reading by scanner. RFID or its equivalent is printed on a package and can be read remotely as the consumer moves out of the retail establishment. The device totals all items acquired, debits the consumer's account, and thus requires no movement of product from the shopping cart. RFID can also automatically change shelf pricing.

1.7 *Food preparation*

In the past 15 years many attempts have been made to integrate food packaging and food heating appliances. Generally, few successful marriages have occurred except in pilot environments.

Microwave reheating is generally not appliance specific and does not accommodate a wide variety of microwave ovens or other appliances, such as toaster ovens, refrigerators, or coffee brewers.

Major differences in microwave ovens include:

- Power levels
- Distribution of energy
- Age of oven
- Position of food in the microwave cavity
- Major differences in energy absorption that depend on the food

Today, the end of microwave heating is now judged by operator, and so automatic control from a package code passive or active signaler can integrate the food, the package, and the oven to optimally heat the contents.

Trends in cooking/hardware today include multimodal (microwave plus forced air convection plus infrared) cooking, which requires consumer education, costs more than conventional, and now requires food processor/packager involvement. Not all foods can be prepared in multimodal ovens such as high-speed convection ovens.

Because new energy source ovens cook differently from conventional, automatic operating controls are required, and they can be directed from package intelligence. For example, conventional recipes and cooking instructions must be converted to multimodal. The new recipes can be downloaded from the Internet since consumers will need directions.

The new package/appliance integers are capable of recording the food intake by the consumer and displaying a summary of food intake, nutritional, and other information. The information can be communicated to an electronic scale for monitoring the weight of the consumer. The device can communicate to the consumer useful suggestions based on the above information. The information may also be transferred to the health care givers' records for action.

1.8 Needs for the future

In conclusion, in the future, food shelf life will be shorter, time in home will be shorter, and food preservation and preparation will integrate with packaging:

- Packaging will be integrated with processing and distribution.
- Package will be a preparation and serving aid.

The future is dictated by the past, the present, and the consumer. Rapid reheating will be sited in strategic home locations to facilitate eating.

We shall need visible inventory communication for refrigerated and frozen foods, e.g., age and time–temperature integrals, coupled with overt signals to the consumer.

Cooperation will be required among food processors and retailers that offer prepared foods in packaging to fit home processing with instruction and linkage by computer sensor read and respond.

Long term, we need accurate information that targets each household member individually for quality such as flavor, mouth feel, and nutrient value. The package will be a communication channel from processor to consumer.

Bibliography

Brody, A.L., Strupinsky, E.R., and Kline, L.R., *Active Packaging for Food Applications*, CRC Press, Boca Raton, FL, 2001.

LaBuza, T.P. and Breene, W.M., Applications of "active packaging" for improvement of shelf-life and nutritional quality of fresh and extended shelf-life foods, *J. Food Process. Preservation*, 13, 1–69, 1989.

Rooney, M.L., *Active Food Packaging*, Blackie, Glasgow, 1995.

chapter two

History of active packaging

Michael L. Rooney

Contents

2.1 Introduction

Active packaging has been developing as a field of scientific and technological interest without a unique historical driver. There have been several independent drivers acting on different timescales (even different centuries), in several countries concurrently and involving a variety of food types, target problems, and many scientific disciplines. The term *active packaging* was introduced as recently as two decades ago,[1] even though the tinplate can has been in use as an example of active packaging for more than a century. Some of the driving forces have been directed toward enhancing a limiting characteristic of the packaging material, while others have brought a novel function to the packaging to allow protection or preservation of the food.

Historically, the interests of the horticultural scientists that have driven them in this field have been based upon the continuing need to overcome seemingly intractable postharvest physiological problems. Besides this scientific curiosity, they have taken up opportunities offered by improvements to the length or consistency of the cold chain, the demands of social change, marketing developments (including global change), and interactions with progress in engineering and science. The latter involved discoveries in fields seemingly unrelated to horticultural produce. Some of these discoveries have allowed minimization of undesirable side effects of other introduced technologies.

Although much of the innovation in active packaging has been directed specifically toward fresh produce, an understanding of the history of active packaging requires an appreciation of the concurrent developments in processed food packaging. Accordingly, where possible, treatment of processed fruits and vegetables is used in this chapter as the basis for understanding the wider picture.

The commercial viability of active packaging is not necessarily derived from an increased distributional life of the produce but may result from the logistical benefits. These include reduced wastage, allowance of mixed loads in shipment, cheaper packaging overall, labor saving, and improved ease of handling. The point in time when any of these becomes important will vary and so affect the history of uptake of the various technologies.

Some of the historical drivers are listed in Figure 2.1 together with some of the research areas that have been generated in response to them. There is not a one-to-one relationship between any one driver and the technologies developed since the latter have been influenced by multiple causes. This is best illustrated by considering each of these drivers in turn.

2.2 *Improved cold chain*

As the control of temperature throughout the cold chain improved in response to ongoing understanding of postharvest physiology and food microbiology, especially in the 1970s and 1980s, it became possible to distribute produce over greater distances. This was facilitated by the introduction of membrane gas separators onto shipping containers and trucks in order to maintain controlled oxygen and carbon dioxide levels. There remained, however, the problem of temperature cycling in refrigerated containers, especially with shipments passing through different climatic zones

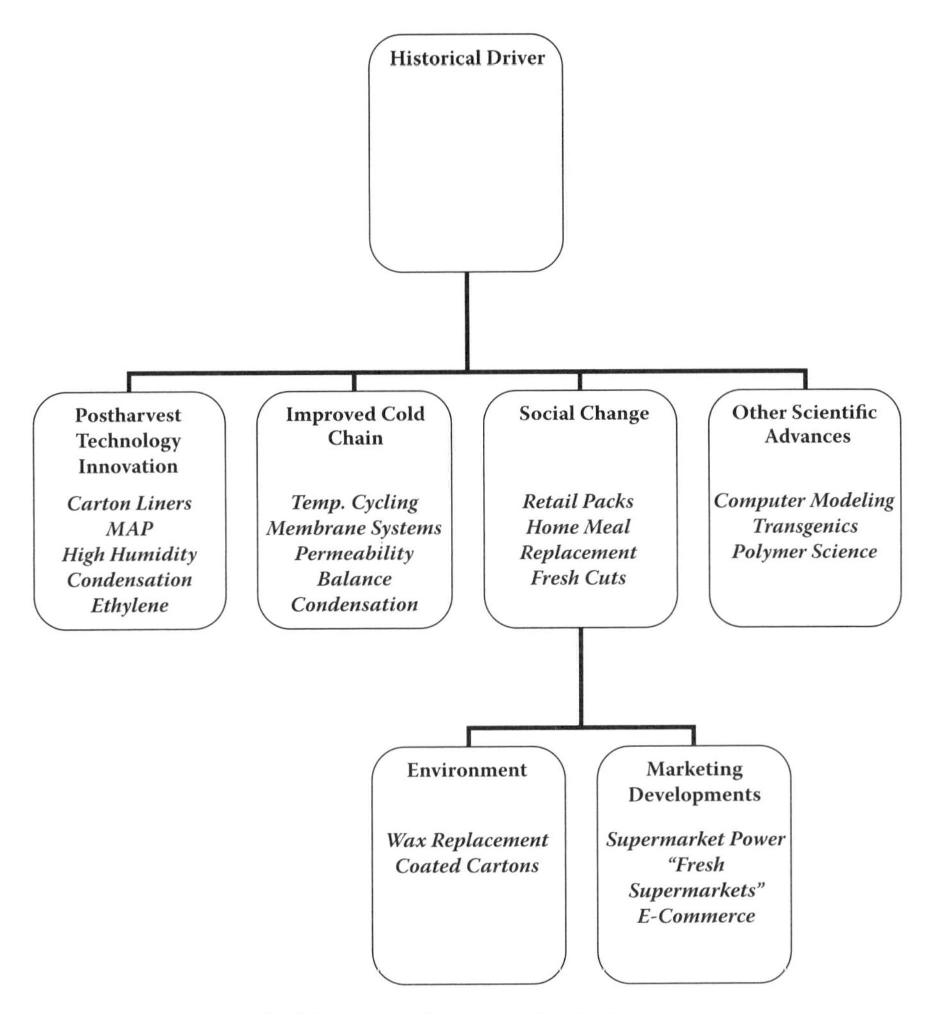

Figure 2.1 Some historical drivers and active technologies.

or when exposed to large diurnal temperature variations. This resulted in condensation of water vapor and consequent dripping onto cargoes or shipments. Some of the consequences of temperature variation *in transit* are set out in Figure 2.2.

One approach to avoiding the cost and logistical problems with the return to the point of origin of gas separators was to attempt to retain modified atmospheres within plastic liners for the cartons. The consequent high relative humidity (RH) values reached led to condensation inside the package with the need for technologies to manage this. Concurrently, it was found that the permselectivity of common plastics did not match the respiration characteristics of most produce, leading to developments in modified atmosphere packaging (MAP), including introduction of some specifically active adjuncts.

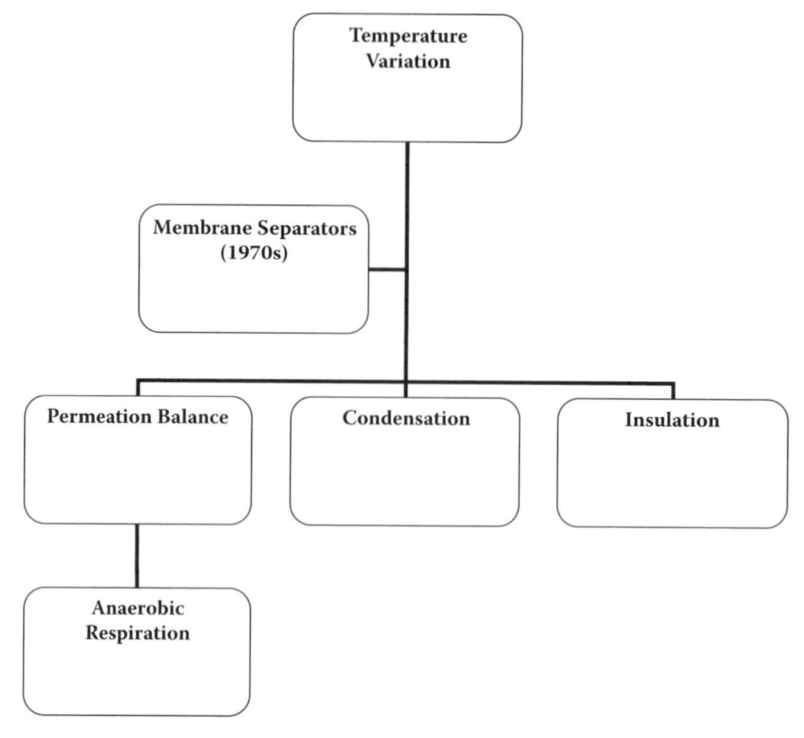

Figure 2.2 Impacts of the cold chain.

2.3 New postharvest technologies

Widespread use of low-density polyethylene in food packaging during the 1960s was soon followed by its adoption as a primary component of wholesale produce packaging during the 1970s. Some implications of its introduction are shown in Figure 2.3. Retail modified atmosphere packages in the late 1980s numbered 500 to 600 million per annum in the U.S. and around 3 billion in Europe.[2] It was during the second half of the 1980s that research into the atmospheres in these packages was matched with the background knowledge of methods of optimization of the storage environment.[3,4] Optimized storage required maintenance of high-humidity atmospheres. Success in achieving this led to localized condensation within plastics packages, thus prompting humidity buffering and condensation control research. Concurrent buildup of ethylene levels that could previously be reduced by sparging of the shipping container with low-oxygen air during transportation left a problem in need of a solution. Earlier physiological research had shown that ripening could be delayed by the correct control of the storage atmosphere.

The introduction of trays with sealed lidding film or with an overwrap has introduced the possibility of using active patches over premade holes to provide localized effects such as gas exchange. Some temperature abuse valves have been proposed or developed to prevent anaerobic conditions

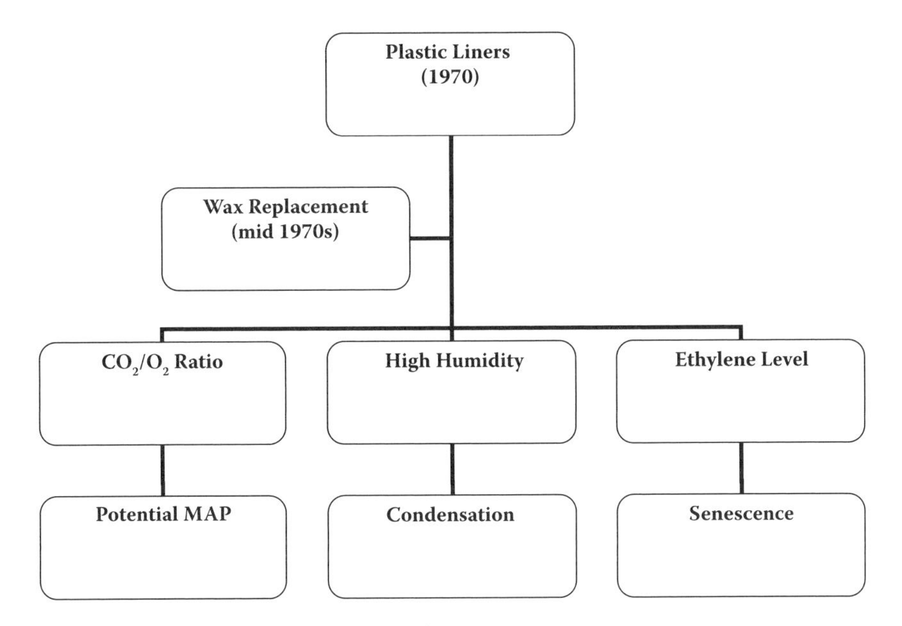

Figure 2.3 Postharvest moisture control.

from developing when the temperature exceeds a critical value. One such development involved the use of waxes with a chosen melting point to block a hole in a patch on a produce tray lid. When the temperature exceeded the melting point of the wax it was wicked away in the liquid form.[5]

The successful maintenance of high relative humidity levels led to localized condensation within the plastics packages, prompting humidity-buffering research. Concurrent accumulation of ethylene, formerly reduced by the increasingly popular ventilation of shipping containers and trucks with nitrogen-enriched air, left a problem awaiting a solution. Earlier physiological research had shown that ripening could be delayed by maintenance of a low-oxygen atmosphere.[6]

2.4 *Social change*

2.4.1 *Time poverty*

The growth of a generation of time-poor individuals and families has also impacted on the potential for use of active packaging, as shown in Figure 2.4. The development of retail markets for cut salad vegetables, "fresh cuts," was first seen in Europe, particularly in France in the mid-1980s.[7] The development of films for the generation of the appropriate equilibrium modified atmosphere (EMA) is a study in itself. Much of the early interest in films for fresh produce was generated in Japan, where films containing a variety of inorganic powders were reported to assist in maintaining freshness or other aspects of shelf life extension. Abe presented a list of products already

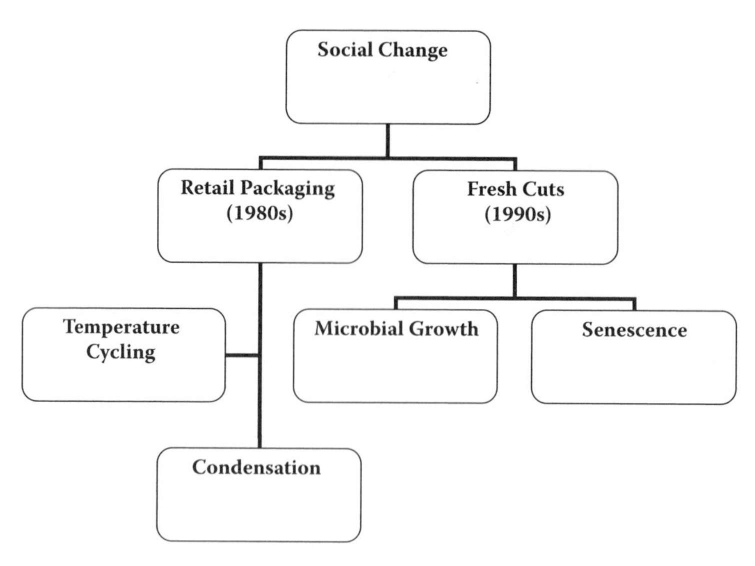

Figure 2.4 Social change impacts.

available in Japan in that year.[8] This commercial interest was not supported by substantial academic research, and the basis has been subject to frequent questioning.

Reproducible perforation of films has been more extensively developed over the years and tested and interpreted. The extensive use of such films today results from well-documented improvement in produce shelf life.[9] There still remain the problems of microbial growth on cut produce and early senescence catalyzed by release of ethylene from damaged tissue.

2.4.2 *An environmental consideration*

The presence of wax in fiberboard cartons for distribution of a large portion of produce produced was long noted as an obstacle to the recycling of the fiberboard. The molten wax was often applied by edgewise curtain coating, which distributed it into the flutes as well as over the surface, with loadings of 25% or more. The lack of recyclability came to a head in the 1970s when some purchasers, such as in Hong Kong, refused to accept produce in such cartons. Research into conventional coatings included both synthetic and biodegradable coatings[10] as well as the use of plastic liners and carton overwraps. The latter interfered with the movement of filled cartons on conveyors and was not widely adopted.

The use of liners created new problems associated with atmospheric compositions not previously encountered. This in turn assisted the development of equilibrium modified atmosphere (EMA) packaging at the carton level. Concurrently, the concept of modified atmosphere distribution was being developed at the shipping container level so that the understanding developed could be applied rapidly. These developments led to advances in

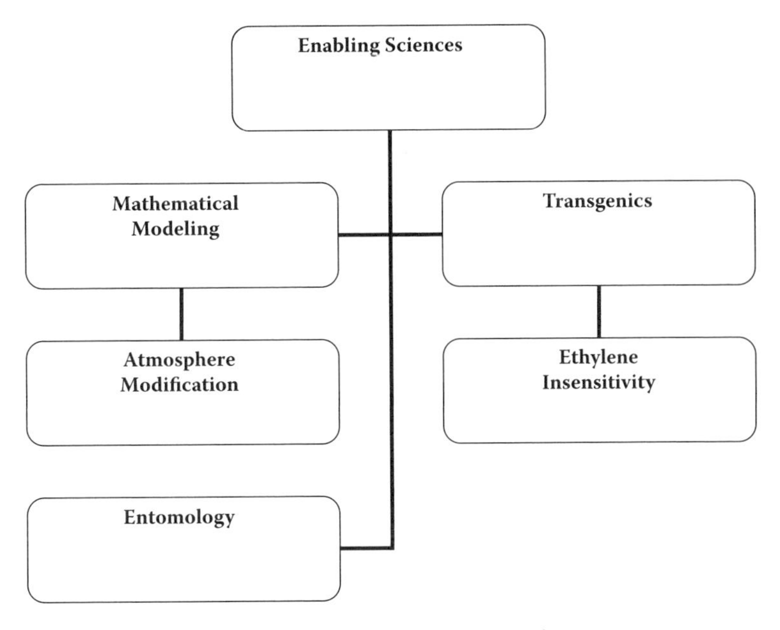

Figure 2.5 Enabling sciences originating outside horticulture.

control of humidity, ethylene level, and gas concentration ratios. However, the initial cooling of the produce and the facile removal of the heat of respiration by means of air circulation remain problems in need of simple solutions, especially in shipping containers.

2.5 Other scientific advances

Progress in active packaging for produce has been leveraged off advances in other sciences, particularly in the field of mathematical modeling, polymer science, and to a lesser extent entomology and transgenics (Figure 2.5). This has the effect of potential users being able to see a wider range of enabling technologies than was available previously.

The use of mathematical modeling has been reviewed frequently with some idea of the historical development listed by Yam and Lee[9] and Solomos.[11] One benefit of this modeling leading to active packaging research has been the appreciation of the limitation of oxygen and carbon dioxide levels that can be achieved using conventional packaging films. This has led to the development of sophisticated perforated films following the early work of investigators such as Mannaperuma and Singh.[12]

Perforations have so far led to limited atmosphere modification in produce packages, especially when the temperature changes, so films with large temperature-induced changes in gas permeability were developed under the Intellimer® trade name of Landec Corporation, in San Francisco.[13] This development utilized an advance in polymer science with the introduction of crystallizable side chains into polymers that are used as patches or whole packages. Economic considerations appear to have restricted these films to the most demanding applications without a comparable development to achieve the same level of effect for cheaper produce.

The difficulty experienced in control of ethylene in packages and whole shipments of produce has led to an alternative: ethylene scavengers. Some research toward development of transgenic varieties not sensitive to ethylene is in progress. This approach will experience the same type of difficulties in market entry as other transgenic crops.

Distribution of produce under conditions conducive to insect attack but with minimal insecticide residues became a subject of serious interest in the 1980s. Produce in plastic sacks and in cartons with bag liners is subject to attack by burrowing insects, and there is a need in developing countries for sacks made from cheap plastics. Inclusion of a low-toxicity insecticide, such as permethrin, in the outer layer of such a sack would offer a cheap solution if several conditions were met. These included lack of inward diffusion and sustained, slow-release, or effective immobilization in the outer layer. Several combinations of insecticide and polymer type were investigated for effectiveness against two insect species, and the order of decreasing effectiveness was as follows: permethrin-treated film, polyester film, polyethylene film, and polypropylene film.[14] This work appears not to have been followed up but might be of value if combined with more recent technologies, such as incorporation of additives into cyclodextrins or other immobilizing agents.

The technologies developed in response to these drivers are addressed in turn, and their development is discussed in relation to the relevant drivers.

2.6 Active packaging technologies

2.6.1 Antimicrobial packaging

The general subject of antimicrobial packaging has been reviewed in some detail.[15–17] The major limitation found in making most research and development relevant to fresh produce distribution has been that most systems rely on intimate contact between the packaging and the food. The most relevant research has generally been restricted to those systems involving a volatile antimicrobial agent.

The release of ethanol from silica gel in a sachet was reported by Shapero et al. and described in the Freund & Company's catalog in 1985.[18,19] The ethanol is absorbed strongly onto the silica gel, but water vapor becomes more strongly bound, thus releasing the ethanol. This product was introduced following successful application of ethanol as a spray on bakery products. There seems to have been no widespread uptake of the sachet product in produce packaging, although it has significant use with processed foods in Japan. The most successful applications are those where the water content of the food is low, so that the ethanol concentration thereon is not substantially reduced.

Several lines of work involved use of the fungicides imazalil and benomyl, although the regulatory status of the latter led Halek and Garg to immobilize it.[20] The effectiveness was determined by placing pieces of the fungicide-grafted film on the surface of an inoculated agar gel. They found that while most of the fungicide was covalently bound to carboxyl groups on the ionomer polymer, some had diffused from the film, creating a halo of inhibition on the gel.

Control of release of *imazalil* for fruits and vegetables was investigated by Miller et al., and the use of shrink wraps on oranges as a source of the same antimycotic was also investigated.[21,22] Weng and Hotchkiss showed that release was effective onto cheese, and the release of food acids as antimycotic agents from edible coatings was also investigated in detail.[23,24] This approach with food-approved antimycotics was part of a more widespread research effort in active edible packaging reviewed by Cuq et al.[25] Given the significant use of coatings on fruits, there is opportunity to extend this approach further.

The separation of grapes from their stems is a common transport problem associated with fungal attack. Large paper pads containing sodium metabisulfite have commonly been used over the past two decades as a source of sulfur dioxide that is released in response to the high relative humidity reached on closing the carton. A problem associated with this type of release agent has been that the sulfur dioxide concentration was too high at first and sometimes results in bleaching part of each berry. During the early 1990s research in CSIRO in Australia showed that plastic film compositions that achieved release of the gas in a more controlled fashion could be prepared.[26]

An antimicrobial plastic film composition has been developed by Bernard Technologies of Chicago under the trade name Microgarde®. The microcapsules in the packaging film can be triggered by light or water vapor to release very low levels of chlorine dioxide, which is an effective antimicrobial gas. It differs from chlorine in that its activity is oxidative only and avoids formation of the common chlorinated hydrocarbons associated with the use of chlorine.

Alternative antimicrobial agents, some of which are volatile, derived from a range of plant sources have been proposed.[17] Allyl isothiocyanate and possibly other components in mixtures derived from horseradish appear to be sufficiently volatile to exert an antimicrobial effect. Sekisui Jushi Co. was marketing such a product as a film to cover lunchboxes in 2002.

This early work has been followed by a much more divergent field of research into antimicrobial packaging in general for applications in food technology. The key feature of such research has been to establish that the agent has not migrated from the packaging material in sufficient quantities to be toxic. Use of nonmigratory antimicrobial agents has become an area of active research.[27]

2.6.2 *Ethylene-scavenging packaging*

The development of packaging that absorbs or actively scavenges ethylene has been derived from research designed to remove it from storage atmospheres for a wide range of produce. Bananas provide an economically important example with their distribution traditionally subject to cycles of over- and undersupply, thus providing an economic driver for packaging innovation. This production problem has been managed by harvesting the fruit green and maintaining it in this state until ripening is desired, for example, in ripening rooms containing 50 to 100 ppm ethylene.[6] The distribution of the unripe fruit is subject to the complication of some fruit having already commenced ripening due to the presence of ambient ethylene levels as high as 0.5 ppm.[28] The addition of potassium permanganate adsorbed in beads held in sachets or dispersed in vermiculite sheets was shown to inhibit ripening.[29,30]

Since these early studies a variety of in-pack removal techniques have been developed commercially.[6,31] The major innovation has been introduction of the scavenging reagent for ethylene into the packaging film.[32] This approach marks the introduction of a packaging material that is a chemical barrier to the diffusion of ethylene into a package of unripe fruit from the external atmosphere. Such an approach is driven by the needs of produce transported in shipping containers containing fruit with a range of maturities and where a high airflow is not available to remove ethylene.

2.6.2.1 *Methylcyclopropene (MCP) exposure*

The recent discovery that MCP may be used to interrupt the process of ethylene-initiated senescence has changed the situation in which fruit distributors find themselves. The effective delivery of low levels of this vapor prior to shipping may reduce the need for ethylene scavenging. The regulatory impact on widespread use will need to be studied, and the potential for release of small quantities within the package may be addressed by active packaging systems.

2.6.3 *Water relations*

The control of water in the environment surrounding packaged produce has lagged behind developments in other areas, largely because of the difficulty in dealing with the high humidity levels typically encountered. Small decreases in temperature at relative humidity values above 95% result in the condensation of liquid water, and thus yeast, fungal, or bacterial activity on the produce surface, especially where bruising, cutting, or skin damage has occurred. Condensation can also lead to fogging of films, which may be only a cosmetic disadvantage in presentation of retail packs. In the early days of plastics packaging (1970s), the problem of managing high-humidity products was sometimes approached by attempting to balance the demands of the product with the water vapor transmission rate (WVTR) of the plastic.

Although it was demonstrated that this could be commercially effective by use of nitrocellulose-coated regenerated cellulose for packaging moist fruitcake, it was still a nonequilibrium approach.[33]

The traditional approach to reducing the relative humidity in packages has been to include a sachet of a desiccant such as clay or silica gel. These have a moisture absorption capacity of 30% or more of their dry weight at a relative humidity of 90%. Use of such materials in horticultural packaging has not been popular since there is not sufficient control of the final humidity. Brody et al. report an innovation in such desiccants that results in a sharp increase in absorbency at the dew point.[16]

The potential for use of compounds such as calcium sulfate, potassium chloride, sodium chloride, xylitol, and sorbitol for controlling relative humidities between 95 and 97% was highlighted by the work of Kader et al.[3] It was shown by Schlimme and Rooney in 1991 that inclusion of 85 g of sodium chloride held in spun-bonded polyethylene sachets in 1.5-kg modified atmosphere packages of peaches was successfully applied to achieve 40 days substantially free of fogging at 1°C.[2] It was shown that an equilibrium relative humidity of around 95% in produce packs can be reduced to around 80% by use of such sachets containing salts like sodium chloride.[34] Dow Chemical supplied Tyvek® (E.I. Dupont de Nemours Co., Wilmington, DE) sachets containing sodium chloride during the early 1990s, but this was not a marketing success.[16]

2.6.3.1 *Humidity-buffering films*

Concurrent with the development of sachets of inorganic salts was an alternative approach of using one or more organic humectants between a duplex plastic film that was highly permeable to water vapor.[35] Examples of this type of product were in use in Japan by 1989, including Pichit®, manufactured by Showa Denko, and Chefkin®.[16] Pichit is described as containing an alcohol, such as propylene glycol, and a carbohydrate,[1] while Chefkin contains a glucose solution.

The desire to achieve these effects without having to use an insert drove research toward devising an active coating for the fiberboard carton that would replace the need for the polyolefin liner or an expensive and nonrecyclable wax coating. Such research was designed to overcome the problem of temperature cycling, which is very difficult to avoid during normal handling of fruits and vegetables. Although this approach is not as compatible with equilibrium modified atmosphere (EMA) packaging as cartons with separate liners, the humidity is buffered at the interface with the fiberboard. An early approach that did not succeed commercially involves:

- An integral water vapor barrier layer on the inner surface of the fiberboard.
- A paper-like material bonded to the barrier and that acts as a wick.

- A nonwettable layer highly permeable to water vapor next to the fruit or vegetable. The latter layer is spot welded to the layer underneath.[5]

This structure is able to take up water vapor when the temperature drops and the RH rises. When the temperature rises, the structure releases water vapor back into the carton in response to a lowering of the RH. The condensation control system therefore acts as an internal water buffer. Despite the passage of a decade and a half, there has been no economic method of manufacture devised for this highly effective and universally applicable concept.

2.6.3.2 *Liquid water removal*

There have also been active packaging approaches to minimizing the impact of water once it has already formed in packaged produce. Temperature cycling of high equilibrium relative humidity (ERH) foods has led most film manufacturers to offer the heat-seal plastics with an antifog additive. This development resulted from the need to present consumer packs of minimally processed produce without the negative impact of fog droplets on the packaging film. The additives are chosen for their amphiphilic nature, with the nonpolar chain in the plastic and the polar end group at the interface. The result is a lowering of the interfacial tension between water condensate and the plastic film. The fog droplets therefore coalesce and form a transparent film on the plastic. This film may even flow on sloped surfaces and gather at the bottom of the pack in extreme circumstances.

Antifog treatments are a cosmetic form of active packaging, assisting the customer in seeing the packaged food clearly. There is no change in the availability of liquid water in the package, and this has the potential to cause produce spoilage unless managed by one of the processes described below. Other active packaging systems go further in removing liquid water from food contact.

Several companies were manufacturing drip-absorbent sheets in the late 1980s, particularly for meat packaging. Pads of these sheets were placed in white polystyrene trays under meat cuts to minimize the negative impact of red drip from the product. They consist of two layers of a microporous or nonwoven polymer, like polyethylene or polypropylene, sandwiching a granular or powdery superabsorbent polymer or highly absorbent paper. The former have not been used for food contact in the U.S. because the polyacrylates used have not had regulatory approval for that purpose. The duplex sheet is sealed at the edges and is normally quilted to allow the water absorbent to be held in place to prevent aggregation toward one edge of the sheet.

Spilling of melted ice in the packaging of seafood for air transportation became a serious insurance risk due to airframe corrosion of the aircraft before the problem was recognized. The superabsorbent polymer used in the Thermarite® sheet manufactured in Australia is capable of absorbing at least 100 times, and possibly as much as 500 times, its own weight of liquid water, depending dramatically upon salinity.[33]

Although the preferred polymers used to absorb the water are polyacrylate salts, graft copolymers of starch can also be used. The interchain bonds between the polysaccharide chains are disrupted, allowing the starch to absorb the water by hydrogen bonding. Such polymers tend to become slimy when swollen with large amounts of water. The swelling of the polymer on hydration results in substantial distortion of the duplex sheet, an effect that is controlled somewhat by the quilted seal pattern.

2.6.4 Postharvest chemical disorders

By the 1960s, observation of storage disorders, such as superficial scald and bitter pit in apples, led to a variety of fruit treatments before packing. In the case of superficial scald, it was found that brown patching of the skin, especially in granny smith apples, resulted from oxidation of α-farnesene present in the fruit. Application of antioxidants was found to be successful in solving the problem, with diphenylamine the best choice.[36] It was shown that this could be applied during storage by packing the fruit with a paper wrap containing the antioxidant.[37] This early application of active packaging appears to have been superseded by improved understanding of postharvest physiology, resulting in the use of low-oxygen atmospheres and restriction of the harvest period.[38] The fact that the condition is still observed in the market suggests that application of an active, antioxidant-laden edible coating might still be of value.

2.7 Processed horticultural produce

Problems associated with the packaging of processed fruits and vegetables have had much in common with the various other food and beverage groups. Consequently, the drivers for introduction of active packaging have included several influences additional to those experienced with fresh produce. Lessons learned from application of these additional technologies may indicate openings where they might be profitably applied with fresh produce also. Table 2.1 lists some examples of processed produce and additional technologies that have either been applied or for which substantial research has been reported.

2.7.1 Debittering of citrus juices

Limonin and naringin in orange and grapefruit juices, respectively, are common bitter principles that are either present in the juice or released from the albedo on pasteurization or during storage. These agents are judged by many consumers to make the juices unpleasantly bitter, thus limiting the source of supply of juice. The rootstocks of some Washington navel crops in

Table 2.1 Active Packaging for Processed Produce

Product	Problem	Activity required
Juices	Oxidation	Oxygen scavenging
Juices	Bitterness	Debittering
Juices	Aroma scalping	Aroma release
Dried fruit	Browning	Oxygen scavenging, sulfur dioxide release
Wine	Browning	Oxygen scavenging, sulfur dioxide release
Retorted vegetables	Discoloration	Oxygen scavenging
Tomato products	Discoloration	Oxygen scavenging

Australia and California were particularly prone to lead to such bitterness in orange juice.[39]

Attempts to process these juices as early as in the 1960s to remove both agents were costly and not always effective. In one successful approach, pasteurized orange juice was passed through columns packed with cellulose acetate/butyrate or nylon 66 beads to lower the limonin concentration to less than 8 to 12 mg/ml.[40] The fact that the limonin is released from the albedo over a period of 24 hours or more created a situation that naturally lent itself to a solution via active packaging. The same authors showed that lining a can with acetylated paper, filling it with juice, and pasteurizing led to sufficient reduction in limonin content after 4 to 13 days of refrigeration for the juice to be deemed by a taste panel to be scarcely bitter. The authors suggested use of a cellulose triacetate absorbent and demonstrated the efficiency of a cellulose acetate butyrate at removing limonin from a bottled juice. This work appears not to have been followed up commercially, as currently sufficient juice from alternate sources is available for blending. In the 1990s, the social movement toward supplying juice squeezed by the consumer at the point of sale would have opened an opportunity for vessels with an absorbent coating, but health issues appear to have closed this opening.

The undesired absorption of juice components by the solution/diffusion mechanism in polymeric packaging has remained a field of significant research since the 1980s.[41–43] This has been developed a step further by incorporation of an enzyme into cellulose triacetate, thus bringing a convergence of packaging polymer research with the juice processing approach of Chandler and Johnson.[40] The naringinase enzyme was dissolved in the polymer and shown to remove an organoleptically acceptable quantity of naringin from grapefruit juice over 14 days.[44] This leaves open the opportunity to utilize the distribution period to actually enhance the quality of the juice rather than merely assisting to retain a level of quality.

2.7.2 *Oxygen scavenging*

The role of oxygen in the loss of quality in processed horticultural produce is consistent with its role in the degradation of the widest range of other foods. It is not surprising that research into methods of removing it from packages can be traced further back than can most other types of active packaging. Oxygen in cans was a key contributor to the oxidation of iron, which led to the application of the internal tin layer. By the 1920s there were attempts to scavenge oxygen by use of ferrous sulfate, and later transition metals were proposed as scavengers by Tallgren.[16,45] Research into use of transition metals continues until the present, but conducting trials on other interesting substrates has been popular concurrently.

The oxidation of hydrogen to produce water was inevitably very attractive, but the catalysts required proved too difficult to produce economically. Palladium-coated steel mesh was attached to the lids of tinplate cans containing milk powder and flushed with a mixture of 7% hydrogen in nitrogen.[46,47] The process was commercialized briefly based on the work of Kuhn et al. involving a laminate-containing palladized alumina sandwiched between an outer gas barrier and a sealant layer permeable to oxygen and hydrogen.[48] The alumina also served as a desiccant to remove the water formed. A pouch vacuum metallized internally with palladium instead of aluminum was patented with application in the storage of oxygen analyzer components.

The most successful commercial products so far have involved use of sachets containing the oxygen-scavenging substance and other adjuncts to provide surface area, hydrogen removal, and water release or binding. The first use of a sachet appears to have been one containing glucose and glucose oxidase.[49] Subsequent scavengers contained in this way include sulfites, ascorbic acid, fatty acids, and reduced iron powder. The development of the popular iron-based scavenger sachets has been reviewed by Smith et al.[19] and Brody et al.[16]

Oxygen-scavenging packaging research, development, and commercialization have made a significant advance with the introduction of scavenging plastics materials. Early approaches involved sandwiching oxidizable materials between two layers of plastic film, as in the use of conventional antioxidants in a solvent, the sulfite solution of Scholle, and the wetting of sulfites Farrell and Tsai.[50–52] The incorporation of oxidizable compounds into a homogeneous solution in a polymer was first reported by Rooney and Holland.[53] Since that time, dozens of patents have described the use of various oxidizable polymers as oxygen-scavenging resins for use as films, cap liners, or bottles.

The susceptibility of ascorbic acid to oxidation is the major shelf-life limiter of orange juice, especially when packaged aseptically. The effects on ascorbic acid loss and color of using oxygen-scavenging films to make pouches for aseptic orange juice have been investigated.[54,55] In both studies, oxygen-scavenging plastic laminates based on totally different chemistries led to a reduction in ascorbic acid loss of around 50% and a reduction in browning of 33 to 50% over 4 months of storage at room temperature. Some oxygen-scavenging compositions are in use commercially in film, bottle, and closure liner form. Their success in the market will depend upon their kinetics of oxygen removal compared with that of the oxidation reactions in the food or beverage. One challenging application is the oxidation of sulfur dioxide in dried fruits and white wine.

2.7.3 *Oxidation of sulfur dioxide*

Sulfured, dried apricots are typical of several varieties of dried fruits that rely on sulfur dioxide to inhibit the Maillard reaction. This fruit is available for only a short season, and year-round availability is a serious management issue. The shelf-life of the product is limited by loss of sulfur dioxide, and the ways in which this loss occurred were not well understood before the early 1970s, at which time legal maximum levels of sulfur dioxide were progressively reduced. Commercially, bulk bins of fruit were sometimes reexposed to the vapor when the level dropped below the legal maximum. Davis et al.[56] investigated the use of various flexible plastics in order to maximize the shelf life and studied several mechanisms via which this vapor might be lost. The results in Table 2.2 show that although four mechanisms could be identified, 44% of the loss was accounted for by reaction with oxygen that was either initially present in the package or that entered by permeation of the packaging film.[56]

During the 1980s there were commercial attempts to remove this oxygen by means of iron-based oxygen-absorbing sachets in flexible packages of the fruit, but the results were inconsistent and indicative of enhanced loss of sulfur dioxide. It may be that the ferrous hydroxide intermediate formed during oxygen scavenging actually reacts with the sulfur dioxide, making the problem worse. An alternative organic-based scavenger is probably called for, and subsequent developments in this field may not interfere in this way, resulting in use of even lower initial levels of sulfur dioxide.

Table 2.2 Contributions of Packaging Variables to Sulfur Dioxide Loss from Dried Apricots

Mechanism of SO_2 loss	Rate constant ($week^{-1} \times 10^2$)
Reaction with fruit constituents	1.82
Oxidation due to oxygen permeability of package	1.52
SO_2 permeation from package	0.85
Oxidation by headspace oxygen	0.60
Combined mechanisms	4.79

2.8 Future opportunities

History so far has been based on several drivers, many of which are outside horticultural science. Continuing application of active concepts for produce packaging will continue to draw upon advances in other disciplines and designed for other products. Economic and regulatory concerns are more likely to be satisfied if the active effect can be achieved using a small area of the package, as with a patch or a perforation cover. Application of the new rules of the European Union for active packaging may be a useful guide here. Linking of activity and intelligence in the one package has been an option for many years but has scarcely ever been used. If improved antimicrobial effects can be achieved within the regulatory framework, a major contribution to the security of fresh-cut produce will result. Application of the many concepts to edible coatings for fruits offers potential for successful outcomes without changing the outer package.

References

1. Labuza, T.P. and Breene, W.M., Applications of "active packaging" for improvement of shelf-life and nutritional quality of fresh and extended shelf-life foods, *J. Food Proc. Preservat.*, 13, 1, 1989.
2. Schlimme, D.V. and Rooney, M.L., Packaging of minimally processed fruits and vegetables, in *Minimally Processed Refrigerated Fruits and Vegetables*, Wiley, R.C., Ed., Chapman & Hall, New York, 1994, chap. 4.
3. Kader, A.A., Zagory, D., and Kerbel, E.L., Modified atmosphere packaging of fruit and vegetables, *Crit. Rev. Food Sci. Nutr.*, 28, 1, 1989.
4. Zagory, D. and Kader, A.A., Modified atmosphere packaging of fresh produce, *Food Technol.*, 42, 70, 1988.
5. Patterson, B.D. and Joyce, D.C., A Package Allowing Cooling and Preservation of Horticultural Produce without Condensation or Desiccants, International Patent Application PCT/AU93/00398, 1993.
6. Zagory, D., Ethylene-removing packaging, in *Active Food Packaging*, Rooney, M.L., Ed., Blackie Academic and Professional, Glasgow, 1995, chap. 2.
7. Carlin, F. and Nguyen, C., Minimally Processed Produce: Microbiological Issues, paper presented at Proceedings International Conference on Fresh-Cut Produce, Chipping Campden, U.K., 1999.
8. Abe, Y., Active packaging: a Japanese perspective, in *Proceedings International Conference on Modified Atmosphere Packaging*, Part 1, Campden and Chorleywood Food Research Association, Chipping Campden, U.K., 1990, p. 15.
9. Yam, K.L. and Lee, D.S., Design of modified atmosphere packaging for fresh produce, in *Active Food Packaging*, Rooney, M.L., Ed., Blackie Academic and Professional, Glasgow, 1995, chap. 3.
10. Rooney, M.L., Interesterification of starch with methyl palmitate, *Polymer*, 17, 555, 1976.
11. Solomos, T., Some biological and physical principles underlying modified atmosphere packaging, in *Minimally Processed Refrigerated Fruits and Vegetables*, Wiley, R.C., Ed., Chapman & Hall, New York, 1994, chap. 5.

12. Mannaperuma, J.D. and Singh, R.P., Design of perforated polymeric packages for the modified atmosphere storage of fresh fruits and vegetables, in *Book of Abstracts, IFT Annual Meeting*, Institute of Food Technologists, Chicago, 1994, p. 53.
13. Stewart, R.F. et al., Temperature-compensating films for modified atmosphere packaging of fresh produce, in *Polymeric Delivery Systems: Properties and Applications*, ACS Symposium Series 520, El Nokaly, M.A., Pratt, D.M., and Charpentier, B.A., Eds., American Chemical Society, Washington, DC, 1994, p. 232.
14. Highland, H.A. and Cline, L.D., Resistance to insect penetration of food pouches made of untreated polyester or permethrin-treated polypropylene film, *J. Econ. Entomol.*, 79, 527, 1986.
15. Hotchkiss, J.H., Safety considerations in active packaging, in *Active Food Packaging*, Rooney, M.L., Ed., Blackie Academic and Professional, Glasgow, 1995, chap. 11.
16. Brody, A.L., Strupinsky, E., and Kline, L.R., *Active Packaging for Food Applications*, Technomic Publishing Co., Lancaster, PA, 2001.
17. Han, J.H., Antimicrobial packaging systems, in *Innovations in Food Packaging*, Han, J.H., Ed., Elsevier Academic Press, San Diego, 2005, chap. 6.
18. Shapero, M., Nelson, D., and Labuza, T.P., Ethanol inhibition of *Staphylococcus aureus* at limited water activity, *J. Food. Sci.*, 43, 1467, 1978.
19. Smith, J.P., Hoshino, J., and Abe, Y., Interactive packaging involving sachet technology, in *Active Food Packaging*, Rooney, M.L., Ed., Blackie Academic and Professional, Glasgow, 1995, chap. 6.
20. Halek, G.W. and Garg, A., Fungal inhibition by a fungicide coupled to an ionomer, *J. Food Safety*, 9, 215, 1989.
21. Miller, W.R., Spalding, D.H., Risse, L.A., and Chew, V., The effects of an imazalil impregnated film with chlorine and imazalil to control decay of bell peppers, *Proc. Fla. State Hort. Soc.*, 97, 108, 1984.
22. Ben Yehoshua, S., Shapiro, B.Y., Gutter, Y., and Barak, E., Comparative effects of applying imazalil by dipping or by incorporation into the plastic film on decay control, distribution and persistence of this fungicide in Shamouti oranges individually seal-packaged, *J. Plastic Film Sheeting*, 3, 9, 1987.
23. Weng, Y.-M. and Hotchkiss, J.H., Inhibition of surface molds on cheese by polyethylene film containing the antimycotic imazalil, *J. Food Prot.*, 9, 29, 1992.
24. Vojdani, F. and Torres, J.A., Potassium sorbate permeability of methylcellulose and hydroxypropyl methylcellulose coatings: effect of fatty acids, *J. Food Sci.*, 55, 841, 1990.
25. Cuq, B., Gontard, N., and Guilbert, S., Edible films and coatings as active layers, in *Active Food Packaging*, Rooney, M.L., Ed., Blackie Academic and Professional, Glasgow, 1995, chap. 5.
26. Jiang, X.Z. and Steele, R.J., Sulfur Dioxide Film, Australian Patent 693424B, 1998.
27. Steven, M.D. and Hotchkiss, J.H., Non-migratory bioactive polymers (NMBP) in food packaging, in *Novel Food Packaging Techniques*, Ahvenainen, R., Ed., Woodhead Publishing Ltd., Cambridge, U.K., 2003, chap. 5.
28. Scott, W.E., Stephens, E.R., Hanst, P.C., and Doerr, R.C., Further developments in the chemistry of the atmosphere, *Proc. Am. Petrol. Inst.*, 37, 171, 1957.
29. Liu, F.W., Storage of bananas in polyethylene bags with an ethylene absorbent, *Hortscience*, 5, 25, 1970.

30. Scott, K.J., McGlasson, B., and Roberts, E.A., Potassium permanganate as an ethylene absorbent in polyethylene bags to delay ripening of bananas during storage, *Aust. J. Exp. Agric. Anim. Husb.*, 10, 237, 1970.
31. Vermeiren, L., Heirlings, L., Devlieghere, F., and Debevere, J., Oxygen, ethylene and other scavengers, in *Novel Food Packaging Techniques*, Ahvenainen, R., Ed., Woodhead Publishing Ltd., Cambridge, U.K., 1993, chap. 3.
32. Holland, R.V., Absorbent Material and Uses Thereof, Australian Patent Application PJ6333, 1992.
33. Rooney, M.L., Active packaging in polymer films, in *Active Food Packaging*, Rooney, M.L., Ed., Blackie Academic and Professional, Glasgow, 1995, chap. 4.
34. Shirazi, A. and Cameron, A.C., Controlling relative humidity in modified atmosphere packages of tomato fruit, *Hortscience*, 13, 565, 1992.
35. Miyake, T., Food Wrap Sheet, Japanese Patent Kokai 65,333/91, March 20, 1991.
36. Anet, E.F.L.J., Superficial scald, *CSIRO Food Res. Q.*, 34, 4, 1974.
37. Heulin, F.E. and Coggiola, I., Superficial scald: a functional disorder of apples. IV. Effect of variety, maturity, oiled wraps and diphenylamine on the concentration of α-farnesene in the fruit, *J. Sci. Food Agric.*, 19, 297, 1968.
38. Erkan, M. and Pekmezci, M., Harvest date influences superficial scald development in granny smith apples during long term storage, *Turk. J. Agric. For.*, 28, 397, 2004.
39. Chandler, B.V., Kefford, J.F., and Ziemilis, G., Removal of limonin from bitter orange juice, *J. Food Sci.*, 19, 83, 1968.
40. Chandler, B.V. and Johnson, R.L., New sorbent gel forms of cellulose esters for debittering citrus juices, *J. Sci. Food Agric.*, 30, 825, 1979.
41. DeLassus, P.T. et al., Transport of apple aromas in polymer films, in *Food and Packaging Interactions*, ACS Symposium Series 365, Hotchkiss, J.H., Ed., American Chemical Society, Washington, DC, 1988, chap. 2.
42. Hirose, K. et al., Sorption of α-limonene by sealant films and effect on mechanical properties, in *Food and Packaging Interactions*, ACS Symposium Series 365, Hotchkiss, J.H., Ed., American Chemical Society, Washington, DC, 1988, chap. 3.
43. Mannheim, C.H., Miltz, J., and Passy, N., Interaction between aseptically filled citrus products and laminated structure, in *Food and Packaging Interactions*, ACS Symposium Series 365, Hotchkiss, J.H., Ed., American Chemical Society, Washington, DC, 1988, chap 6.
44. Soares, N.F.F. and Hotchkiss, J.H., Naringinase immobilization in packaging films for reducing naringin concentration in grapefruit juice, *J. Food Sci.*, 63, 61, 1998.
45. Tallgren, H., Keeping Food in Closed Containers with Water Carrier and Oxidizable Agents Such as Zn Dust, Fe Powder, MN dust etc., British Patent 496935, 1938.
46. King. J., Catalytic removal of oxygen from food containers, *Food Manuf.*, 30, 441, 1955.
47. Abbott, J., Waite, R., and Hearne, J.F., Gas packing of milk with a mixture of nitrogen and hydrogen in the presence of a palladium catalyst, *J. Dairy Res.*, 28, 285, 1961.
48. Kuhn, P.E., Weinke, K.F., and Zimmerman, P.L., paper presented at Proceedings of the 9th Milk Concentrates Conference, Pennsylvania State University, State College, 1970.

49. Scott, D. and Hammer, F., Oxygen-scavenging packet for in-package deoxygenation, *Food Technol.*, 15, 99, 1961.
50. Cook, J.M., Flexible Film Wrapper, U.S. Patent 3429717, 1969.
51. Scholle, W.R., Multiple Wall Packaging Material Containing Sulfite Compound, U.S. Patent 4041209, 1976.
52. Farrell, C.J. and Tsai, B.C., Oxygen Scavenger, U.S. Patent 4536409, 1985.
53. Rooney, M.L. and Holland, R.V., Singlet oxygen: an intermediate in the inhibition of oxygen permeation through polymer films, *Chem. Ind.*, 900, 1979.
54. Rogers, B.D. and Compton, L., New polymeric oxygen scavenging system for coextruded packaging structures, in *Proceedings of Oxygen Absorbers: 2001 and Beyond*, George O. Schroeder and Associates, Madison, WI, 2000, 11 pp.
55. Zerdin, K., Rooney, M.L., and Vermüe, J., Packaging of orange juice in an oxygen scavenging film, *Food Chem.*, 82, 387, 2003.
56. Davis, E.G., McBean, D.McG., and Rooney, M.L., Packaging of foods that contain sulfur dioxide, *CSIRO Food Res. Q.*, 35, 57, 1975.

chapter three

MAP as a basis for active packaging

Randolph Beaudry

Contents

The imposition of modified levels of oxygen and carbon dioxide partial pressures can alter the physiology of harvested fruits and vegetables in a desirable manner, resulting in an improvement in quality maintenance relative to air storage. Gas modification technologies can be segregated into two classes based on the manner in which the atmospheres are generated and maintained. One class of technologies is referred to as controlled atmosphere (CA) storage, in which the atmosphere is either manually or mechanically controlled to achieve target headspace gas concentrations. In CA storages, O_2 and CO_2 concentrations can be modulated independently from one another. The second class of technologies is modified atmosphere packaging (MAP), in which a package possessing a film or foil barrier passively limits gas exchange by the living produce enclosed in the package, thereby altering the headspace atmosphere. In MAP, both oxygen and carbon dioxide are modified simultaneously and their concentrations at steady state are a function of one another. In MAP, the primary route of gas exchange may be through gas-permeable films, perforations in films, or both. In what is referred to as active or intelligent packaging techniques, packages may be flushed with specific gas mixtures designed to obtain a desired initial atmospheric composition, gases may be actively released or scavenged in the package, a partial vacuum can be imposed, biologically active materials can be incorporated in the package, sensors may be used to respond to the product or package conditions, and so on.

No matter what the packaging technology employed, however, the capacity of active packaging to deliver and maintain functional and beneficial atmospheres is partly or wholly dependent upon the factors that control the passively modified atmospheres in MAP. MAP usually depends upon the respiratory activity of the enclosed product as a driving force for atmosphere modification and the permeability of the packaging material to maintain atmospheres within desired limits. It is the continued depletion of O_2 or the release of CO_2 (and water vapor) by the product that enables the modified atmosphere to persist after flushing and sealing. For the success of any packaging approach, factors that must be controlled or incorporated include film permeability, film area, film thickness, temperature, and the respiratory behavior (responses to O_2, CO_2, and temperature) of the product.

The aim of MAP (passive, active, or intelligent in design) is to take advantage of physiological responses of the enclosed plant material or plant or human pathogens to the respiratory gases O_2 and CO_2. Presumably, MAP use is intended to maintain product quality, thereby ensuring appropriate value to the consumer and adequate cash flow back through the marketing and handling chain such that the production and marketing system is sustainable (Figure 3.1). Knowledge of the physiological responses to atmosphere modification is beneficial in terms of anticipating improved quality retention as a result of technology investment. This chapter will describe factors that influence the decision to modify atmospheres using MAP, the generation of target atmospheres, and the design of the MAP system to achieve target atmospheres.

3.1 Historical development of MAP

The common perception that modified atmospheres are useful for improving storability has significant historical precedent. The written history of the use of modified atmospheres can actually be traced back at least 2000 years to

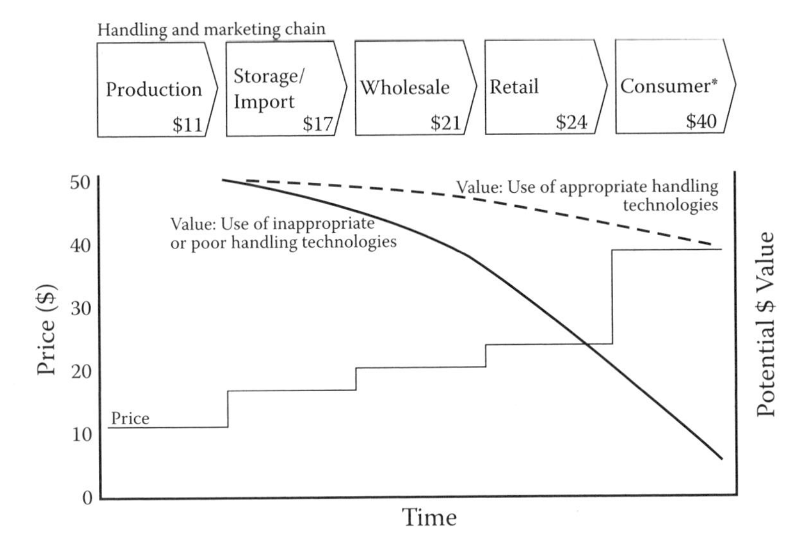

Figure 3.1 Diagram representing the flow of product through the handling and marketing chain (upper portion of figure) and its value, based on changing quality over time, and the concomitant reverse flux of cash (price) from the consumer to the various links in the market chain (lower portion of the figure). *Consumer decisions drive the flow of cash to all members in the supply chain; the amount of money the consumer is willing to pay will reflect perceived value in the product.

the use of underground, sealed silos, or *syri*, for grain storage (Owen, 1800). In this case, atmosphere modification was detected as "foul air" that was dangerous to enter and was likely a result of O_2 depletion and CO_2 accumulation due to the respiratory activity of the grain. The modified atmosphere was unintentional, although probably beneficial. The foul air in the storages would presumably control rodent and insect pests, thereby acting to preserve the quality and storage life of the grain. The potential for a positive impact from alteration in the respiratory gases O_2 and CO_2 became increasingly known through the early research of Berard (1821), Mangin (1896), Kidd and West (1914, 1927, 1945), and Blackman and Parija (1928). Through their work and the contributions of numerous other researchers to follow, CA storage technology was initiated and successfully implemented (Dilley, 1990).

The success in CA set the stage for the application of modified atmospheres in packages soon after. However, the utility of MAP was not immediately realized. Rather, the need to understand the relationship between a packaging film and the atmosphere generated within the package grew out of problems associated with the use of films to minimize water loss from fresh produce in the 1940s. Evidently, spoilage due to an insufficient supply of O_2 could be traced to those instances in which packages were tightly sealed (Platenius, 1946). The lack of more frequent problems with fermentation at the time was partially explained by the fact that most packages leaked. It was recognized that diffusion of O_2 and CO_2 through the film could predictably affect the atmosphere obtained within sealed packages. Platenius (1946), using permeability data from Davis (1946), determined that in all but a few instances the diffusion rate of O_2 through transparent films available at the time was inadequate to meet respiratory demands of the packaged produce. Allen and Allen (1950) further noted that MAP suppressed ripening in tomatoes, which at the time was considered undesirable. It was suggested that polymers then in use needed to be perforated if sealed or that polymers with higher permeability to O_2 were needed (Allen and Allen, 1950; Platenius, 1946).

Polyethylene, a polymer with very high O_2 and CO_2 permeability, and perhaps the most widely used polymer in MAP applications today, was first synthesized in an industrially practical way by Imperial Chemical Industries (ICI) in Britain in the mid- to late 1930s. However, the discovery was suppressed until after 1945 for national security reasons. Ryall and Uota (1955) and Workman (1957, 1959) were among the first to report on beneficial effects of MAP using polyethylene for suppressing apple ripening. Additional reports soon followed on other commodities that underwent ripening, including banana (Scott and Roberts, 1966). An early observation was that control of the package atmosphere was variable, which limited MAP applicability (Tomkins, 1962; Workman, 1959). For this reason, mathematical models were constructed to try to understand how various package, product, and environmental factors affected the atmosphere in MA packages.

Early work on modeling to predict steady-state package headspace O_2 levels at a single temperature was accomplished by Jurin and Karel (1963). Factors investigated include film permeability, film thickness and area, product respiration, and product type and quantity. A dynamic model including package volume was useful in predicting the O_2 partial pressure in the package over time (Henig and Gilbert, 1975). This was probably the first study that employed computers for modeling in MAP, taking advantage of the computational power needed for numerical solution of differential equations. One of the first studies incorporating temperature into the prediction of steady-state O_2 concentrations was accomplished by Cameron et al. (1994) for blueberry fruit. Subsequently, Hertog et al. (1998) combined temperature responsiveness with a dynamic model that is able to predict the impact of changing temperature throughout the marketing chain on package O_2 and CO_2.

As our understanding of the factors controlling the atmosphere in MAP improved and as polymer diversity and technology expanded, the application of MAP also expanded. It is telling, however, that the greatest expansion came in the form of MAP of lettuce and salad preparations, rather than for fruit-ripening control, which was identified early on as a useful application. In the 1990s in the U.S., total sales for lightly processed (fresh-cut) produce increased from near zero to approximately $3.9 billion by 2003 (Anon., 2003). Of these sales, approximately $2.4 billion was for fresh-cut salads and lettuce, a significant proportion of which employ modified atmospheres. In 2005, fresh-cut salads were estimated to be valued at approximately $3 billion in the U.S. (Calderon, 2006). Ironically, Platenius (1946) identified a need for investigating the keeping quality of chopped vegetables in packages 60 years before this widespread adoption. Despite early recognition of the potential of fresh cuts in packages, development of this market sector required improvements in refrigeration, transportation capabilities, food safety, and packaging technology, as well as an evolving lifestyle that put a premium on convenience.

3.2 *Physiological bases of responses to O_2 and CO_2*

The determination of the extent to which oxygen or carbon dioxide modification is appropriate is dependent on the biology of the harvested plant organ and those components of physiology and pathology comprising limiting factors for quality. In the postharvest arena, as we attempt to improve quality retention using tools like MAP, it should be with a strategy that takes advantage of our fullest knowledge of the biology of the commodity. The biology of every commodity is different, and even cultivar differences can impact our decision-making. For perspective, it is instructional to recognize that the proteomes of some crop species contain approximately 50,000 enzymes and structural proteins (Brendel et al., 2002). The large number of proteins suggests that opportunities for metabolic targets must exist, but those opportunities exist amid immense complexity and interrelatedness. Identifying a successful metabolic target revolves around the degree to which the pathway of interest can be isolated from the remainder of the proteome. In some cases, the goal may be not only to impact the metabolism of the crop plant, but also to influence the metabolism of accompanying pathogens or beneficial organisms.

Metabolic targets might be broken down into two broad classes: specific metabolic targets and global metabolism. A specific target would include a pathway that can be impacted with some measure of isolation from the remainder of metabolic activities. Affecting a metabolic process in complete isolation is not practically possible, but there are many opportunities to impact a process in a beneficial manner while other, nontarget processes proceed with only minimal changes. A well-documented specific metabolic target is the ethylene response pathway, which can be inhibited with great benefit using modified atmospheres and various ethylene production or action inhibitors (Watkins, 2002). In addition to, or in place of, controlling specific metabolic targets, strategies exist to impact global metabolism. Probably the most powerful tool to control global metabolism postharvest is through appropriate temperature regulation. While temperature control is not specifically related to MAP, it is essential for effective use of MAP.

The physiological processes typically targeted for control via atmosphere modification in the postharvest sector are listed in Table 3.1. Of these, the impact of low O_2 on ethylene biology is probably the most economically important application of atmosphere modification and is commonly applied using CA facilities; relatively little MAP is used to regulate ripening. As

Table 3.1 Responses of Plant Organs to Ranges of O_2 and CO_2 and Examples of Commodities Benefiting from Modification of These Responses

Gas	Range (kPa)	Response	Examples of benefited commodities
O_2	0.5–5	Reduced ethylene perception	Apple, pear, banana, kiwifruit, cabbage
	0.0–1.0	Reduced oxidative browning of cut surfaces	Fresh-cut lettuce
	0.5–5	Suppressed respiratory and associated metabolic activity	Apple, pear, banana, kiwifruit, cabbage
	3	Suppressed meristematic activity	Onion
CO_2	1–5	Reduced ethylene perception	Apple, banana, kiwifruit, cabbage
	3–5	Reduced chlorophyll degradation	Apple, banana, kiwifruit, asparagus, cabbage, broccoli, spinach
	5–20	Suppressed decay (fungal sporulation or growth)	Blueberry, blackberry, cherry, raspberry, strawberry, onion

Derived from Ben-Yehoshua, S. et al., in *Environmentally Friendly Technologies for Agricultural Produce Quality*, Ben-Yehoshua, S., Ed., CRC Press, Boca Raton, FL, 2005, pp. 63–67; Hardenburg, R.E. et al., *The Commercial Storage of Fruits, Vegetables, and Florist and Nursery Stocks*, Agriculture Handbook 66, revised, U.S. Department of Agriculture, 1986; Smittle, D.A., in *Proceedings of the 5th International Controlled Atmosphere Research Conference*, Wenatchee, WA, 1989, Vol. 2, pp. 171–177; Smyth, A.B. et al., *J. Agric. Food Chem.*, 46, 4556–4562, 1998.

noted, the use of low O_2 to prevent browning of cut surfaces in salad products has a large economic importance, and this application is entirely via MAP. A less common, but still important, application is to suppress decay in berry crops, typically by highly elevated CO_2 levels. The suppression of global metabolic activity through respiratory inhibition may be beneficial for some tissues, but this has not been clearly demonstrated except under hypobaric conditions (Burg, 2004).

3.2.1 *Targeting ethylene biology*

As noted previously, inhibition of ethylene production and action has been a successful approach to improve the storability of a number of ethylene-sensitive commodities. Kidd and West (1914, 1927, 1945), Blackman and Parija (1928), and others originally demonstrated the usefulness of atmosphere modification to slow ripening of apple. However, while it was initially believed that the impact was via respiratory suppression, it was later conclusively demonstrated that the primary benefit of reduced O_2 and elevated CO_2 was the suppression of ethylene perception and production (Burg and Burg, 1965, 1967). Inhibition of ethylene action in climacteric crops (i.e., those that require and produce ethylene for normal ripening) by the application of low O_2 is most effective when instituted prior to the onset of endogenous ethylene production (Beaudry et al., 2006). Once ethylene action has begun, for many climacteric crops, there is a loss in the capacity to control ripening through ethylene action inhibition (Ekman et al., 2004). Interestingly, tomato ripening can be arrested at any stage by 1-methylcyclopropene (1-MCP) (Mir et al., 2004), reinforcing the importance of knowledge of the specifics of the biology of the plant material evaluated. For tomato, a potential opportunity exists to temporarily arrest the ripening of still firm half-ripe fruit using MAP, which may be useful for slowing ripening during shipment.

The extreme effectiveness of low O_2 and elevated CO_2 atmospheres and 1-MCP treatment on ripening inhibition has led researchers to evaluate the potential for these strategies to preserve the quality of many different commodities. However, the primary successes of these technologies have been on those commodities where the target metabolic process is ethylene driven. Commercial successes for low O_2 CA include apple, pear, banana (during transport only), kiwifruit, and cabbage storage. Low O_2 and elevated CO_2 do not suppress the ripening of nonclimacteric fruit such as blueberry, raspberry, watermelon, and strawberry, nor do they markedly impact ethylene-independent senescence found in most vegetable tissues.

3.2.2 *Targeting oxidative browning*

The control of oxidative browning of fresh-cut produce can be achieved by inhibiting the action of polyphenyl oxidase (PPO), a highly specific metabolic target that initiates a series of reactions ultimately responsible for the formation of brown pigmentation on the cut surfaces of many plant tissues. While cut surface browning can be quickly inhibited using processing aids such as sulfating agents like sulfur dioxide gases and meta-bisulfite salts, derivatives of ascorbic acid, hexyl resorcinol and derivatives, sodium dehydroacetic acid, citric acid, and cysteine, long-term protection is achieved for some products using low O_2. For lettuce, low O_2 inhibits browning by PPO at roughly the same concentrations that inhibit respiration. In order to ensure that the O_2 concentrations in commercial packages of cut lettuce and salad products are effective, commercial MA packages are designed to achieve O_2 levels that are often below the fermentation threshold (Cameron et al., 1995). However, the fermentation of lettuce, if not severe, produces only a minor amount of off-flavors (Smyth et al., 1998). The tolerance of lettuce quality to low O_2 again demonstrates the need for a thorough knowledge of the product. Unlike lettuce, most fresh-cut products will not endure exposure to fermentative levels of O_2 for extended periods. It is interesting to note that the inhibition of this one enzymatic reaction likely accounts for the bulk of MAP use and has a value in the billions of U.S. dollars, based on the sales of fresh-cut salads noted previously.

3.2.3 Targeting respiratory metabolism

Respiratory metabolism can be readily reduced using low O_2 partial pressures, elevated CO_2, or respiratory inhibitors. Respiration impacts all metabolic pathways dependent on sustained energy transfer and carbon skeleton formation. Inhibition of the respiratory pathway cannot, therefore, be isolated from impacts on the remainder of metabolism and, using the criteria discussed previously, would therefore represent a global metabolic target.

Kidd and West (1914), from their work on mustard seeds, developed the concept that restricting O_2 availability and increasing the concentration of the respiratory product CO_2 would suppress respiration and attendant metabolic processes enough to improve retention of metabolic materials such as sugars and acids to prolong storability. They expanded their work from mustard seeds to apple fruit and found that low O_2 and elevated CO_2 effectively increased storability. However, the central thesis of Kidd and West has been demonstrated to have only limited value; unbeknownst to them, they were primarily successful because of their ability to impair ethylene action. A careful analysis of O_2 responses for apple fruit reveals that levels insufficient to inhibit respiration in preclimacteric fruit (e.g., 3 to 6 kPa) are quite effective at inhibiting ethylene action and consequent ripening (Sfakiotakis and Dilley, 1973).

Beaudry (2000) indicated that under steady-state hypoxic conditions, a respiratory suppression of 50% could induce fermentation in some commodities. If one supposes that a 50% inhibition of respiratory/metabolic activity were desirable from a standpoint of deriving commercial benefit, the induction of fermentation would not typically permit the beneficial use of atmospheres modified in O_2 for respiratory suppression. Under such circumstances, Beaudry (2000) indicated there was no "safe working atmosphere." However, for actively ripening climacteric fruit, 50% inhibition of respiration was easily achievable at O_2 levels well above the fermentation threshold. The implication is that metabolic suppression may be desirable for suppressing ethylene-related respiratory activity and may, in part, explain the effectiveness of CA imposed even after the onset of ripening for apple. However, for the vast majority of stored crops, endogenous ethylene has little impact and the benefit of respiratory suppression by O_2 limitation is likely to be minor or nonexistent. An exception to this generalization may be offered by low-pressure storage, wherein nonclimacteric tissues have been shown to benefit by low O_2 (Burg, 2004).

To date, the use of low O_2 or elevated CO_2 to suppress global metabolic activity via respiratory inhibition has not been widely realized, despite innumerable research efforts to use this approach to prolong storability. One example where metabolic suppression may indeed impact a nonclimacteric commodity is in sprout suppression in bulbs of the liliaceae (Legnani et al., 2002; Smittle, 1989). An O_2 concentration of 3 kPa provides good control of sprout formation in Vidalia onions (Smittle, 1989) and has resulted in the commercial use of CA storage in this industry. Nevertheless, the mechanism for this suppression is not known; it may be the inhibition is due to effects on respiration and thereby on global energy metabolism, but there is also the possibility that the inhibition of sprouting is via an O_2-dependent process needed for meristem development or sprout elongation.

Elevated CO_2 has also been reported to suppress respiration via inhibition of the respiratory pathway, although the magnitude of the effect is much less than that of O_2. It therefore may have even less potential for improving storability through the mechanism of respiratory suppression than O_2. Kubo et al. (1990) indicated that suppression of respiration by CO_2 may be related to its impact on ethylene biology. Interestingly, high CO_2, like low O_2, can induce fermentation (Beaudry, 1993). Beaudry (1993) found little impact of CO_2 on O_2 uptake by blueberry, a nonclimacteric fruit, even at CO_2 levels as high as 60 kPa. CO_2 is also recognized as a useful tool for suppressing the sporulation and growth of fungal decay organisms.

3.2.4 *Targeting decay control*

Decay is an important cause of storage losses, but modified atmospheres offer a viable means of pathogen control. Brown (1922) determined that CO_2 concentrations of 10% or more can control the sporulation and growth of numerous fungal decay organisms, while O_2 has a very minor effect on pathogen activity or survival at levels above the fermentation threshold of most commodities. Under hypobaric conditions, extremely low O_2 partial pressures have been shown to be effective in decay control (Burg, 2004). The effect of superatmospheric O_2 (>21 kPa) on pathogens varies with the species treated, but it is usually less effective than CO_2 (Kader and Ben-Yehoshua, 2000).

Decay is reduced by elevated CO_2 via its effect on the fungal organism, rather than on the product. However, as noted previously, CO_2 has an impact on the product as well. For most commodities, the levels of CO_2 needed to control decay adversely affect quality. Nevertheless, there are numerous products that are less sensitive to CO_2 and for which decay tends to be the limiting factor in storability. These crops include strawberry, blueberry, blackberry, raspberry, and cherry; all can be stored successfully under an atmosphere of 10 to 20% CO_2. Even among these crops, elevated CO_2 can prove to be damaging, often inducing fermentation (Zhang and Watkins, 2005) or making the product more sensitive to low-O_2-induced fermentation (Beaudry, 1993).

Most of the applications of elevated CO_2 for the purpose of controlling decay employ MAP systems for pallets or smaller bulk systems. For blueberries, high-CO_2 CA storage is sometimes used. As will be discussed, perforated MA packages attain higher levels of CO_2 than nonperforated packages for a given O_2 level, easily achieving CO_2 partial pressures of 8 to 15 kPa without developing fermentation-inducing levels of O_2. Thus, MAP designs incorporating perforated films would likely be appropriate for establishing atmospheres that control decay of berry fruits. Another approach to decay prevention was by introducing a volatile fungicide into the MAP storage environment. Song et al. (1996) used the natural volatile hexanal to control decay. A MAP system could be used as a micro-fumigation chamber (Suppakul et al., 2003).

3.2.5 Temperature regulation

Plant metabolism encompasses the entirety of enzymatic reactions, tens of thousands of which are occurring simultaneously. Enzymatic catalysis is fundamentally temperature-sensitive. The energy of activation (E_a) characteristic of catalysis renders all metabolic reactions sensitive to temperature. Temperature control therefore targets global metabolism.

It is interesting to note that many of the reactions catalyzed by the proteome exhibit a wide range in sensitivity to temperature. For instance, the energy of activation (E_a) of catalase is approximately 8.4 kJ/mol, which is equivalent to a Q_{10} of about 1.1 or an increase in activity of approximately 10% for every 10°C increase in temperature. At the higher end, the E_a of pectin methyl esterase is approximately 100 kJ/mol (Q_{10} = 4.5). Most enzymatic reactions have an E_a somewhere near the middle; that of alpha-amylase, for instance, is between 30 and 50 kJ/mol (Q_{10} of approximately 2). Despite this wide diversity in temperature sensitivity, many harvested plant organs can survive without obvious negative effects across a broad range in temperatures. For this reason, temperature regulation is typically the most powerful tool in postharvest quality maintenance of fresh produce.

A snapshot of the impact of temperature on global metabolism can be taken by measuring respiratory rate. Respiration integrates metabolic activity by virtue of its central position in supplying energy for the many reactions that collectively comprise metabolism. Assuming that many different processes affect respiratory demand, one might suppose that the temperature sensitivities of respiration for most commodities would be similar. To a certain extent, this is true, with the apparent E_a for respiration being between 50 and 80 kJ/mol (Q_{10} between 2.2 and 3.1) for most products (Figure 3.2) (Hardenburg et al., 1986). However, some (~25%) have temperature sensitivities outside of this range. This evidence suggests that some of the regulatory processes for global metabolism are quite sensitive to temperature, while others are relatively insensitive. The existence of very high and very low temperature sensitivities might mean that a low number of processes regulate metabolic activity in some plant species. The practical implication is that some commodities have a greater potential for benefit from temperature regulation than others. This is an oversimplification; individual tolerances of the commodity to low temperature need to be superimposed on the benefits of reduced temperature storage, and factors that limit quality may have a temperature sensitivity that markedly differs from that of global metabolic activity. Nevertheless, these simple numbers help us develop a mechanistic framework for developing storage rationales upon which we

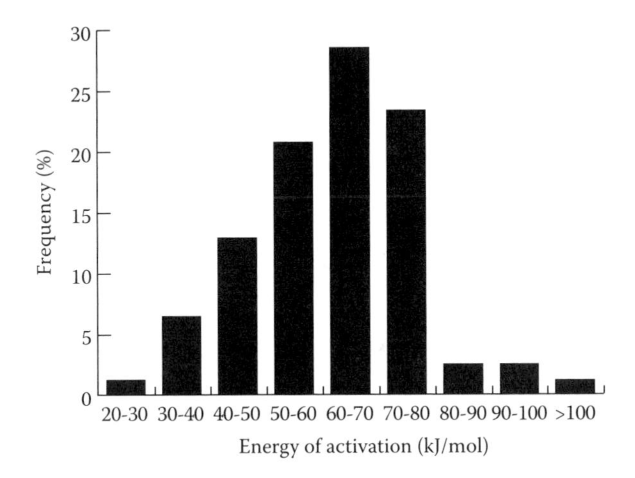

Figure 3.2 Distribution of the energy of activation for respiration calculated for 77 commodities listed in Handbook 66. (From Hardenburg, R.E. et al., *The Commercial Storage of Fruits, Vegetables, and Florist and Nursery Stocks*, Agriculture Handbook 66, revised, U.S. Department of Agriculture, 1986.)

can continue to build. The development of mathematical models that incorporate the temperature response of global metabolism as integrated by respiratory activity has proven useful in predicting how a package behaves in response to temperature.

3.3 MAP modeling to achieve target atmospheres

Atmosphere modification in a package requires a barrier through which gas exchange is restricted. Enclosing an actively respiring product within a package composed all or in part of a film that acts as a gas barrier reduces O_2 and increases CO_2, creating gradients across the film barrier. These gradients provide the driving force for gas flux into or out of the package. In passive MAP, a package always tends toward steady state, in which O_2 and CO_2 levels are constant and O_2 uptake and CO_2 production by the product are equal to those permeating through the package, a situation that exists only when the respiratory rate is constant or nearly so. The steady-state levels of O_2 and CO_2 within a package are dependent on the interaction of respiration of the product and the permeability properties of packaging film or microperforations (Beaudry et al., 1992; Cameron et al., 1989; Jurin and Karel, 1963; Kader et al., 1989).

Depending on the rates of respiration and transmission through the package, atmosphere modification can be rapid or relatively slow. At low temperatures, atmosphere modification can take several days, such that some package systems would not achieve steady state prior to the end of their shelf-life. In many cases, package atmosphere purging with CO_2, N_2, or a combination of gases (i.e., active packaging) is often desirable at the time of package filling and sealing to rapidly obtain the maximum benefits of modified atmospheres, as in the case of fresh-cut cantaloupe melon (Bai et al., 2001).

3.3.1 *Gas exchange properties of continuous and perforated films*

Two strategies for creating film barriers exist. The first employs continuous (nonperforated) films that control movement of O_2 and CO_2 into or out of the package. The second uses perforated films with small holes or microperforations as the primary route of gas exchange; essentially, O_2 andCO_2 flow around the packaging film. The relative flux of O_2 and CO_2 differ substantially between the two packaging strategies, resulting in considerable differences in gas exchange behavior.

In packages composed of continuous films, the permeability of the film to CO_2 is usually two to eight times the permeability of the film to O_2. If the rate of O_2 uptake is roughly the same as CO_2 production (RQ-1), as is often the case unless fermentation is taking place, the CO_2 gradient for continuous film packages will be a fraction of the O_2 gradient (Figure 3.3). For example, low-density polyethylene (LDPE) has a permeability to CO_2 that is approximately four times that to O_2. In a LDPE package having a 10-kPa O_2 atmosphere at steady state, the CO_2 level would be calculated to be (21 kPa – 10 kPa)/4, or 2.75 kPa CO_2. For perforated films, the permeability to CO_2 is only about 0.8 times that of O_2 (Emond et al., 1991; Fonseca et al., 2002). As a result, in a package relying on perforations for gas exchange, CO_2 levels climb to roughly the same extent that O_2 levels decline (i.e., the gradients are nearly equal), such that the sum of O_2 and CO_2 partial pressures is usually in the range of 19 to 22 kPa. For any given O_2 level, therefore, the perforated package will have a considerably higher CO_2 partial pressure than the continuous film package (Figure 3.3). To extend the example above, if a perforated package were designed to generate 10 kPa O_2 using perforations as

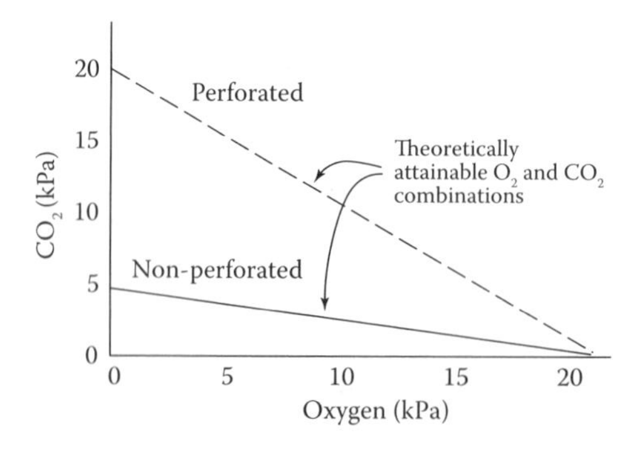

Figure 3.3 Relationship between the attainable O_2 and CO_2 partial pressure combinations in the headspace of perforated and nonperforated packages. The line for the nonperforated package is representative of low-density polyethylene, for which the permeability to CO_2 is approximately four-fold higher than the permeability to O_2.

the route of gas exchange, the CO_2 level would be 0.77 × (21 kPa – 10 kPa), or 8.8 kPa, which is approximately three times greater than that achieved in the continuous film package with the same O_2 level. The relative elevation of CO_2 in perforated film packages may be critically important in package design if CO_2 is needed to help control decay or degreening. It is possible to engineer a package to have permeation take place through both perforations and the film at similar rates. However, as noted later, a relatively small pore provides a very high capacity for gas exchange such that it would be difficult to design a package that would reliably obtain a consistent partition between the two routes of gas movement. If engineered properly, such combined MA packages would have features of both systems and the attainable atmosphere combinations would be in between those of packages dependent on permeation only and those dependent on diffusion through perforations only. These packages would also have a temperature sensitivity for gas diffusion somewhere between those for perforated packages and those for continuous packages.

3.3.2 *Continuous films*

According to Fick's first law of mass transfer, the flux of O_2 and CO_2 is usually directly proportional to the partial pressure or concentration gradient of the gas. For packages composed of continuous films, the rate of O_2 or CO_2 flux is

$$F_{O_2} = \frac{P_{O_2} A\left(p_{o,O_2} - p_{i,O_2}\right)}{l} \quad or \quad F_{CO_2} = \frac{P_{CO_2} A\left(p_{i,CO_2} - p_{o,CO_2}\right)}{l} \tag{3.1}$$

where F_{O_2} and F_{CO_2} are the flux (mol·sec^{-1}) of O_2 and CO_2, respectively; p_{o,O_2} is the partial pressure of O_2 external to the package (21,000 Pa, 21 kPa, or ~21% of an atmosphere); p_{o,CO_2} is the partial pressure of CO_2 external to the package (0.04 kPa or 0.04% of an atmosphere); p_{i,O_2} and p_{i,CO_2} are the partial pressures of O_2 and CO_2, respectively, in the package atmosphere; A is the area (m^2) of the film available for gas transfer; l is the thickness (m) of the packaging film; and P_{O_2} and P_{CO_2} are the O_2 and CO_2 permeabilities (mol·m·m^{-2}·Pa^{-1}·sec^{-1}), respectively, for the film.

As noted, the flux of O_2 or CO_2 through the film increases linearly with the increase in gradient for the gas (Figure 3.4). As the produce in a package uses up the O_2 in the package, the O_2 level declines, reducing respiration, and the gradient increases, increasing O_2 transmission into the package until the two fluxes are equal. At this point, steady state is reached (provided respiration does not change) and O_2 uptake is essentially the same as transmission through the film. Dividing the flux by the weight of the produce in the package yields the rate of respiration for a given O_2 level. Beaudry et al. (1992) and Gran and Beaudry (1993) used this relationship to determine the rates of respiration for O_2 and CO_2 for MA packaged blueberry fruit after

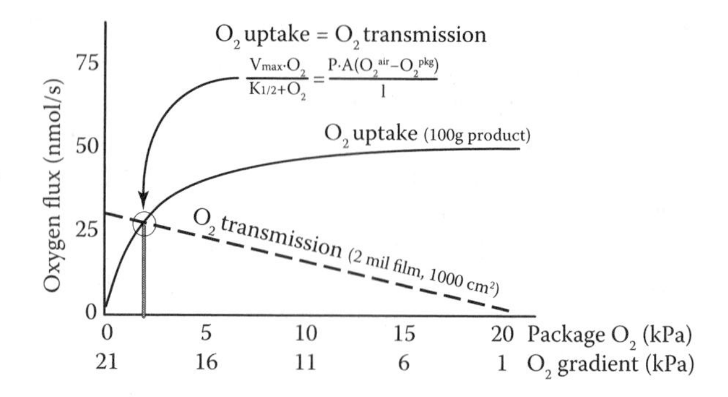

Figure 3.4 Oxygen flux into hypothetical produce (i.e., O_2 uptake or respiration) as a function of oxygen content in the headspace of the package (solid line) and O_2 flux into a hypothetical package (i.e., O_2 transmission) as a function of the oxygen gradient between the headspace of the package and the outside air (dashed line). As the produce in a package uses up the O_2 in the package, the O_2 level declines, reducing respiration, and the gradient increases, increasing O_2 transmission into the package until the two fluxes are equal. At this point, steady state is reached (provided respiration does not change) and the equation shown applies. The O_2 uptake data and O_2 transmission curves are similar to those reported for blueberry fruit and LDPE, respectively. (Data are taken from Cameron, A.C. et al., *J. Am. Soc. Hort. Sci.*, 119, 534–539, 1994.)

steady state was reached. Since the rates of flux of both O_2 and CO_2 are easily determined, the ratio of CO_2 production to O_2 uptake can be used to estimate the respiratory quotient and identify the O_2 level below which fermentation is induced. In Figure 3.4, steady state is reached at approximately 2 kPa O_2. If the O_2 transmission rate of the package is altered by changing the area or thickness of the film, the steady-state O_2 level will change accordingly (Figure 3.5). Importantly, the range of O_2 levels obtained for any package and product

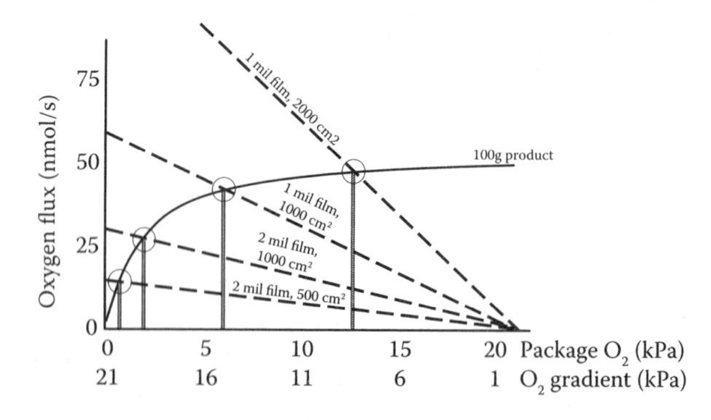

Figure 3.5 The dependence of package oxygen on the oxygen transmission rate for a hypothetical package containing a hypothetical product. The O_2 uptake data and O_2 transmission data are similar to those reported for blueberry fruit and LDPE, respectively. (Data are taken from Cameron, A.C. et al., *J. Am. Soc. Hort. Sci.*, 119, 534–539, 1994.)

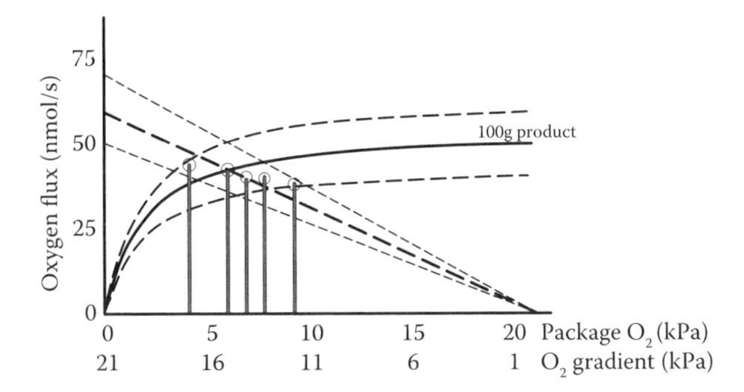

Figure 3.6 The variability in package oxygen is a function of variation (±20% shown) in the O_2 transmission rate of the package and O_2 uptake by the produce. The O_2 uptake data and O_2 transmission data are similar to those reported for blueberry fruit and LDPE, respectively, at 20°C. (Data are taken from Cameron, A.C. et al., *J. Am. Soc. Hort. Sci.*, 119, 534–539, 1994.)

combination can vary significantly if there is variation in the transmission rate of O_2 or in the O_2 uptake of the product (Figure 3.6) (Talasila et al., 1994). In the example given, a package designed to yield a 20% variation in O_2 transmission for a product having a 20% variation in respiratory rate will yield a marked variation in package O_2 levels. While the target O_2 level is 6 kPa O_2, the range of O_2 partial pressure anticipated will be approximately 2.5 to 9.5 kPa. If the package were designed to protect against ethylene action, only those O_2 levels below approximately 6 kPa would generate the desired response (Burg and Burg, 1967).

Cameron et al. (1993) calculated the risk of the package O_2 falling below the lower O_2 limit tolerated by any particular product and resulting in fermentation. They suggested that packages be designed to generate O_2 levels well above the lower O_2 limit in order to ensure aerobic conditions. An interesting phenomenon was also observed for packages that are designed to achieve oxygen levels low enough to retard respiration significantly: as the target O_2 level declined, the absolute ranges of variation in the O_2 and CO_2 levels in the package headspace also declined (Talasila and Cameron, 1995). This relationship can be used to advantage in some situations, as in the case of cut lettuce, for which the optimal O_2 level to minimize fermentation while retarding browning lies between 0.25 and 1% (Cameron, 2003; Smyth et al., 1998).

Permeability varies with temperature. As temperature increases, the O_2 and CO_2 permeability of many packaging films increases markedly, such that the transmission rate of O_2 and CO_2 increases (Figure 3.7). The relationship between temperature and the permeability of a continuous film is depicted by the Arrhenius equation:

$$P_{O_2} = P_{0,O_2} e^{\left(\frac{-E_a^{P_{O_2}}}{RT}\right)} \tag{3.2}$$

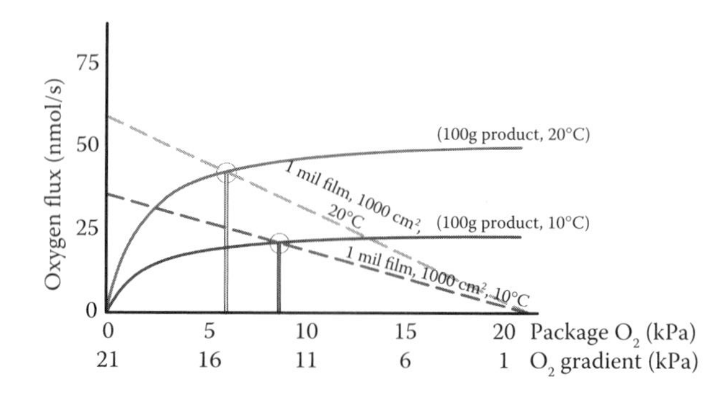

Figure 3.7 Effect of temperature on O_2 flux into hypothetical produce as a function of oxygen content in the headspace of the package (solid lines) and O_2 flux into a hypothetical package as a function of the oxygen gradient between the headspace of the package and the outside air (dashed lines). As temperature declines, the respiratory demand for O_2 declines to a greater extent than does O_2 transmission, resulting in an increase in the steady-state package O_2. Conversely, as temperature increases, the O_2 partial pressure declines. The O_2 uptake data and O_2 transmission curves are similar to those reported for blueberry fruit and LDPE, respectively. (Data are taken from Cameron, A.C. et al., *J. Am. Soc. Hort. Sci.*, 119, 534–539, 1994.)

where p_{0,O_2} is the Arrhenius constant (mol·m·m^{-2}·Pa^{-1}·sec^{-1}) for O_2; T is the temperature in degrees Kelvin (K, where K = 273.15 + °C); R is the gas constant (8.3144 J·mol^{-1}·K^{-1}); and $E_a^{PO_2}$ is the activation energy (J·mol^{-1}) for the permeation of O_2. As noted for respiration, the energy of activation describes the temperature sensitivity of the process. The E_a values of permeation of O_2 and other gases through films have units that are often expressed in kilojoules (kJ) per mole of molecules undergoing the interaction. The $E_a^{PO_2}$ for LDPE is approximately 38 to 42 kJmol^{-1} (Cameron et al., 1994), which yields a 2.5-fold higher permeability as temperature increases from 0 to 15°C. A higher $E_a^{PO_2}$ would relate to a greater change in O_2 permeability over the temperature range, and a lower $E_a^{PO_2}$ indicates a lower temperature responsiveness. For most films, the E_a of O_2 permeation is lower than the apparent E_a of O_2 uptake for most plant products. The impact of this disparity is that package O_2 levels usually increase as temperature declines below the target for the package system and decline as temperature increases above the target temperature (Figure 3.7).

3.3.3 *Perforated films*

In packages with perforation-mediated gas exchange, the gas transmission through the package is a sum of gas diffusion through the perforation and gas permeation through the polymeric film. Generally, gas diffusion through the film is negligible relative to the total gas flow through the perforations. Mathematically, gas transmission can be explained in a fashion similar to that of nonperforated packages; the perforation is modeled as a short tube through which gas diffusion occurs. Gas flux, as in the case of continuous films, is proportional to the differences in partial pressures between the outside and inside of the package (Fishman et al., 1996). For O_2,

$$F_{O_2} = \frac{D_{O_2} A_h \left(C_{o,O_2} - C_{i,O_2}\right)}{L_h} \tag{3.3}$$

where D_{O_2} is the diffusion coefficient of the gas in air ($m^2 \cdot sec^{-1}$); A_h is the combined area of the perforations (m^2) (with $A_h = R_h^2 N$, where = 3.1415, R_h is the radius (m) of the perforation, and N is the total number of perforations); C_{o,O_2} is the concentration ($mol \cdot m^3$) of O_2 external to the package (with $C_{O_2} = p_{O_2}/RT$, where T is the temperature in degrees Kelvin and R is the gas constant [8.3144 $J \cdot mol^{-1} \cdot K^{-1}$]); and L_h is the effective length of the pore (with $L_h = l + R_h$, where l is the thickness [m] of the perforated film). The diffusion coefficient is dependent on the type of gas and the temperature of the gas (Table 3.2).

The rate of diffusion through a pore is also sensitive to temperature. In contrast to permeation through continuous films, the transmission of gas through perforations has an extremely low temperature sensitivity (Nobel, 1983). The diffusivity of gases through a perforation depends upon temperature (K) by $T^{1.8}$, which is approximately equivalent to an E_a of 4.3 $kJ \cdot mol^{-1}$.

Table 3.2 Diffusion Coefficients (D) for Gases in Air at 1 atm of Pressure (1 atm = 101.3 kPa)

Gas	Temperature (°C)	Diffusion coefficient ($m^2 \cdot sec^{-1}$)
CO_2	0	1.33×10^{-5}
	10	1.42×10^{-5}
	20	1.51×10^{-5}
	30	1.60×10^{-5}
O_2	0	1.72×10^{-5}
	10	1.83×10^{-5}
	20	1.95×10^{-5}
	30	2.07×10^{-5}
H_2O	0	2.13×10^{-5}
	10	2.27×10^{-5}
	20	2.42×10^{-5}
	30	2.57×10^{-5}

Note: To calculate the rate of flux for perforated films, substitute the data for the appropriate gas and temperature into Equation 3.3.

Source: Adapted from Nobel, P.S., *Biophysical Plant Physiology and Ecology,* W.H. Freeman and Co., New York, 1983.

The low temperature sensitivity of diffusion through perforations yields only an 11% increase in gas flux as temperature increases from 0 to 15°C. As previously mentioned, the respiration of fruits and vegetables usually has a temperature sensitivity significantly higher than that of O_2 diffusion, leading to a relative deficiency in O_2 transmission as temperature increases (Beaudry et al., 1992).

The absolute rates of gas exchange for perforated films can be much greater than for continuous films because of the much higher gas transmission rates for diffusion through a pore than for a film. On a per area basis, a pore permits 4 to 20 million times the flux of O_2 or CO_2 compared to LDPE, depending on temperature, film thickness, and the diameter of the perforation. A 100-μm-diameter perforation in a 50-μm-thick LDPE film at 20°C has a diffusion capacity roughly equivalent to that of a package of the same film with a surface area of approximately 400 cm^2. Therefore, even the smallest pinhole in a package that is supposed to be nonperforated has a dramatic effect on package atmosphere at steady state.

3.3.4 *Respiratory parameters*

The respiratory response of plant material to O_2 concentration can be mathematically described using saturation-type curves. A standard means of expressing the dependence of a reaction (e.g., O_2 uptake) on a substrate (e.g., O_2) is the Michaelis–Menton model, which is primarily applied to specific enzymatic reactions. When this model is adapted for use in packaging, it is expressed as follows:

$$r_{O_2} = \frac{V_{\max} p_{i,O_2}}{K_{1/2} + p_{i,O_2}} \tag{3.4}$$

where r_{O_2} is the O_2 uptake (mol·kg^{-1}·sec^{-1}), V_{max} (mol·kg^{-1}·sec^{-1}) is the maximal rate of O_2 uptake, p_{i,O_2} is the partial pressure (Pa) of O_2 in the package, and $K_{1/2}$ is the O_2 partial pressure in the package at 50% of V_{max}. This expression differs from a true Michaelis–Menton model since it represents a system, rather than a single enzyme, and for this reason the term *apparent* K_m or $K_{1/2}$ is often substituted for K_m (Cameron et al., 1994, 1995). While other empirical mathematical models have been used in various analyses (Fishman et al., 1996), this model has the advantage of being broadly interpretable in terms of the physiology and biochemistry of the plant material.

Both V_{max} and $K_{1/2}$ have been shown to have temperature dependence (Cameron et al., 1994; Lakakul et al., 1999). Where $K_{1/2}$ changes with temperature, it has been described using linear or curvilinear equations. The simplest representation is a linear equation:

$$K_{1/2} = xT + y \tag{3.5}$$

where x and y are variables determined by fitting data and T is temperature (K). Typically, $K_{1/2}$ is between 0.1 and 3 kPa O_2 at 0°C, and it increases with temperature at a rate between 0.05 and 0.5 kPa per degree.

As mentioned previously, the maximal rate of respiration increases with increasing temperature. The temperature dependence of V_{max} can be expressed in a manner similar to that in the permeation process (Lakakul et al., 1999) using the following form of the Arrhenius expression:

$$V_{\max} = V_{\max_0} e^{\left(\frac{-E_a^{rO_2}}{RT}\right)} \tag{3.6}$$

where V_{max_0} is the equivalent of the Arrhenius constant, R is the gas constant (8.3144 J·mol^{-1}·K^{-1}), T is temperature (K), and $E_a^{rO_2}$ is equivalent to the energy of activation for the maximal rate of O_2 uptake and can also be termed the apparent E_a. This form of describing temperature sensitivity has the advantage that the units are the same as those for $E_a^{PO_2}$. The maximal rate of respiration for most fruit and vegetable products has an $E_a^{rO_2}$ between 60 and 70 kJ·mol^{-1} (Figure 3.2), undergoing a four- to six-fold increase from 0 to 15°C (Beaudry et al., 1992; Cameron et al., 1994, 1995; Lakakul et al., 1999). This means that product respiration increases at 2 or 3 times the rate of LDPE permeability and 30 times the rate of perforation permeability with increasing temperature. As noted, when respiratory demand for O_2 increases faster than O_2 permeation as temperature increases, O_2 levels decline and may pose a risk to product quality. This limits the usefulness of MAP in some situations. The temperature problem caused by the imbalance in the temperature sensitivity of respiration and permeation was recognized by early workers in the field (Workman, 1959; Tomkins, 1962).

Safe levels of exposure to O_2 and CO_2 are important criteria for package design and the interpretation of the package O_2 models. A lower O_2 limit has been associated with the onset of fermentation and accumulation of the fermentation products ethanol and acetaldehyde (Beaudry et al., 1992; Dostal-Lange et al., 1991; Gran and Beaudry, 1993). The lower O_2 limit increases with temperature. Lower oxygen limits vary from 0.15 to 5 kPa, varying by temperature, commodity, and even cultivar.

3.3.5 *Integrating package, product, and environment*

Once information regarding the respiratory response to O_2 (and, in some cases, CO_2), film permeability to O_2, CO_2, and H_2O, the lower O_2 limit, and the upper CO_2 limit is acquired, the information can be integrated into mathematical models (Beaudry et al., 1992; Cameron et al., 1994; Fishman et al., 1996; Hertog et al., 1998; Lakakul, 1999). These models enable the reasonably accurate prediction of package performance (i.e., O_2, CO_2, and H_2O content in the package headspace) under a variety of environmental conditions prior to construction of the package. Additionally, they permit the identification of limiting features of the film, package, product, or environment. Models typically include temperature dependency, but can also be developed to predict effects of package volume, resistance to heat flow, and developmental changes in product physiology on headspace gases.

A model can be developed to predict the steady-state concentration of O_2 in the package headspace by setting Equation 3.3 equal to Equation 3.6 and solving for p_{i,O_2} as follows:

$$p_{i,O_2} = \frac{1}{2}\left(\left(\left(K_{1/2} + \frac{Wl}{P_{O_2}A}V_{\max} - p_{o,O_2}\right)^2 + 4P_{o,O_2}K_{1/2}\right)^{1/2} - \left(K_{1/2} + \frac{Wl}{P_{O_2}A}V_{\max} - p_{o,O_2}\right)\right) \tag{3.7}$$

When the temperature-dependent expressions for P_{O2} (or diffusion), $K_{1/2}$, and V_{max} (Equations 3.2, 3.5, and 3.6, respectively) are substituted into Equation 3.7, the model becomes temperature compensating and can be used to predict steady-state O_2 partial pressure as a function of temperature, fruit weight, surface area, film thickness, or other variable, such as $E_a^{PO_2}$. Models incorporating temperature effects on respiration and permeability have been published for whole apple fruit, apple slices (Lakakul et al., 1999), blueberry fruit (Beaudry et al., 1992; Cameron et al., 1994), chicory leaves (Hertog et al., 1998), broccoli florets (Cameron et al., 1995), lettuce leaves (Cameron et al., 1995), strawberry (Joles, 1993), tomato (Hertog et al., 1998), and raspberry (Joles, 1993; Joles et al., 1994). In addition, more complex dynamic models have been developed to account for temporal changes in package volume, product respiration, and the humidity and temperature of the environment (Fishman et al., 1996; Hertog et al., 1998).

3.3.6 *Integrating the model and metabolism*

An example of the full knowledge of limiting factors for quality and limitations imposed by plant responses to atmosphere manipulation can be provided using blueberry fruit. Research by Beaudry (1993) indicated that as CO_2 partial pressure increases, the tolerance of blueberry to low oxygen declines such that for 20, 15, 10, and 5 kPa CO_2 needs to remain above approximately 17, 7, 4, and 1.5 kPa, respectively (Figure 3.8A). These represent the lower limits for O_2 under these CO_2 conditions. However, the limiting factor for blueberry storage is often decay, which can be controlled by elevated CO_2 concentrations between 8 and 17 kPa without damaging fruit. Our knowledge of the behavior of perforated and nonperforated film packages indicates that only perforated films will generate the required CO_2 levels (Figure 3.8B). The lower O_2 limit for perforated packages is imposed by CO_2

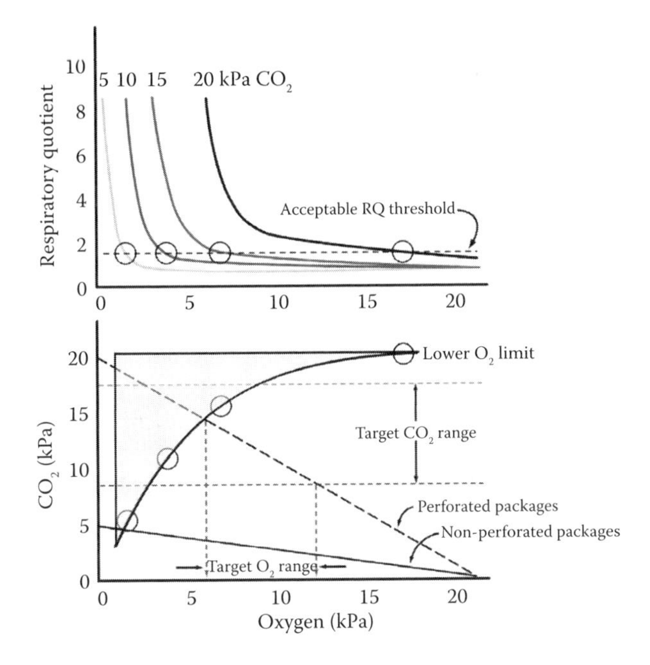

Figure 3.8 The respiratory response of blueberry to O_2 and CO_2 can be used to identify oxygen levels below which fermentation results (upper graph) for various applied CO_2 partial pressures. These threshold values for O_2 and CO_2 can be combined with knowledge about CO_2 partial pressures needed for decay control (8 to 17 kPa) and thereby identify suitable O_2 levels (based on the gas exchange characteristics of the film, e.g., perforated or nonperforated) package for designing an MA blueberry. Data for the respiratory quotient are taken from Beaudry (1993).

sensitivity, and the upper limit is imposed by the lower 8-kPa CO_2 limit. Using the models previously described and respiratory data ($K_{1/2}$, V_{max}) for blueberry, one can design a package to achieve target O_2 levels by adjusting pore area and number using equations in this chapter.

3.4 Conclusions

An understanding of the metabolic factors controlling quality loss and their capacity to be manipulated is fundamental to the application of postharvest agricultural technologies. Global metabolic manipulation is possible by regulating temperature and O_2 levels. A degree of more specific control over some metabolic processes such as the ethylene response, degreening (chlorophyll loss), oxidative browning, and other postharvest phenomena can also be obtained. Understanding the limits of the commodity with respect to the application of these technologies and coupling this with knowledge of the limits of the existing handling and marketing chain allow us to design an appropriate storage strategy for each commodity.

Modification of the oxygen and carbon dioxide partial pressures in the atmospheres can alter the physiology of harvested fruits and vegetables in a desirable manner, resulting in an improvement in quality maintenance relative to air storage. The determination whether oxygen or carbon dioxide modification is appropriate is dependent on the biology of the harvested plant organ and those components of physiology and pathology that comprise biological limiting factors. The potential for atmosphere modification to impair product quality must also feature in the decision-making process; desirable or optimal gas concentrations can exist in a fairly narrow range.

References

Allen, A.S. and N. Allen. 1950. Tomato-film findings. *Modern Packaging* 23:123–126, 180.

Anon. 2003. Fresh-Cut Sales of Retail Produce Approaching $4 Billion a Year. *Fresh Cut*, November 2003, http://www.freshcut.com.

Bai, J.-H., R.A. Saftner, A.E. Watada, and Y.S. Lee. 2001. Modified atmosphere maintains quality of fresh-cut cantaloupe (*Cucumis melo* L.). *J. Food Sci.* 66:1207–1211.

Beaudry, R.M. 1993. Effect of carbon dioxide partial pressure on blueberry fruit respiration and respiratory quotient. *Postharvest Biol. Technol.* 3:249–258.

Beaudry, R.M. 2000. Responses of horticultural commodities to low oxygen: limits to the expanded use of modified atmosphere packaging. *HortTechnology* 10:491–500.

Beaudry, R.M., A.C. Cameron, A. Shirazi, and D.L. Dostal-Lange. 1992. Modified-atmosphere packaging of blueberry fruit: effect of temperature on package O_2 and CO_2. *J. Am. Soc. Hort. Sci.* 117:436–441.

Beaudry, R., V. Luckanatinvong, and T. Solomos. 2006. Maintaining quality with CA and MAP. *Acta Hort.* 712:245–252.

Ben-Yehoshua, S., R.M. Beaudry, S. Fishman, S. Jayanty, and N. Mir. 2005. Modified atmosphere packaging and controlled atmosphere storage. In *Environmentally Friendly Technologies for Agricultural Produce Quality*, S. Ben-Yehoshua (Ed.). CRC Press, Boca Raton, FL, 534 pp. 61–112.

Berard, J.E. 1821. Memoire sur la maturation des fruits. *Ann. Chim. Phys.* 16:152–183, 225–251.

Blackman, F.F. and P. Parija. 1928. Analytic studies in plant respiration. I. The respiration of a population of senescent ripening apples. *Proc. R. Soc. London B* 103:412–445.

Brendel, V., S. Kurtz, and V. Walbot. 2002. Comparative genomics of Arabidopsis and maize: prospects and limitations. *Genome Biol.* 3:1005.1–1005.6.

Brown, W. 1922. On the germination and growth of fungi at various temperatures and in various concentrations of oxygen and carbon dioxide. *Ann. Bot.* 36:257–283.

Burg, S.P. 2004. *Postharvest Physiology and Hypobaric Storage of Fresh Produce.* CAB International, Wallingford, U.K.

Burg, S.P. and E.A. Burg. 1965. Ethylene action and the ripening of fruits. *Science* 148:1190–1196.

Burg, S.P. and E.A. Burg. 1967. Molecular requirements for the biological activity of ethylene. *Plant Physiol.* 42:114–152.

Calderone, V. 2006. Valley's Agriculture Is Ever Evolving. The Salinas Californian. *Ag Outlook*, June 12.

Cameron, A.C. 2003. Modified-atmosphere packaging of perishable horticultural commodities can be risky business. 8th International Controlled Atmosphere Research Conference. *Acta Hort.* 600:305–310.

Cameron, A.C., R.M. Beaudry, N.H. Banks, and M.V. Yelanich. 1994. Modified-atmosphere packaging of blueberry fruit: modeling respiration and package oxygen partial pressures as a function of temperature. *J. Am. Soc. Hort. Sci.* 119:534–539.

Cameron, A.C., W. Boylan-Pett, and J. Lee. 1989. Design of modified atmosphere packaging systems: modelling oxygen concentrations within sealed packages of tomato fruits. *J. Food Sci.* 54:1413–1416, 1421.

Cameron, A.C., B.D. Patterson, P.C. Talasila, and D.W. Joles. 1993. Modeling the risk in modified-atmosphere packaging: a case for sense-and-respond packaging. In *Proceedings of the Sixth International Controlled Atmosphere Research Conference*, NRAES-71, G. Blanpied, J. Bartsch, and J. Hicks (Eds.). Cornell University, Ithaca, NY, pp. 95–102.

Cameron, A.C., P.C. Talasila, and D.J. Joles. 1995. Predicting the film permeability needs for modified-atmosphere packaging of lightly processed fruits and vegetables. *HortScience* 30:25–34.

Davis, D.W. 1946. Gas permeability: an isostatic test method. *Modern Packaging* 19:145–149.

Dilley, D.R. 1990. Historical aspects and perspectives of controlled atmosphere storage. In *Food Preservation by Modified Atmospheres*, M. Calderon and R. Barkai-Golan (Eds.). CRC Press, Boca Raton, FL, pp. 187–196.

Dostal-Lange, D. and Beaudry, R.M. 1991. The effects of modified atmosphere packaging and temperature on postharvest storage life of three high bush blueberry cultivars. *HortScience* 23:742.

Ekman, J.H., M. Clayton, W.V. Biasi, and E.J. Mitcham. 2004. Interactions between 1-MCP concentration, treatment interval and storage time for 'Bartlett' pears. *Postharvest Biol. Technol.* 31:127–136.

Emond, J.P., F. Castaigne, C.J. Toupin, and D. Desilets. 1991. Mathematical modeling of gas exchange in modified atmosphere packaging. *Trans. ASAE.* 34:239–245.

Fishman, S., V. Rodov, and S. Ben-Yehoshua. 1996. Mathematical model for perforation effect on oxygen and water vapor dynamics in modified-atmosphere packages. *J. Food Sci.* 61:956–961.

Fonseca, S.C., F.A.R. Oliveira, I.B.M. Lino, J.K. Brecht, and K.V. Chau. 2002. Modelling O_2 and CO_2 exchange for development of perforation-mediated modified atmosphere packaging. *J. Food Eng.* 52:99–119.

Gran, C.D. and R.M. Beaudry. 1993. Determination of the low oxygen limit for several commercial apple cultivars by respiratory quotient breakpoint. *Postharvest Biol. Technol.* 3:259–267.

Hardenburg, R.E., A.E. Watada, and C.Y. Wang. 1986. *The Commercial Storage of Fruits, Vegetables, and Florist and Nursery Stocks*, Agriculture Handbook 66, revised. U.S. Department of Agriculture, 136 pp.

Henig, Y.S. and S.G. Gilbert. 1975. Computer analysis of the variables affecting respiration and quality of produce package in polymeric films. *J. Food Sci.* 40:1033–1035.

Hertog, M.L.A.T.M., H.W. Peppelenbos, R.G. Evelo, and L.M.M. Tijskens. 1998. A dynamic and generic model of gas exchange of respiring produce: the effects of oxygen, carbon dioxide and temperature. *Postharvest Biol. Technol.* 14:335–349.

Hirata, T., Y. Makino, Y. Ishikawa, S. Katusara, and Hasegawa. 1996. A theoretical model for designing a modified atmosphere packaging with a perforation. *Trans. ASAE.* 39:1499–1504.

Joles, D.W. 1993. Modified-Atmosphere Packaging of Raspberry and Strawberry Fruit: Characterizing the Respiratory Response to Reduced O2, Elevated CO2 and Changes in Temperature. M.S. thesis, Michigan State University, East Lansing.

Joles, D.W., A.C. Cameron, A. Shirazi, P.D. Petracek, and R.M. Beaudry. 1994. Modified-atmosphere packaging of 'Heritage' red raspberry fruit: respiratory response to reduced oxygen, enhanced carbon dioxide, and temperature. *J. Am. Soc. Hort. Sci.* 119:540–545.

Jurin, V. and M. Karel. 1963. Studies on control of respiration of McIntosh apples by packaging methods. *Food Technol.* 17:104–108.

Kader, A.A. and S. Ben-Yehoshua. 2000. Effects of superatmospheric oxygen levels on postharvest physiology and quality of fresh fruits and vegetables. *Postharvest Biol. Technol.* 20:1–13.

Kader, A.A., D. Zagory, and E.L. Kerbel. 1989. Modified atmosphere packaging of fruits and vegetables. *CRC Crit. Rev. Food Sci. Nutr.* 28:1–30.

Kidd, F. and C. West. 1914. The controlling influence of carbon dioxide in the maturation, dormancy, and germination of seeds. *Proc. R. Soc. Lond. B* 87:408–421.

Kidd, F. and C. West. 1927. A relation between the concentration of oxygen and carbon dioxide in the atmosphere, rate of respiration, and length of storage of apples. In *Food Investigation Board Report.* London, pp. 41–42.

Kidd, F. and C. West. 1945. Respiratory activity and duration of life of apples gathered at different stages of development and subsequently maintained at constant temperature. *Plant Physiol.* 20:467–504.

Kubo, Y., A. Inaba, and R. Nakamura. 1990. Respiration and C_2H_4 production in various harvested crops held in CO_2-enriched atmospheres. *J. Am. Soc. Hort. Sci.* 115:975–978.

Lakakul, R., R.M. Beaudry, and R.J. Hernandez. 1999. Modeling respiration of apple slices in modified-atmosphere packages. *J. Food Sci.* 64:105–110.

Legnani, G., C.B. Watkins, and W.B. Miller. 2002. Use of low-oxygen atmospheres to inhibit sprout elongation of dry-sale Asiatic lily bulbs. *Acta Hort.* 570:183–189.

Mangin, L. Sur la végétation dan une atmosphère viciée par la respiration. *Comp. Rend. Acad. Sci.* 122:747–749.

Mir, N., M. Cañoles, R. Beaudry, E. Baldwin, and C. Mehla. 2004. Inhibition of tomato ripening by 1-methylcyclopropene. *J. Am. Soc. Hort. Sci.* 129:112–120.

Nobel, P.S. 1983. *Biophysical Plant Physiology and Ecology.* W.H. Freeman and Co., New York.

Owen, T. 1800. *The Three Books of M. Terentius Varro Concerning Agriculture.* Oxford, 257 pp. (translated).

Platenius, H. 1946. Films for produce: their physical characteristics and requirements. *Modern Packaging* 20:139–143, 170.

Ryall, A.L. and M. Uota. 1955. Effect of sealed polyethylene liners on the storage life of Watsonville Yellow Newtown apples. *Proc. Am. Soc. Hort. Sci.* 65:203–210.

Scott, K.J. and E.A. Roberts. 1966. Polyethylene bags to delay ripening of bananas during transport and storage. *Aust. J. Exp. Agric. Animal Husb.* 6:197–199.
Sfakiotakis, E.M. and D.R. Dilley. 1973. Induction of autocatalytic ethylene production in apple fruits by propylene in relation to maturity and oxygen. *J. Am. Soc. Hort. Sci.* 98:504–508.
Smittle, D.A. 1989. Controlled atmosphere storage of Vidalia onions. In *Proceedings of the 5th International Controlled Atmospheric Research Conference*, Wenatchee, WA, Vol. 2, pp. 171–177.
Smyth, A.B., J. Song, and A.C. Cameron. 1998. Modified-atmosphere packaged cut iceberg lettuce: effect of temperature and O_2 partial pressure on respiration and quality. *J. Agric. Food Chem.* 46:4556–4562.
Song, J., R. Leepipattanawit, W. Deng, and R.M. Beaudry. 1996. Hexanal vapor is a natural, metabolizable fungicide: inhibition of fungal activity and enhancement of aroma in apple slices. *J. Am. Soc. Hort. Sci.* 121:937–942.
Suppakul, P., J. Miltz, K. Sonneveld, and S.W. Bigger. 2003. Active packaging technologies with an emphasis on antimicrobial packaging and applications. *J. Food Sci.* 68:408–420.
Talasila, P.C. and A.C. Cameron. 1995. Modeling distribution of steady-state oxygen levels in modified-atmosphere packages: an approach for designing safe packages. *J. Food Process Eng.* 18:199–217.
Talasila, P.C., A.C. Cameron, and D.W. Joles. 1994. Frequency distribution of steady-state oxygen partial pressures in modified-atmosphere packages of cut broccoli. *J. Am. Soc. Hort. Sci.* 119:556–562.
Tomkins, R.G. 1962. Film packaging of fresh fruit and vegetables: the influence of permeability. In *Institute of Packaging Conference Guide 1961*. Larkfield, Maidstone, Kent, England, pp. 64–69.
Watkins, C.B. 2002. Ethylene synthesis, mode of action, consequences and control. In *Fruit Quality and Its Biological Basis*, M. Knee (Ed.). CRC Press, Boca Raton, FL, 279 pp.
Workman, M. 1957. *A Progress Report on the Use of Polyethylene Film Box Liners for Apple Storage*, HO-50-4. Purdue University Agriculture Extension Service, 8 pp.
Workman, M. 1959. The status of polyethylene film liners to provide modified atmosphere for the storage of apples. *East. Fruit Grower* 23:6, 10–14.
Zhang, J. and C.B. Watkins. 2005. Fruit quality, fermentation products, and activities of associated enzymes during elevated CO_2 treatment of strawberry fruit at high and low temperatures. *J. Am. Soc. Hort. Sci.* 130:124–130.

chapter four

Active packaging for fruits and vegetables

A.D. Scully and M.A. Horsham

Contents

4.1 Introduction

Active food packaging technologies in the form of sachets or inserts capable of scavenging oxygen or absorbing water vapor have been commercially available for more than two decades. However, the intensification in research and development activity relating to plastic-based active food packaging technologies in the past 10 years has been spectacular. In this time, a plethora of active food packaging materials have been reported and reviewed,[1–4] all designed to counteract a wide range of deleterious quality- and safety-limiting effects, including rancidity, color loss/change, nutrient loss, dehydration, microbial proliferation, senescence, gas buildup, and off odors, among others.

The focus of this chapter is to provide an overview of the active packaging technologies that have the potential to enhance the preservation of packaged fruits and vegetables in the distribution chain, with a particular emphasis on emerging plastics-based materials.

4.2 Ethylene control

Ethylene (C_2H_4) is a gaseous natural plant growth hormone and has long been known to have a detrimental impact, even at low concentrations, on the product quality and shelf-life of many fruits and vegetables during storage and distribution. Ethylene is sometimes referred to as the ripening or death hormone because it induces fruit ripening and accelerates fruit softening and senescence (aging). Ethylene can also cause a range of postharvest physiological disorders such as russet spotting on lettuce and scald on apples. Although ethylene is produced by all plants, the principal sources of the low background levels of ethylene in the atmosphere are climacteric fruits (fruits that ripen after harvest and are characterized by an increase in respiration rate and a burst of ethylene production as they ripen), damaged or rotten produce, and exhaust gases from petrol combustion engines.

There have been a variety of methods employed by the horticultural industry to minimize the impact of ethylene during storage and distribution, including the use of low temperatures and controlled atmospheres, as well as the introduction of filters/scrubbers to remove the ethylene from the atmosphere around the stored produce. Low-temperature storage reduces the production of ethylene by lowering the respiration and metabolic rates of the produce. Controlled atmosphere storage is based on the use of low oxygen and high carbon dioxide concentrations to suppress respiration rates and render the produce less sensitive to the effects of ethylene.

Packaging technologies designed to scavenge or absorb ethylene from the surrounding environment of packaged produce have also been developed. The most widely used ethylene-scavenging packaging technology today is in the form of a sachet containing potassium permanganate immobilized on an inert porous support, such as alumina and silica, at a level of about 5% w/w. The ethylene is scavenged through an oxidation reaction with the potassium permanganate to form carbon dioxide and water. Although these permanganate-based ethylene-scavenging sachets are effective at removing ethylene, their use is sometimes accompanied by undesirable effects. These include possible migration of the potassium permanganate from the sachet onto the produce, lack of specificity to ethylene resulting in desirable aromas being scalped, and a general lack of user enthusiasm for the use of sachets. Other types of sachet-based ethylene-scavenging technologies utilize activated carbon with a metal catalyst (for example, palladium), such as SendoMate® from Mitsubishi Chemicals and Neupalon™ from Sekisui Jushi Ltd.[5]

In recent years, a number of ethylene-removing plastic film-based products consisting of polyethylene impregnated with finely dispersed minerals, such as clays, zeolites, and carbon, have become available. Examples of commercially available plastic-based ethylene-removing materials include Evert-Fresh (Evert-Fresh Co., U.S.), Peakfresh™ (Peakfresh Products, Australia), Orega™ (Chang Yo, Korea),[6] and Bio-fresh™ (Grofit Plastics, Israel). Although bags made using these materials have been demonstrated to provide an extension of shelf-life compared with standard polyethylene bags, it is thought that this benefit arises from the enhanced ethylene permeability of the film containing the minerals, thereby allowing the ethylene to escape from the inside of the package.

A novel plastic-based ethylene-scavenging technology that has the capability to be incorporated into a variety of packaging structures, including films, trays, and fiberboard cartons, is currently being developed at Food Science Australia.[7,8] This technology is based on the irreversible and specific reaction between electron-deficient dienes, such as tetrazines, and ethylene in what is known as an inverse electron-demand Diels–Alder reaction (Figure 4.1). Tetrazines are generally highly colored compounds (usually red), whereas the product formed on reaction with ethylene, dihydropyridazine, is essentially colorless. Consequently, a color change from red/pink to colorless occurs upon reaction with ethylene, and this feature can be used to indicate residual ethylene-scavenging capacity in the packaging material.

One of the early limitations of this technology was the sensitivity to moisture of some tetrazine derivatives under the high relative humidity (RH) conditions experienced in the storage and distribution of horticultural produce, but this limitation was overcome recently through the development of highly moisture stable tetrazine derivatives.[8] A comparison of the moisture stability of the ethylene-scavenging capacity of films containing the new moisture stable compounds to that of films containing a tetrazine compound used in earlier work is shown in Figure 4.2. It can be seen from these results that the scavenging capacity of films containing the earlier compound is reduced by 90% within 24 hours when stored at 93% RH, whereas films containing the latest compound are completely stable, within experimental uncertainty, under these storage conditions.

An alternative approach to minimizing the effects of ethylene is the use of ethylene inhibitors such as 1-methylcylcopropene (1-MCP). 1-MCP is a

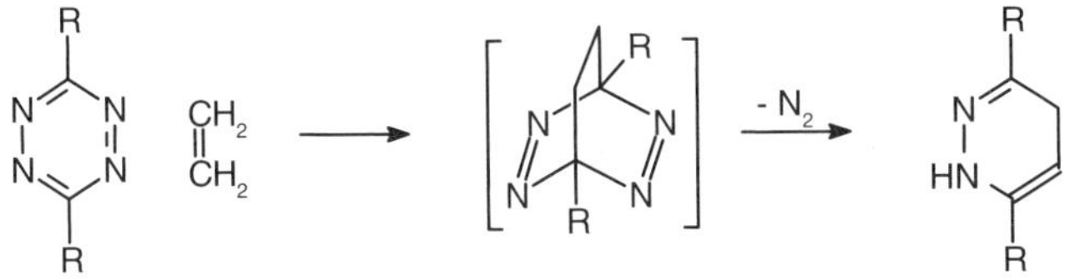

R N N N N R + $CH_2{=}CH_2$ → [bicyclic intermediate, R, N, N, N, N, R] $\xrightarrow{-N_2}$ R N HN R

Figure 4.1 Reaction scheme for Food Science Australia's ethylene-scavenging technology.

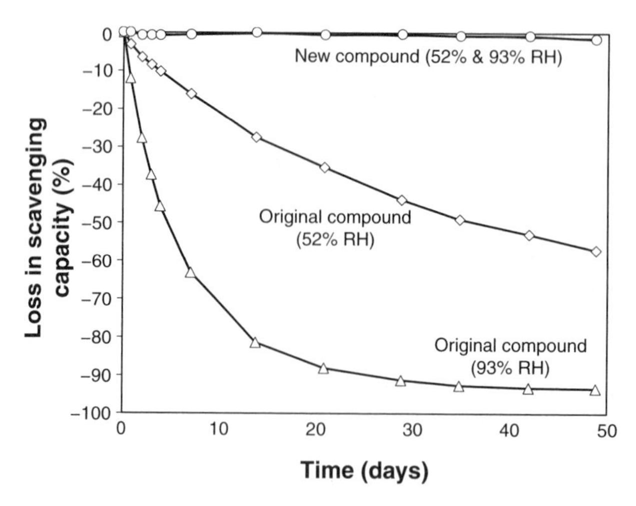

Figure 4.2 Comparison of the moisture stability of Food Science Australia's latest ethylene-scavenging compound (new compound) in a polymer film with the tetrazine compound used in earlier studies (original compound).

gas that binds to the ethylene receptors in plant tissue and, as a result, prevents the hormonal action of ethylene. Research has shown that the shelf-life of many fruits, vegetables, and flowers can be extended by the application of low concentrations of 1-MCP.[9] 1-MCP is marketed by AgroFresh as EthylBloc™ (formulated for floral use only) and SmartFresh™ (approved for food use) and is registered for use in more than 25 countries, including the U.S., Europe, and Japan. It is supplied in the form of a white powder where the 1-MCP is encapsulated in a molecular encapsulating agent such as alpha-cyclodextrin,[10] and the controlled release of the 1-MCP is facilitated through addition of water to this complex.

One of the drawbacks in the use of 1-MCP is the need for a dedicated fumigation room or chamber to treat the produce. For fruit that typically passes through a ripening room, such as bananas and tomatoes, this may not be a problem, but for other horticultural produce there will be a need for some capital investment to provide such rooms. In an attempt to overcome this limitation, Joseph Hotchkiss and his team at Cornell University are developing a packaging film for the controlled release of an ethylene inhibitor during storage and transportation. Hotchkiss reported that an increased rate of release of 3-methylcyclopropene (3-MCP), an isomer of 1-MCP, results when the packaging film is exposed to increasing RH.[11] Although 3-MCP can act as an ethylene receptor blocking agent, it has been reported that, at a given concentration, it is less effective than 1-MCP.[12]

4.3 Microbial control

Antimicrobial packaging technologies that require direct contact with the packaged product have been investigated for their potential use with fruits and vegetables,[13] but technologies that release volatile or gaseous microbial control agents are preferred due to the typically limited contact of the produce with the package surfaces.

4.3.1 Sulfur dioxide release

Sulfur dioxide (SO_2) is an effective gaseous microbial agent and has been used for more than 80 years to control the growth of the fungus *Botrytis cinerea*, which causes grey mold on table grapes. Without fumigation with sulfur dioxide, long-term storage of table grapes would not be possible. Fumigation of table grapes is often conducted in cold storage rooms, but can also be effected *in situ* after the grapes have been packed into a fiberboard carton through the use of sulfur dioxide-releasing pads. An example of these pads is the OSKU Grape Guard pads marketed by Quimica Osku S.A. in Chile.[14] The mechanism of sulfur dioxide release from these pads is based on the reaction of water with the sodium metabisulfite incorporated into paper or plastic film layers, or sachets. So-called single-releasing (slow or fast) or dual-releasing (slow and fast) pads are available depending on the length of storage required, and according to Mustonen,[15] a storage life for table grapes of 8 to 10 weeks at 0°C is achievable using the Grape Guard pads.

The rate of release of sulfur dioxide is highly dependent on the RH, and one of the major drawbacks with the use of these pads is the potential for a sudden catastrophic release of sulfur dioxide as a result of condensation occurring upon a sudden decrease in temperature during storage and distribution. High levels of sulfur dioxide can result in undesirable bleaching of the fruit, making them unacceptable for sale. Furthermore, in 1989, a residue tolerance level of 10 ppm for sulfur dioxide was introduced by the U.S. Environmental Protection Agency (EPA) in response to the findings that certain individuals, in particular severe asthmatics, can be hypersensitive to sulfite residues, such as those resulting from sulfur dioxide fumigation.

Several approaches to developing a plastic-based packaging film for the controlled release of sulfur dioxide have been reported.[16,17] In the technology developed by Steele and Zhou at Food Science Australia,[16] the packaging film typically comprises two laminated films, an outer layer containing the sulfur dioxide-releasing compound, calcium sulfite, and an inner layer containing an organic acid such as citric acid or stearic acid. Moisture from the produce is absorbed by the inner film containing the acid, which in turn causes the migration of hydrogen ions from the acid compound to the outer layer containing calcium sulfite. Reaction of the acid with the calcium sulfite results in the liberation of sulfur dioxide, and it was shown[16] that the rate of sulfur dioxide release can be controlled through appropriate selection of the pK_a of the acid and the moisture permeability of the inner film. The chemical reaction sequence used in this approach is summarized in Figure 4.3.

The system developed by Corrigan[17] is based on the action of water on sodium metabisulfite blended into the polymer, and the rate of release of

$$\text{Organic acid} + H_2O \longrightarrow H_3O^+ + Org^-$$

$$H_3O^+ \text{(inner film)} \longrightarrow H_3O^+ \text{(outer film)}$$

$$Ca_2SO_3 + H_3O^+ \longrightarrow Ca_2^+ + H_2SO_3$$

$$H_2SO_3 \longrightarrow H_2O + SO_2\uparrow$$

Figure 4.3 Chemical reaction sequence for controlled release of sulfur dioxide from packaging.

sulfur dioxide from the film is controlled by selecting polymer blends with appropriate water vapor transmission rates.

4.3.2 Chlorine dioxide release

Chlorine dioxide (ClO_2) is a broad-spectrum biocidal gas that is used commonly today for disinfection in both the medical and food industries. For example, it has recently been approved for use in the washing medium for cleaning fruits and vegetables and is used worldwide for the treatment of drinking water. The mode of microbial inactivation of chlorine dioxide is primarily through disruption of cell membranes, which prevents transportation of nutrients, and there is no known resistance to the effects of this agent by mutant strains.

MicroActive Corporation's Microsphere® technology is designed to enable the sustained release of chlorine dioxide and is composed of an encapsulated core material containing anions such as sodium chlorite that are capable of reacting to form chlorine dioxide gas.[18] This core is surrounded by a hydrophobic layer containing an acid-releasing component. In a fashion similar to that of the sulfur dioxide system discussed above,[16] the release of chlorine dioxide is triggered by the movement of hydrated protons (formed on hydrolysis of the acid component) from the outer layer to the inner core. The rate of acid release is controlled by temperature and humidity levels, and this in turn controls the rate of chlorine dioxide release. As well as being triggered by moisture, a light-triggerable version has also been developed. The Microsphere system can be incorporated into films, coatings, and adhesives, or used as powders, and has approval for food contact applications.

4.4 Active MAP

The atmosphere inside a sealed package containing a respiring product changes with time due to the consumption of oxygen and production of carbon dioxide that occurs as a result of the respiration process, until eventually an equilibrium concentration of these gases is established in the atmosphere within the package. The equilibrium concentrations of these gases in the package headspace are governed by a complex function of the intrinsic respiration rate of the product, the weight of produce in the package, and the permeability characteristics of the package to these gases. The intrinsic respiration rate of the produce and the permeability characteristics of the package increase with temperature, and so can potentially vary considerably over the storage life of the product, depending on the integrity of the cold chain during storage and distribution of the packaged product.

The respiration rate of the packaged product decreases as the oxygen in the headspace is consumed, but to achieve the specific oxygen and carbon dioxide levels required in the package headspace to provide the optimal shelf-life for the particular fruit and vegetable, it is necessary to use a packaging material that has the necessary permeability characteristics for these optimal gas concentrations to be developed and maintained inside the package, without the onset of anaerobiosis, which will result in spoilage and reduced shelf-life. This technique of utilizing a combination of packaging permeability and respiration rate to generate optimal headspace concentrations of oxygen and carbon dioxide is referred to as equilibrium or passive modified atmosphere packaging (MAP).

A number of innovative passive packaging technologies, such as the incorporation of microsized inorganic particles or microperforations into materials such as polyethylene, have been developed to enable the production of materials having gas permeability characteristics that can be customized for particular products. The porous inorganic particles and perforations provide a facile route for gas transmission through the packaging film, and so materials having a wide range of gas permeability characteristics can be conveniently produced by simply varying the amount and size of the particles or perforations. A shortcoming of this type of approach is the inability of these materials to deal with the variability in respiration rate of the produce that invariably occurs due to temperature fluctuations in the distribution chain.

One method of overcoming this limitation for packaged respiring products is through the use of packages having holes or pores over which are positioned substances that undergo a phase change (for example, solid to liquid) at a given threshold temperature.[19] In this way, the overall permeability of the package to oxygen and carbon dioxide increases with temperature, thereby assisting in maintaining the optimal headspace gas concentrations inside the package. A more sophisticated version of this concept has been developed in which temperature-responsive plastic membranes made from side chain crystallizable polymers are affixed over holes in packages.[20] An example of the commercial uptake of this technology is the recent announcement by Chiquita Brands International of the use of Landec Corp.'s Intellipac® membrane technology in packaging for distribution of single-serve bananas to convenience stores in the U.S.

However, a significant deficiency in the use of these equilibrium or passive MAP techniques for optimization of shelf-life is the delay after packaging in establishing the optimal headspace gas composition at the low temperatures typically required for storage of many types of respiring produce. Although the length of this delay is dependent on factors such as initial package headspace and the respiration rate of the produce, it has been reported,[21] for example, that the time required to attain the equilibrium headspace gas composition in packages of broccoli stored at 1.5°C is around 1 week, and for blueberries stored in the temperature range of 0 to 5°C, this delay has been predicted to be as long as 2 to 3 weeks.[22] During this period the product is exposed to suboptimal gas concentrations, and this offsets the benefits of this approach to some extent.

So-called active MAP refers to the use of physical or chemical means to rapidly replace the initial headspace gas with a gas composition that has a concentration of oxygen and carbon dioxide that is closer to the optimal levels (2 to 5% oxygen/3 to 10% carbon dioxide) in order to avoid prolonged exposure of the product to suboptimal headspace gas concentrations. This can be achieved by simply flushing the package headspace with a suitable gas mixture; however, more recently chemical approaches have been reported that have the commercial advantage of potentially providing a means of avoiding the slow and costly gas-flush step.

4.4.1 Active scavenging systems

Oxygen-scavenging sachets have been used commercially for more than 25 years, and their use for extending the shelf-life of a wide range of packaged foods is well documented. Although the potential for use of active oxygen-scavenging systems for enhancing the shelf-life of fruits and vegetables has been recognized for some time,[23] it was not until recently that the potential use of this approach was investigated experimentally. Results reported by Charles and coworkers[24,25] demonstrate that incorporation of iron-based oxygen-scavenging sachets can produce a rapid depletion of the oxygen level in the headspace of tomatoes and mushrooms, thereby resulting in a substantial reduction in the time required to achieve equilibrium oxygen and carbon dioxide concentration. In their work, it is reported that incorporation of an oxygen-scavenging sachet into packages containing tomatoes or mushrooms results in a 50% reduction in the time to reach equilibrium oxygen and carbon dioxide concentrations. Furthermore, they reported that the harmful transient peak in carbon dioxide concentration observed in the absence of the sachet was almost completely eliminated though incorporation of a sachet in the package. Clearly, careful matching of oxygen-scavenging capacity and rate with the respiration characteristics of the produce and the permeation properties of the packaging material is required to avoid onset of anaerobiosis.

The results of investigations into the potential use of carbon dioxide-absorbing sachets, such as those containing calcium hydroxide, for extending the shelf-life of packaged carbon dioxide-sensitive produce such as strawberries and mushrooms have been reported by Day and Brydon[26] and Charles and coworkers.[25] It was concluded from these studies that the use of carbon dioxide-scavenging sachets alone appears to have little impact on the storage life of these products, although substantial storage life extension was observed for packaged strawberries when the sachets were used in combination with high initial oxygen levels.[26]

4.4.2 *Combination active MAP*

It has been reported by Day[27] that replacement of the air in the headspace of packages of respiring produce by physically flushing with a gas mixture comprising oxygen concentrations in the range of 70 to 100%, sometimes referred to as active high-oxygen MAP, can provide enhanced microbiological and nutritional benefits compared with low-oxygen MAP systems, which typically contain oxygen concentrations in the range of 2 to 5%, without the risk of development of hazardous anaerobic conditions that would result in undesirable fermentation reactions and potential proliferation of pathogenic microorganisms.

This use of high initial levels of oxygen is often accompanied by concomitant buildup of substantially higher levels of carbon dioxide in the package than observed for equilibrium or low-oxygen MAP, which could be potentially deleterious to the shelf-life of carbon dioxide-sensitive produce. EMCO Fresh Technologies Ltd. has pioneered the development of combined oxygen-emitting/carbon dioxide-scavenging sachets, the use of which has been reported[26,28,29] to be effective in extending the beneficial high oxygen levels in high-oxygen MA packages while minimizing the accumulation of excessive carbon dioxide levels. However, the overall extension in shelf-life beyond that achieved with high-oxygen MAP alone, as determined on the basis of sensory and microbiological evaluations, was found to be highly dependent on the type of produce.[26]

4.4.3 *Plastics-based active MAP*

Intense technical development of oxygen-scavenging plastics for flexible and rigid food packaging applications has been in progress for more than a decade, and a number of products based on these technologies are now becoming available commercially. In principle, these materials should provide the same potential shelf-life benefits as the active MAP techniques based on the use of sachets or physical headspace gas replacement, although to the best of our knowledge no reports have been published to indicate that this is being investigated. The advantages of plastic-based active packaging over sachets for optimizing the storage life of a wide range of foodstuffs have been well documented,[30] and many of these advantages apply to fruits and vegetable products.

One distinct advantage of plastic-based materials compared with sachets is that they can be used to prolong the shelf-life of fruit- and vegetable-based beverages. Several oxygen-scavenging technologies suitable for use in polyethylene terephthalate (PET) containers have been developed, some of which are available commercially. Food Science Australia's generic oxygen-scavenging technology is based on the reactions of reduced organic compounds with oxygen[31] and is suitable for use in a wide range of conventional packaging materials and applications. Incorporation of this technology into PET containers, either in the PET[32] or in a buried layer of a conventional gas barrier material such as MXD6,[33] can provide a glass-like oxygen barrier for a substantial period, where the duration of this barrier enhancement depends on the level of oxygen scavenger incorporated into the wall of the container. The use of oxygen-scavenging packaging materials is playing a key role in enabling the current trend away from the use of glass and metals toward plastic containers for the packaging of oxygen-sensitive fruit- and vegetable-based beverages.

Plastic materials having combined oxygen-scavenging/carbon dioxide-releasing capabilities have also been explored at Food Science Australia.[34] In this case, ethyl cellulose was used as the base substrate that contained the active agents. The oxygen-scavenging functionality of the polymer film was achieved using the visible light-driven photooxidation of a sensitizer molecule, tetraphenylporphine, to produce short-lived and highly reactive singlet-oxygen molecules. The singlet-oxygen molecules produced in this reaction then proceed to react with furoic acid also incorporated in the film containing the sensitizer, leading to the formation of an endoperoxide derivative of furoic acid that releases carbon dioxide upon rearrangement. As shown in Figure 4.4, a feature of this system is that, on exposure to light, the oxygen in the package headspace can be replaced rapidly with carbon dioxide, with the total headspace volume of the package remaining constant. In effect, the package respires in competition with the produce.

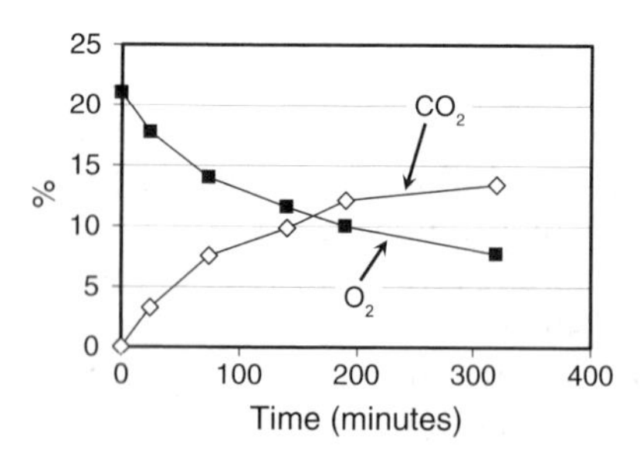

Figure 4.4 Rapid oxygen scavenging/carbon dioxide release using a plastic film containing a singlet-oxygen-sensitizing dye and furoic acid.

4.5 Humidity and condensation control

The storage life of fruits and vegetables can be significantly reduced if the control of moisture levels within the package is inadequate. Moisture loss from packaged produce through evaporation from the surface of the produce followed by transmission of the water vapor through the package wall can result in desiccation of the product. Furthermore, the development of elevated relative humidity inside a package due to respiration or use of materials having low permeability to water vapor can cause condensation, which can then lead to reduced quality and safety of the produce due to microbial proliferation. A number of commercial products are available for humidity control for a range of applications, including Grace Davison's Condensation-Gard® sachets and Pitchit Film from Showa Denko.

One of the earliest approaches to developing a plastic-based active packaging material for this purpose was that of Patterson and Joyce,[35] which involves the use of a multilayer package wall comprising a layer of moisture-absorbent material, such as PVOH, or a cellulosic fiber-based material like paper, sandwiched between an outer layer that is impermeable to water vapor and liquid water, such as polyethylene, and an inner layer that is hydrophobic but permeable to water vapor and that is spot-welded to the moisture-absorbent layer. A schematic representation of this technology in the form of a liner bag for use in a fiberboard carton is shown in Figure 4.5. This structure prevents condensation by allowing the absorbent layer to take up water vapor when the relative humidity increases as a result of reduced temperature and prevents dehydration of the produce by releasing water vapor into the headspace of the package when the relative humidity

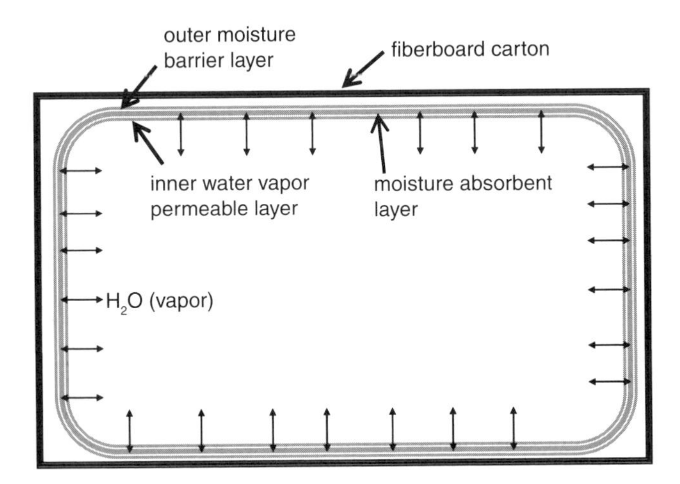

Figure 4.5 Schematic representation of CSIRO's MCT technology as a plastic carton liner.

decreases in response to increasing temperature. This approach has undergone further recent improvement[36] and forms the basis of CSIRO's Moisture Control Technology (MCT), which has been reported[37] to provide substantial extension to the storage life of oranges and cauliflower in trial shipments from Australia to the U.S. and Singapore, respectively.

4.6 Odor control

A number of active packaging technologies have been developed to remove odors associated with the volatile compounds that can accumulate inside a package as a result of food degradation, such as aldehydes, amines, and sulfides, and these technologies have been reviewed thoroughly by Brody et al.[2]

However, some produce, such as onions and potatoes, have a naturally pungent odor. In these cases, a primary concern is the prevention of cross-contamination during storage and transportation of mixed loads containing these commodities. Durian has a particularly strong odor and has been banned on all international airlines and in virtually all mixed storerooms, which has severely restricted the distribution of this fruit despite its being highly prized in Southeast Asia for its exceptional flavor. One of the problems in designing an odor-proof package for fresh produce like durian is that the produce needs to respire, so the package must allow movement of respiratory gases but restrict the transport of the volatile odor compounds. Morris[38] designed an odor-proof package for the storage and transportation of durian in particular in which the package is comprised of an odor-impermeable plastic, such as polyethylene terephalate (PET) or polyethylene of a suitable thickness, to prevent the transmission of odor, together with a port to allow for the passage of respiratory gases. Over this port is affixed a sachet containing an odor-absorbing material made from a mixture of charcoal and about 10% w/w nickel. This approach was found to be successful in laboratory trials in preventing the escape of principal odor components of durian, namely propanethiol, ethyl disulfide, and methyl-methylbutyrate, and performed well in overall odor-proofing of packaged durian.

4.7 The future

Because postharvest produce is living and highly perishable, packaging of it presents unique challenges compared with many other food products. As a consequence, matching the properties of the package to the needs of the produce has often been difficult to achieve and has involved compromises. The package must not only act as an inert barrier to the external environment, but also take into consideration respiration issues. The demands on package performance continue to increase, influenced in particular by market and social changes, and there is a need for packaging innovations such as active packaging that can further enhance the performance of the package such that the properties of the package more closely match the needs of the produce. For instance, the growing demand by consumers for packaged fresh-cuts introduces further challenges, such as the need to deal with elevated microbial loads, as well as exacerbating condensation and respiration issues. The demand for longer shelf-lives in order to access distant markets and the increasing demand by major retailers for produce in shelf-ready packaging are other examples of future pressures on packaging.

It is likely that the trend of integrating active packaging technologies into the packaging material itself will continue, eventually culminating in the development of a truly interactive package that responds directly to the needs of the produce. The feasibility of such interactive packaging is approaching rapidly with the recent advances in the development of flexible electronic components, such as batteries and sensors. Incorporation of active packaging technologies into edible coatings is another relatively unexplored area that has the potential to grow in the future.

Although a number of sachet-based active packaging technologies are used commercially for the preservation of fruits and vegetables, further uptake of active packaging technologies (especially plastic-based materials) will be dependent on the demonstration of a clear benefit to the horticultural industry through increased shelf-life and quality to justify any cost associated with their implementation. This is a major challenge in an industry where there are typically low profit margins.

References

1. Rooney, M.L., *Active Food Packaging*, Blackie, Glasgow, 1995.
2. Brody, A.L., Strupinsky, E.R., and Kline, L.R., *Active Packaging for Food Applications*, Technomic, Lancaster, PA, 2001.
3. Rooney, M.L., Introduction to active food packaging technologies, in *Innovations in Food Packaging*, Han, J.H., Ed., Elsevier Academic Press, San Diego, 2005, chap. 5.
4. Vermeiren, L., Devlieghere, F., van Beest, M., de Kruijf, N., and Debevere, J., Developments in the active packaging of foods, *Trends in Food Science and Technology*, 10, 77, 1999.
5. Takashi, H., Japanese Patent 2113849, 1990.
6. Matsui, M., U.S. Patent 4,847,145, 1989.
7. Holland, R.V., International Patent Application WO 91/04292, 1991.
8. Horsham, M.A., Murphy, J.K.G., and Santangelo, R., International Patent Application WO 2004/07654, 2004.
9. Blankenship, S.M. and Dole, J.M., 1-Methylcyclopropene: a review, *Postharvest Biology and Technology*, 28, 1, 2003.
10. Daly, J. and Kourelis, B., U.S. Patent 6,017,849, 2000.
11. Hotchkiss, J., Innovative Approaches to Active and Intelligent Packaging, paper presented at Proceedings of the BARD-sponsored International Workshop on Active and Intelligent Packaging for Fruits and Vegetables, Shepherdstown, West Virginia, September 2004.
12. Sisler, E.C., Serek, M., Dupille, E., and Goren, R., Inhibition of ethylene responses by 1-methylcyclopropene and 3-methylcyclopropene, *Plant Growth Regulation*, 27, 105, 1999.

13. Shetty, K. and Dwelle, R., Disease and sprout control in individual film wrapped potatoes, *American Potato Journal*, 67, 705, 1990.
14. Quimica Osku S.A., http://www.osku.cl/oskuen/portal/jsp/productos/generadores/index.jsp?subsec=Grape%20Guards, accessed June 5, 2006.
15. Mustonen, H.M., The efficacy of a range of sulfur dioxide generating pads against *Botrytis cinerea* infection and on out-turn quality of Calmeria table grapes, *Australian Journal of Experimental Agriculture*, 32, 389, 1992.
16. Steele, R.J. and Zhou, J.X., International Patent Application WO 94/10233, 1994.
17. Corrigan, P., International Patent Application WO 00/03930, 2000.
18. MicroActive Corporation, http://www.mac.barriersafe.com/Patents.asp, accessed June 20, 2006, and patents listed therein.
19. Patterson, B.D. and Cameron, A., International Patent Application WO 92/21588, 1992.
20. Clarke, R., Intelligent Packaging for Safeguarding Product Quality, paper presented at Proceedings of the 2nd International Conference on Active and Intelligent Packaging, Chipping Campden, U.K., September 2002.
21. Christie, G.B.Y., Macdiarmid, J.I., Schliephake, K., and Tomkins, R.B., Determination of film requirements and respiratory behavior of fresh produce in modified atmosphere packaging, *Postharvest Biology and Technology* 6, 41, 1995.
22. Cameron, A.C., Chowdry Talasila, P., and Joles, D.W., Predicting film permeability needs for modified-atmosphere packaging of lightly processed fruits and vegetables, *HortScience*, 30, 25, 1995.
23. Rooney, M.L., Active and Intelligent Packaging of Fruits and Vegetables, paper presented at Proceedings of the International Conference on Fresh-Cut Produce, Chipping Campden, U.K., September 1999.
24. Charles, F., Sanchez, J., and Gontard, N., Active modified atmosphere packaging of fresh fruits and vegetables: modeling with tomatoes and oxygen absorber, *Journal of Food Science*, 68, 1736, 2003.
25. Charles, F., Sanchez, J., and Gontard, N., Modeling of active modified atmosphere packaging of endives exposed to several postharvest temperatures, *Journal of Food Science*, 70, E443, 2005.
26. Day, B.P.F. and Brydon, L., *Active Packaging for Chilled Fresh Prepared Produce and Combination Products*, R&D Report 156, Campden and Chorleywood Food Research Association, Chipping Campden, U.K., 2002.
27. Day, B.P.F., High Oxygen MAP for Fresh Prepared Produce and Combination Products, paper presented at Proceedings of the International Conference on Fresh-Cut Produce, Chipping Campden, U.K., September 1999.
28. Parker, N., Innovative Oxygen Emitting/Carbon Dioxide Scavenging Technology for Fresh Produce Applications, paper presented at Proceedings of the 2nd International Conference on Active and Intelligent Packaging, Chipping Campden, U.K., September 2002.
29. Brydon, L., *Active Packaging for Chilled Fresh Prepared Produce and Combination Products: Part 2*, R&D Report 201, Campden and Chorleywood Food Research Association, Chipping Campden, U.K., 2004.
30. Scully, A. and Horsham, M., Emerging packaging technologies for enhanced food preservation, *Food Science and Technology*, 20, 16, 2006.
31. Rooney, M.L., International Patent Application WO 94/12590, 1994.

32. Scully, A., Barrier performance of a new triggerable oxygen scavenging PET, in *Nova-Pack Asia 2004 Proceedings*, Shanghai, China, Schotland Business Research, 2004, p. 67.
33. Horsham, M.A, Scully, A.D., Murphy, J.K.G., Santangelo, R.A., and McNally, M., International Patent Application WO 2006/000055, 2006.
34. Zerdin, K. and Rooney, M., unpublished results, 2000.
35. Patterson, B.D. and Joyce, D.C., International Patent Application WO 94/03329, 1994.
36. Gibberd, M.R. and Symons, P.J., International Patent Application WO 05/053955, 2005.
37. CSIRO Plant Industry Communication Group, http://www.csiro.au/files/files/p2if.pdf, accessed June 2006.
38. Morris, S., Odour-Proof Package, WO 99/25625, 1999.

chapter five

Modified atmosphere packaging for vegetable crops using high-water-vapor-permeable films

Nehemia Aharoni, Victor Rodov, Elazar Fallik, Uzi Afek, Daniel Chalupowicz, Zion Aharon, Dalia Maurer, and Janeta Orenstein

Contents

Abstract

This review summarizes more than 10 years of research and development in the area of combined modified atmosphere and modified humidity packaging of fresh produce performed in Israel by collaborative efforts of academic and industrial research teams.

In many cases, the commercialization of modified atmosphere packaging (MAP) is limited due to the accumulation of condensed water in the package causing increased pathological and physiological disorders. In a joint venture between the Agricultural Research Organization–The Volcani Center and StePac L.A. we have developed a series of plastic films having a higher permeability to water vapor than most commercially available MAP products. A desired in-pack relative humidity was obtained using Xtend® films manufactured from various proprietary blends of polyamides with other polymeric and nonpolymeric compounds. A beneficial modified atmosphere was achieved in the package by microperforation of the film. Film composition and extent of microperforation were tailored in accordance with the respiratory activity and weight of the produce packaged, anticipated temperature fluctuations during storage and shipment, and expected physiological and pathological responses of the produce to CO_2/O_2 concentrations and humidity levels inside the package.

The effectiveness of Xtend films in maintaining quality of MA-packed fresh produce was greater than that of other commercially available films such as polyethylene, polypropylene, and polyvinyl chloride. The microperforated Xtend packaging allowed the formation of a desirable modified atmosphere, retarding ripening and senescence of the produce. Additional beneficial effects of Xtend films included reduction of decay, chilling injury, leaf elongation, leaf sprouting, tissue discoloration, peel blemishes, and formation of off-odors, and inhibition of bacterial growth on the produce surface. Most of the technologies based on the Xtend packaging reviewed in this chapter have been successfully implemented in postharvest practice in Israel, the U.S., and other countries.

5.1 *Benefits and hazards of MAP*

Although the principles of modified atmosphere packaging (MAP) of fresh produce are well known, the technology is still applied mainly in retail packages characterized by a high ratio between the film surface and produce weight. The method is still limited in bulk packaging, especially with highly respiring products, because of the difficulties in controlling in-package atmosphere and humidity during commercial shipments. These problems very often occur when the produce is exposed to temperature fluctuations during storage or shipment.

The beneficial and hazardous effects of MAP in a range of fruits and vegetables have been extensively reviewed (Barkai-Golan, 1990; Church and Parsons, 1995; Day, 1993; El-Goorani and Sommer, 1981; Geeson, 1988; Isenberg, 1979; Kader, 1986; Kader et al., 1989; Kader and Saltveit, 2003; Mir and Beaudry, 2004; Zagory and Kader, 1988).

A reduced level of O_2 or an elevated level of CO_2 can reduce respiration, decrease ethylene production and action, retard tissue ripening and softening, retard chlorophyll degradation and biosynthesis of carotenoids and anthocyanins, reduce enzymatic browning, alleviate physiological disorders and chilling injury, retard development of decay, and maintain nutritional quality of fresh produce. The effect of decreased O_2 and increased CO_2 levels on senescence and ripening processes are additive and can be synergistic (Kader et al., 1989; Kader and Saltveit, 2003; Zagory and Kader, 1988). Elevated CO_2 in MAP can nullify the enhancing effect of ethylene on ripening and senescence processes (Aharoni et al., 1986; Herner, 1987; Kenigsbuch et al., 2007; Rodov et al., 2002). Additionally, the beneficial effect of CO_2 in some physiological processes is independent of ethylene action (Philosoph-Hadas et al., 1993). In leaves, elevated CO_2 concentrations, which effectively retarded chlorophyll loss, were found to promote ethylene production through blocking the autoinhibitory effect of ethylene in the first phase of senescence (Aharoni et al., 1979).

In vegetative tissues optimal MAP can also reduce leaf regrowth (green onion and leek), stem toughening (asparagus), and leaf sprouting and rooting in root vegetables (parsnip, radishes). The delay of ripening and senescence of fruits and vegetables also reduces their susceptibility to pathogens (Barkai-Golan, 1990; El-Goorani and Sommer, 1981).

It seems that in many cases the inhibition of pathogenesis in MAP is not due to a direct suppression of the pathogens since it would require O_2 levels below 1% or CO_2 levels above 10%. Most fruits and vegetables do not tolerate such concentrations of CO_2 and O_2. Therefore, the beneficial effect of MAP in suppressing pathogen development at lower concentrations of CO_2 and higher concentrations of O_2 is indirect, through retardation of ripening and senescence of the host (Barkai-Golan, 1990).

Nevertheless, some fresh fruits and vegetables, such as strawberry, sweet corn, asparagus, cantaloupes, mushrooms, spinach, lychee, cherry, blueberry, and blackberry, are tolerant to CO_2 concentrations within the fungistatic range (>10%) (Kader and Saltveit, 2003). Elevated CO_2 concentrations make plants more sensitive to a low level of O_2 (Beaudry, 1999, 2000). On the other hand, it is plausible that with an increase in O_2 concentration the plant or plant organ will be more tolerant to high CO_2. Therefore, packaging having high levels of CO_2 and medium levels of O_2 (>5%) could be useful in the retardation of both physiological and pathological processes.

In those fruits and vegetables that tolerate high concentrations of CO_2, levels above 10% in MAP can inhibit spore germination or mycelial growth of certain fungi (Barkai-Golan, 1990; Brown, 1922). The bacterial and fungal growth inhibition of CO_2 is dependent upon its dissolution in the packaged product. The solubility of CO_2 is inversely proportional to storage temperatures, and its effect therefore increases as temperature decreases (Brown, 1922; Church and Parsons, 1995; El-Goorani and Sommer, 1981; Phillips, 1996; Werner and Hotchkiss, 2006).

While MAP can improve the storability of fruits and vegetables, the potential hazards of inducing undesirable effects should be taken into consideration. Packing of fresh produce in an impermeable film may result in depletion of O_2 levels lower than 2 to 4% and accumulation of CO_2 to undesirable levels. The depletion of O_2, especially when accompanied with a high level of CO_2, results in anaerobic respiration with the accumulation of acetaldehyde, ethanol, ethyl acetate, and lactic acid, all products of fermentation, contributing to the development of off-odors, off-flavors, and tissue deterioration (Kays, 1997; Mattheis and Felman, 2000).

Extreme changes in O_2 and CO_2 to intolerable levels often occur when fresh produce packed in improper MAP is exposed to higher temperatures during shipment and distribution. Other undesirable effects of improper MAP are initiation or aggravation of certain physiological disorders, irregular fruit ripening, and increased susceptibility to decay following physiological damage (Kader and Saltveit, 2003).

5.2 *Polymeric films for MAP*

The most commercially used polymeric films for fresh produce are low-density polyethylene (LDPE), polypropylene (PP; cast or oriented), and polyvinyl chloride (PVC). The permeability to CO_2 in these films is usually in three to six times greater than that of O_2. These films are suitable for fruit and vegetables with low respiration rates that are packed in consumer packaging and kept at low temperatures. Under these conditions, due to the high ratio between film surface area and produce weight, a steady state of low O_2 level and nondetrimental level of CO_2 can be achieved. However, after transfer of the packaged produce to higher temperature, anaerobic respiration is likely to occur. This situation is typical for produce with a medium–high respiration rate (Hardenburg, 1971; Zagory and Kader, 1988). For these fresh produce items, most of the commercially available films used for MAP do not adequately provide gas fluxes or selectivity for achieving optimal conditions, especially in bulk packaging (Exama et al., 1993).

Increased permeability of packaging was achieved by using a composite film comprising ethylene vinyl acetate (EVA), LDPE, and oriented polypropylene (OPP). This packaging is suitable for fresh-cut produce (Ahvenainen, 1996). Another way of increasing gas exchange in plastic packaging is by mixing the polymeric film with inert inorganic material such as $CaCO_3$ and SiO_2, thereby generating microporous films. Additional technologies for increasing the oxygen transmission rate (OTR) are detailed in other chapters dealing with active and intelligent packaging.

5.3 Controlling in-pack atmosphere by microperforation

The most common way for increasing the OTR of plastic films, used in a wide range of fresh produce, is the addition of microperforations by electrostatic discharge, laser technology, or mechanically by small needles. The diameter of the microperforation generally ranges from 40 to 250 μm. It is therefore possible to tailor a specific microperforated polymeric film (number and size of the holes) in accordance with the rate of respiration of the produce, permeability of the film, film surface area as related to produce weight, void volume of the packaging, storage temperature, and anticipated fluctuation in temperature during shipment. This technology is also suitable for high-respiring fresh produce such as strawberry, asparagus, broccoli, leek, fresh herbs, sweet corn, mushrooms, and others, that are packed in either consumer or bulk packaging and can tolerate medium–high levels of CO_2 (Aharoni et al., 1993a, 1996b, 1997; Aharoni and Richardson, 1997; Emond et al., 1991; Geeson, 1989, 1990; Geeson et al., 1988; Rodov et al., 2000, 2002; Rooney, 1995). For such vegetables, due to microperforation technology, it is possible to achieve high levels of CO_2 and safe levels of O_2, thereby preventing anaerobic respiration.

When using microperforation technology, the OTR of most polymeric films is not as important as it is in nonperforated film, since total gas flow through the holes is much greater than movement through the film. Gas transmission through microperforated and perforated films has been modeled (Cameron et al., 1995; Emond et al., 1991; Fishman et al., 1996; Lee and Renault, 1998; Mannapperuma and Singh, 1994; Renault et al., 1994a, 1994b).

In contrast to continuous films having several-fold higher permeability to CO_2 than to oxygen, the barrier properties of perforations are roughly similar for both gases. Therefore, after steady state is achieved, the gradients of CO_2 and O_2 between the in-package atmosphere and the surrounding air are about the same, on condition that the respiratory quotient of the produce is close to 1 (normal case of aerobic respiration). In this situation, the sum of O_2 and CO_2 concentrations in the package is expected to be not far from 21% (Beaudry, 1999). Accordingly, if the requested O_2 concentration for a given product is 5 to 10%, then the expected steady-state CO_2 range in a microperforated package may be about 11 to 16%. Hence, MAP in microperforated film is suitable for those products that benefit from medium–high levels of CO_2 but not for those requiring low O_2 and low CO_2.

5.3.1 The coriander model

The storability of fresh green herbs is limited due to high metabolic activity expressed by high respiration rate and accelerated leaf senescence and decay. These processes are further enhanced following harvesting, handling, and marketing. Many of the fresh herbs are also sensitive to the senescence-enhancing effect of ethylene (Aharoni, 1994; Aharoni et al., 1989; Cantwell and Reid, 1993; Kenigsbuch et al., 2007; Meir et al., 1992; Philosoph-Hadas, 1989).

Packing yellowing-susceptible herbs such as coriander, dill, chervil, parsley, watercress, chives, and others in nonperforated PE bags resulted in the development of a modified atmosphere capable of retarding yellowing and decay (Aharoni et al., 1993a; Hruschka and Wang, 1979). Elevated concentrations of CO_2 inhibited the senescence-inducing effect of accumulated ethylene in the package, especially when combined with a decreased level of O_2 (Aharoni et al., 1989).

In experiments in which moderate and nonfluctuating temperatures were maintained during both storage and shelf life (6°C and 12°C, respectively) anaerobic conditions did not occur (Aharoni et al., 1993a). However, when herbs are exposed to extreme fluctuations in storage temperature, as is likely to occur during air freight, anaerobic respiration may be induced.

The typical pattern of change in respiratory gases in coriander packed in a PE-lined carton is shown in Figure 5.1. The depletion of O_2 and the accumulation of CO_2 in nonperforated PE packages to intolerable levels resulted in anaerobic conditions characterized by a burst of ethanol, development of objectionable odor, and tissue deterioration. When microperforated PE film was used, anaerobic respiration was prevented. The CO_2 concentration in the microperforated packaging did not exceed 10%, and O_2 concentrations did not drop below 5%. Consequently, the keeping quality of fresh herbs packed in microperforated packaging was much better than that for herbs packed in nonperforated film. The beneficial effect of MAP was also retained after repacking of the herbs for shelf life in ventilated consumer packages (Aharoni et al., 1993a). Loaiza and Cantwell (1997) found that CO_2 concentrations of 5 to 9% (in air) effectively retarded yellowing of coriander stored at 7.5°C for 14 days. However, extending storage duration by an additional 4 days resulted in dark spots on the leaves due to CO_2 injury.

5.4 Controlling humidity and condensation in MAP

Most of the commercially available plastic films used for MAP have very low permeability to water vapors, and consequently, packages are able to maintain high in-pack humidity, even when they are kept at very low

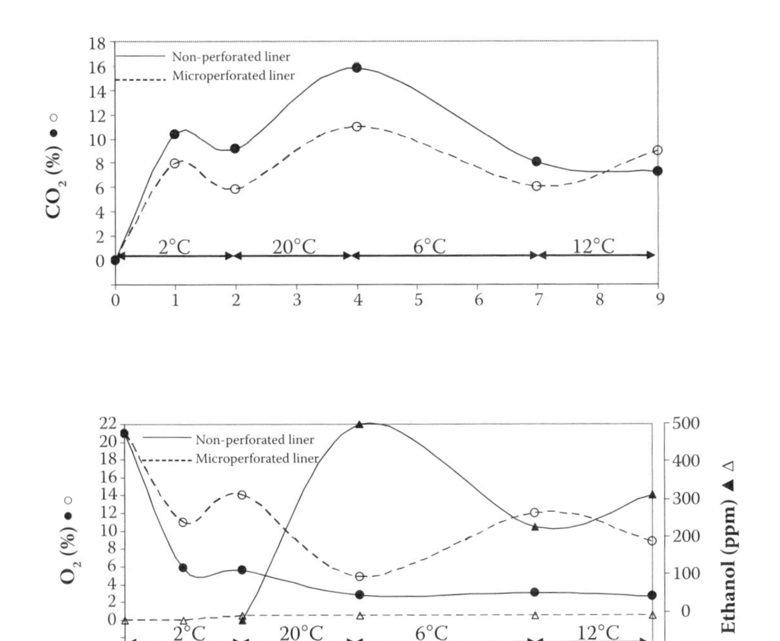

Figure 5.1 Changes in concentrations of CO_2, O_2, and ethanol in the atmosphere of coriander packed in nonperforated or microperforated polyethylene and held under extreme fluctuation in storage temperatures. The percentage of film perforation area was 0.00023%. Data are means of three cartons weighing 1 kg each. (Redrawn from Aharoni, N., *Israel Agresearch*, 7, 35–59, 1994.)

ambient relative humidity (RH). The high humidity in MAP results in reduced transpiration of water from the produce, thereby reducing wilting, shriveling, and loss of firmness. The accumulation of water vapor in the packaging depends on the rate of water loss from the product, its surface area, the water vapor transmission rate (WVTR) of the film, and the external environment temperature (Geeson, 1989). Films with high-water-vapor barriers are essential for commodities with greater exposed surface per unit weight, such as leafy vegetables and prepared cut fruits and vegetables, which have a faster rate of water loss (Day, 1993; Mir and Beaudry, 2004; Zagory and Kader, 1988).

The RH in most sealed packaging is near the saturation level. Therefore, even very small fluctuations in temperature during storage or shipment may result in water condensation on the surface of both film and produce. Condensed water on the produce surface may adversely affect the gas exchange, leading to an unfavorable internal atmosphere (Burton and Wigginton, 1970; Cameron et al., 1995). However, the most deleterious effect of in-pack condensation is enhancement of growth of many plant pathogens, resulting in produce decay. In addition, a combination of high humidity and low O_2 is favorable to the development of human pathogens (Brackett, 1994; Hao et al., 1999; Hintlian and Hotchkiss, 1986; Zagory and Kader, 1988).

Many polymeric films contain antifog additives that prevent visible condensation but do not affect considerably in-pack RH. In antifog films the condensed water may accumulate in the bottom of the package. This can be eliminated by adding moisture-absorbing pads.

Perforation of plastic film is the most common and simple approach to reduce in-pack moisture and condensation. However, perforation that is adequate to achieve effective concentration of CO_2 and O_2 in the package does not significantly affect moisture loss (Ben-Yehoshua et al., 1996; Hardenburg, 1971; Rodov et al., 1995; 1998a). Fishman et al. (1996) found in a model system imitating LDPE packaging that 40 perforations, each 2 mm in diameter, were required to reduce in-pack RH from about 100% to about 95% (70% RH and 20°C in ambient), while only four perforations increased O_2 to the ambient level nullifying the MA effect.

Other methods have been tried in order to achieve optimal levels of both atmosphere and humidity in packaging of fruits and vegetables. These approaches, some of them tested only in laboratory-scale experiments, consist of adding water-absorbing materials either as separate microporous sachets or placed between two layers of plastic polymers having high permeability to water (Barmor, 1987; Patterson et al., 1993; Rooney, 1995; Shirazi and Cameron, 1992). More details on these methods are described in chapters in this book dealing with intelligent and active packaging.

5.5 Controlling of atmosphere and humidity by using hydrophilic films

Some polymeric films having high WVTR are available and may be used for achieving desired modified atmosphere and humidity. Among them are ethylene vinyl alcohol (EVOH), ethylene vinyl acetate (EVA), highly plasticized polyvinyl chloride (PVC), oriented polystyrene (OPS), and various polyamides (PAs; nylon). Some of them are not used as a single film but as laminates or coextruded mixed polymers to improve OTR and WVTR.

Barron et al. (2002) have experimentally tested MAP of mushrooms with two hydrophilic films: biodegradable wheat gluten-based material and polyether polyamide copolymer film (Pebax®, Elf Atochem, France). The best results (low levels of both CO_2 and O_2) were achieved with the wheat gluten hydrophilic film, but its poor mechanical and sealing properties limit commercial application in vegetables. The use of Pebax film resulted in detrimental levels of CO_2 and O_2. Some hydrophilic films have disadvantages, especially when used as a single polymer. Among the disadvantages are excess weight loss when the permeability to water vapor is too high, less mechanical strength, and inability to be used with automatic packaging machinery because of poor sealing.

The most important factors that determine the type of polymeric film to be used are storage and shipment temperature, product respiration rate and quotient, response to levels of CO_2, O_2, and humidity, and product weight. Hence, film packaging that is adequate for consumer packages is not always suitable for bulk packaging and vice versa. So far, no published data show a wide commercial use of hydrophilic films in MAP of fresh produce in bulk packages.

5.6 Controlling physiological and pathological processes by modified atmosphere and modified humidity packaging

5.6.1 Properties of the hydrophilic film Xtend for MAP

Xtend (XF), a hydrophilic plastic packaging, was developed by StePac L.A. Ltd, Tefen, Israel, in cooperation with the Agricultural Research Organization, Ministry of Agriculture, Israel (Israel Patent 112151 and U.S. Patent 6190710). The plastic packaging was manufactured by co-extrusion of proprietary blends consisting of different polyamides with other polymeric and nonpolymeric compounds. The different blends allow manufacturing plastic films with different WVTR, according to the requested RH in a given produce packaging.

In the experiments described in the following sections, three types of XF films (20 µm thick) were used — XF10, XF12, and XF14 — having the following permeance to water vapor: 18×10^{-10}, 6×10^{-10}, and 24×10^{-10} mol S^{-1} m^{-2} Pa^{-1}, respectively. Oxygen permeance of XF10 was 48×10^{-14} and of XF12 and XF14, 24×10^{-14} mol S^{-1} m^{-2} Pa^{-1}. Due to the very low gas transmission rate of the XF films to both O_2 and CO_2, the actual gas transmission characteristics of the packages were determined by microperforations. According to Day (1993) the O_2 transmission rate of microperforated film is $>70.5 \times 10^{-12}$ mol S^{-1} m^{-2} Pa^{-1} at 23°C.

In most cases, similar O_2 and CO_2 transmission rates may be expected through microperforated films because their permselectivity is between 0.72 and 1.0 (Mir and Beaudry, 2004; Oliveira et al., 1998). The degree of perforation in the various packages described in this chapter, expressed as percentage of total perforation area related to film surface, was in the range of 0.00012 to 0.0012% and 0.19% for microperforation and macroperforation, respectively. In most of the experiments, LDPE packaging was used as a control to Xtend packaging. Water vapor and O_2 permeance of nonperforated LDPE film were about 11.8×10^{-11} and 35×10^{-12} mol S^{-1} m^{-2} Pa^{-1}, respectively. The permeability data were provided by the manufacturers and converted into SI units according to Banks et al. (1995).

The unique behavior of the polyamides used is characterized by interaction between water molecules and specific polar amide groups of the polymer that affect the sorption–desorption cycles. The rate of water sorption is greatly increased as water activity in the package and the temperature increase (Hernandez, 1994; Sfirakis and Rogers, 1980). The presence of water within the polymeric matrix reduces the barrier properties for O_2 and for polar organic volatiles such as ethanol (Sfirakis and Rogers, 1980). Therefore, when highly respiring fresh produce is transferred to a higher temperature, these properties may help to maintain O_2 level above the fermentation threshold.

5.6.2 *Retardation of yellowing, decay, and off-odors*

5.6.2.1 *Broccoli as a model*

Broccoli, a highly respiring crop, serves as a good model for studying MAP because the beneficial and hazardous factors in the system can be determined. Therefore, the following review on the effects of MAP on broccoli is described and discussed more widely than the other crops.

Broccoli is a highly perishable vegetable and its storability is very low, especially at temperatures above 5°C (Lieberman and Hardenburg, 1954). Respiration rate and ethylene production, both of which are more intensive in the florets, increase in a climacteric-like pattern and the inflorescence turns yellow due to opening of the flower buds and degradation of chlorophyll in the green tissues. This may be followed by drop of the flowers. In advanced stages of senescence, decay, caused mainly by *Alternaria alternata*, *Botrytis cinerea*, and *Erwinia carotovora*, starts to develop. The flower buds are extremely sensitive to ethylene; therefore, blocking ethylene biosynthesis or action results in retardation of senescence processes (Aharoni et al., 1985; Hyodo et al., 1995; King and Morris, 1992; Tian et al., 1994; Wang, 1977; Yamauchi and Watada, 1998).

Many studies have shown the senescence retarding effect of high CO_2 or low O_2 in CA and MAP of broccoli (Aharoni et al., 1985; Anelli et al., 1984; Barth and Zhuang, 1996; Gillies et al., 1997; Isenberg, 1979; Lebermann et al., 1968; Lieberman and Hardenburg, 1954; Lipton and Harris, 1974; Makhlouf et al., 1989a, 1989b; Wang, 1979; Wang and Hruschka, 1977; Zhuang et al., 1994). However, the commercial implementation of these technologies has very often failed. Packing broccoli in improper plastic film or inadequate ventilation in the storage environment resulted in the development of strong offensive odors and off-flavors (Ballantyne et al., 1988; Kasmire et al., 1974; Lipton and Harris, 1974; Makhlouf et al., 1989a; Wang and Hruschka, 1977). Forney et al. (1991) considered that under anaerobic conditions the sulfurous compound methanethiol (MT) is primarily responsible for the off-odor in broccoli. Hansen et al. (1992) considered that in addition to MT, dimethyl trisulfide and β-ionone could also be major contributors to the offensive off-odor. All three of these metabolites, which are induced by 0.5% O_2, were inhibited by 20% CO_2. Obenland et al. (1994) have found that elevated CO_2 concentrations up to 26.5% inhibited low-O_2-induced MT production. Izumi et al. (1996) have also shown that 10% CO_2 reduced undesirable odors at all storage temperatures tested (0 to 10°C).

Since the undesirable odors have been reported to occur when O_2 was below 1% and CO_2 above 10%, the recommended gas compositions for CA storage at 0 to 5°C are 1 to 2% O_2 and 5 to 10% CO_2 (Cantwell and Suslow, 1999; Makhlouf et al., 1989a; Ryall and Lipton, 1979; Saltveit, 1997). Even though broccoli is relatively tolerant to high CO_2 concentrations, levels of 10% or more, under certain circumstances, may result in the development of off-odors (Kasmire et al., 1974; Makhlouf et al., 1989a). A combination of 6% CO_2 and 2.5% O_2 was found to be adequate for 6 weeks storage at 1°C (Makhlouf et al., 1989a). When fluctuations in temperature are likely to occur during shipment, then the proper MAP seems to be an ideal method for keeping quality of broccoli. It has been found that elevated CO_2 was more effective than reduced O_2 in retarding yellowing (Kasmire et al., 1974; Lebermann et al., 1968; Lipton and Harris, 1974; Makhlouf et al., 1989a; Wang, 1979).

Broccoli heads, when packed in proper polymeric films, can be preserved in good quality for several weeks in cold storage (Aharoni et al., 1985; DeEll and Toivenen, 2000; Forney et al., 1989; Joyce, 1988; Serrano et al., 2006; Wang and Hruschka, 1977). Steady-state concentrations of CO_2 and O_2 depend mainly on storage temperature, produce respiration, its weight, and permeability of the polymeric packaging.

5.6.2.2 Broccoli in retail packaging

In most studies retail film packages for various produce items have been examined under low–moderate temperatures (1 to 10°C) without subsequent exposure to higher temperatures, such as 20°C. Adequate steady-state concentrations of O_2/CO_2 established in the packaging at low–moderate temperatures are not likely to be maintained at higher temperatures. We designed an experiment in which broccoli in retail sealed bags was held in prolonged cold storage (1°C) followed by shelf life at 20°C (Table 5.1 and Figurc 5.2).

Broccoli heads (500 to 600 g) were packed in a nonperforated PVC-wrapped (12-μm-thick) commercial microperforated OPP bag for broccoli (SP/240, "P-Plus," obtained from Sidlaw Packaging Co., England) and microperforated polyethylene (PE) and microperforated Xtend (XF12) bags. The bags were closed with rubber bands. The control consisted of unpacked heads.

As expected, the fastest quality loss was observed in unpackaged heads. The weight loss percentage and yellowing index of the unpacked broccoli heads after 30 days at 1°C followed by an additional 4 days at 20°C were 25.7% and 5.0 (maximum index), respectively (Table 5.1). The extent of yellowing in the PVC-wrapped broccoli was high (index = 3.7) because of the excessive gas permeability of the polymeric film. The CO_2 and O_2 concentrations in the PVC packages during shelf life leveled off at 4 and 11%, respectively (Figure 5.2).

The remarkable reduction of yellowing observed in the three other types of bags is mainly accounted for by the high level of CO_2 during shelf life

Table 5.1 Effect of Various Consumer Packages of Broccoli on Water Condensation, Produce Weight Loss, Yellowing, Decay, and General Appearance[a,b]

Consumer packaging	Condensation index (1–5)[c]	Weight loss (%)	Yellowing index (1–5)[c]	Decay index (1–5)[c]	Appearance index (1–5)[c]
Naked	—	25.7 a	5.0 a	2.5 b	1.0 c
PVC[d] wrapping	1.5	3.6 b	3.7 b	2.4 b	1.5 c
Microperforated SP/240[e]	1.9	0.7 c	2.2 c	1.9 c	2.6 b
Microperforated PE[f]	3.7	0.5 c	2.0 c	2.8 a	2.4 b
Microperforated XF12[f]	2.0	4.0 b	1.5 d	1.3 d	3.2 a

[a] Broccoli heads (cv. Marathon) packed in plastic bags (500 to 600 g each) were checked after 30 days at 1°C followed by 4 days at 20°C.

[b] Data are presented as the mean of 3 cartons, with 10 heads sampled from each. Mean values that are followed by the same letter are not significantly different at $p < 0.05$ according to Duncan's multiple range test.

[c] Indices: 1 = the lowest level; 5 = the highest level. Appearance indices of 2.5 are marketable.

[d] PVC = nonperforated polyvinyl chloride.

[e] SP/240 = Microperforated oriented polypropylene (OPP) "P-Plus," Sidlaw Packaging Co., England.

[f] Total perforation area of PE and XF12 was 0.00023%.

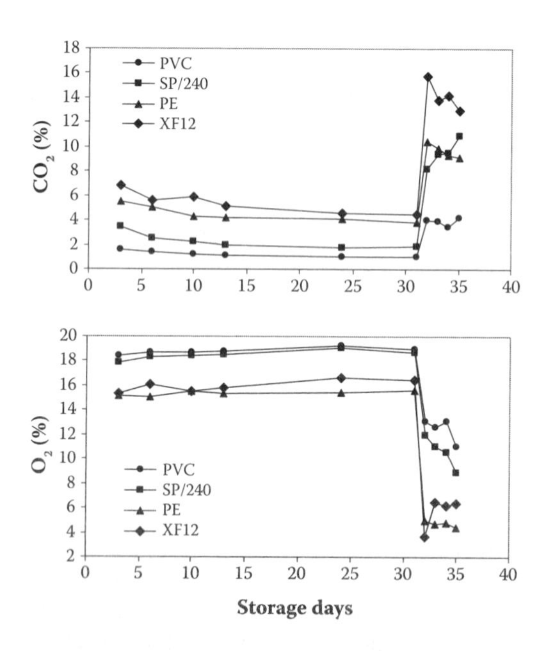

Figure 5.2 Changes in concentrations of CO_2 and O_2 in consumer packages of broccoli during 30 days at 1°C followed by 4 days at 20°C. Heads (500 to 600 g each) were packed in nonperforated polyvinyl chloride (PVC) wrapping and microperforated bags of SP/240 ("P-Plus," Sidlaw Packaging Co., England), polyethylene (PE), and Xtend (XF12). Data are means of three gas samples taken from each of three cartons. Percentages of perforation area of the films are given in Table 5.1.

(9.5 to 16%). However, O_2 concentrations of 4.0 to 6.5% at the end of shelf life, and CO_2 concentrations of 4.0 to 4.5% at the end of cold storage, found in the microperforated PE and XF12 bags, contributed also to the reduction in yellowing. Oxygen levels during cold storage were in the range of 15 to 19% in the various packages.

The lowest level of yellowing and decay was found in the XF12 bags and is attributed to the beneficial concentrations of CO_2 during cold storage (4.5 to 7.0%) and shelf life. CO_2 concentration in the first day at 20°C was 16% and decreased to 13% on the fourth day. In spite of the higher levels of CO_2 in the XF12 bags, which were much above the recommended 10%, no off-odors were perceived. This may be related to the increase of O_2 concentration from 3% in the first day at 20°C to 6% on the fourth day. The increase of O_2 at 20°C could be related to the unique property of the polyamide-based packaging (see Section 5.6.1).

The high humidity and condensation in the PE bags (index = 3.7) resulted in a relatively high level of decay (index = 2.8), whereas the hydrophilic characteristic of the XF packaging, which reduced in-pack humidity, contributed to the lowest level of decay (index = 1.3). In the XF packaging, the weight loss after the 4-day shelf life was 4.0%, which is the maximum permissible weight loss for broccoli (Kays, 1997), but still, the appearance index was significantly higher than in the other packages. StePac L.A. now manufactures a range of hydrophilic films with lower WVTR values that still have reduced condensation. Therefore, weight loss in these bags is expected to also be reduced.

5.6.2.3 Broccoli in bulk packaging

The problem of O_2 depletion and CO_2 accumulation to intolerable levels is more severe in bulk packaging than in retail packaging, since the ratio between film surface area and produce weight in bulk packaging is smaller. Inadequate levels of O_2 and CO_2 might occur very often in bulk packaging during prolonged cold storage or following transfer of the closed packages for distribution at higher temperature, even for a very short time.

Results of an experiment designed to compare several types of bulk packaging are depicted in Table 5.2 and Figure 5.3. Broccoli heads (10 kg) in a plastic-lined carton were packed in a commercial microporous packaging (mineral-impregnated LDPE), microperforated polyethylene (PE), and Xtend (XF12). The sealed bags were stored for 19 days at 1°C followed by 1 day at 20°C. The gas transmission rate of the microporous packaging was improper; after 2 weeks at 1°C, the O_2 level was stabilized at around 4%, but CO_2 concentration gradually increased to 14.5% (Figure 5.3). At this stage, slight anaerobic respiration began, as determined by ethanol production, and became stronger during the next 5 days. After 1 day at 20°C, concentrations of O_2, CO_2, and ethanol in the sealed packaging were 2.7%, 19.5%, and 190 ppm, respectively. At this stage, the broccoli heads remained green without decay (Table 5.2) but developed offensive odors that persisted even after removal of the heads to ventilated conditions.

Table 5.2 Effect of Various Bulk Packages (10 kg) of Broccoli on Weight Loss, Yellowing, Decay, General Appearance, and Development of Off-Odor[a,b]

Packaging	O_2 (%)	CO_2 (%)	Weight loss (%)	Yellowing index (1–5)[c]	Decay index (1–5)[c]	Off-odor	Appearance index (1–5)[c]
Microperforated PE[d]	4.6	14.0	1.1 b	2.6 a	2.1 a	Slight	2.4 a
Microperforated XF12[d]	2.2	18.0	2.0 a	1.9 b	1.4 b	None	3.1 b
Microporous PE[e]	2.7	19.5	1.1 b	2.0 b	1.2 b	Severe	—

[a] Broccoli heads (cv. Stolto) packed in 10-kg plastic-lined cartons were checked after 19 days at 1°C followed by 1 day at 20°C. Details of treatments as in Figure 5.3.

[b] Data are presented as the means of 3 cartons, with 10 heads sampled from each. Mean values that are followed by the same letter are not significantly different at $p < 0.05$ according to Duncan's multiple range test

[c] Indices: 1 = the lowest level; 5 = the highest level. Appearance indices of 2.5 are marketable.

[d] Total microperforation area of PE and XF12 was 0.0007%.

[e] Microporous PE is a mineral-impregnated film obtained from a commercial company.

In the microperforated PE packaging stored at 1°C, O_2 and CO_2 concentrations after 19 days were 17 and 4.2%, respectively. After 1 day at 20°C, O_2 decreased to 4.6% and CO_2 increased to 14.0% and the broccoli developed slight off-odors that dissipated with subsequent ventilation.

In the microperforated XF12 packages stored at 1°C, O_2 and CO_2 concentrations equilibrated at 13.7 and 8.4%, respectively. After 1 day at 20°C, O_2 decreased to 2.2% and CO_2 increased to 18%. No ethanol or off-odors were detected. The broccoli remained fresh and green, weight loss was only 2.0%, and levels of yellowing and decay were significantly lower than for broccoli packed in the PE packaging. The XF12-packed broccoli also remained in a salable state (appearance index = 2.5) after 3 additional days at 20°C in ventilated packaging (data not shown), whereas the PE-packed heads were rated 2.4 after only 1 day at 20°C in the sealed packaging (Table 5.2).

The results obtained with both consumer and bulk packaging (Table 5.1 and Table 5.2; Figure 5.2 and Figure 5.3) may suggest that in hydrophilic packaging characterized by a slight water loss, the broccoli (and maybe some other crops) can tolerate lower levels of O_2 and higher levels of CO_2 without shifting to anaerobic respiration and developing undesirable odors. This suggestion could be supported by recent findings of DeEll et al. (2006), which show that addition of sorbitol, a water-absorbent compound, to sealed packaging of broccoli resulted in increased weight loss by 1.3% and decreased production of the anaerobic products ethanol and acetaldehyde as well as undesirable odor.

Toivonen (1997a) showed that addition of clay, known as water absorbent, to a package of diced onions resulted in mild water loss and a better odor score. He suggested that transpired water vapors can be carriers of many volatiles in the water stream. In addition, the clay composite may also

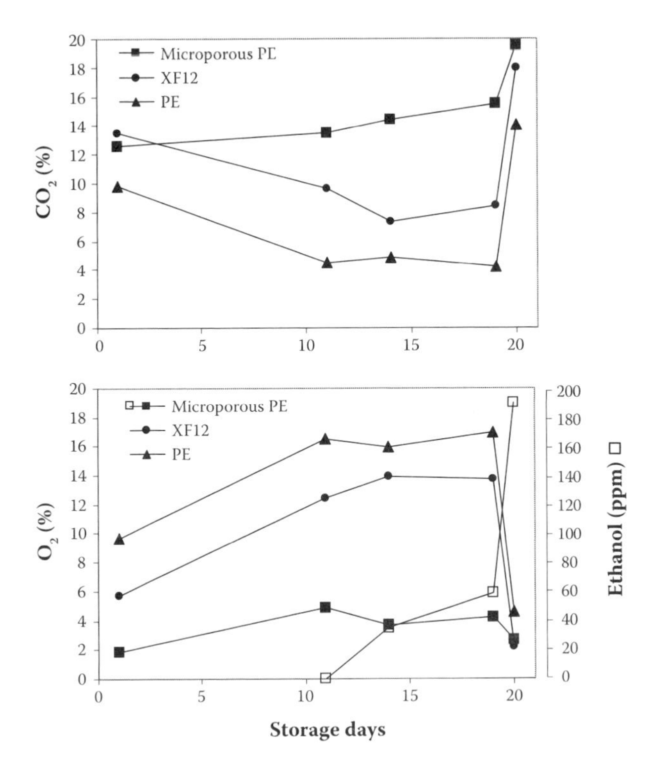

Figure 5.3 Changes in concentrations of CO_2, O_2, and ethanol in bulk packages (10 kg) of broccoli during 19 days at 1°C followed by 1 day at 20°C. Heads (500 to 600 g each) were packed in liners of microporous polyethylene (PE), microperforated polyethylene (PE), and microperforated Xtend (XF12). Data are means of three gas samples, one sample from each carton. Percentages of perforation area of the films are given in Table 5.2.

adsorb some organic volatiles. Studies of Wills (1968) and Wills and McGlasson (1970) revealed that losses of specific volatiles in stored apples can be enhanced by water loss. Addition of desiccants to packaging of tomato and mushrooms improved their keeping quality (Shirazi and Cameron, 1992; Roy et al., 1996). Toivonen (1997b) concluded that water loss from the packed fresh produce resulted in an increased vapor pressure deficit, which helps drive the volatiles from the tissue along with water vapor. In addition, undesirable odors may also be contributed by microbial growth, the process that is inhibited under lower RH.

On the basis of the findings and assumptions mentioned above, it appears that the good results obtained with the hydrophilic Xtend packaging are due to the high permeability of the film to water vapors in combination with microperforation. In addition to the reduction of decay due to the reduced in-pack humidity, the increased evapotranspiration may also help the removal of undesirable volatiles from the produce. We assume that the hydrophilic characteristic of the Xtend films facilitates the specific modification of atmosphere in the packaging, causing stronger retardation of yellowing, decay, objectionable odors, and some other undesirable processes. The hydrophilic characteristic of the packaging is attributed to the polyamides that are an important constituent of the mix with other polymeric and nonpolymeric compounds.

5.6.3 *Retardation of yellowing, decay, and leaf elongation*

5.6.3.1 *Green onions*

Green onion, like broccoli, is also a highly perishable crop. Bunched onions can be stored for 3 to 4 weeks at 0°C and 95 to 100% RH. However, when kept at 5°C, the storage life sharply decreases to only 1 week. Storage at higher temperatures results in severe yellowing, sprig curvature, and decay (Hruschka, 1974; Hardenburg et al., 1986; Adamicki, 2004). Since green onion is very tolerant to low O_2, its storability can be extended to 6 to 9 weeks when held at 0°C in 1% O_2 and 5% CO_2 (Hruschka, 1974). For commercial practice, storage at 0°C in CA conditions of 2 to 4% O_2 plus 10 to 20% CO_2 is recommended (Suslow and Cantwell, 1998a).

Good quality of both intact and minimally processed green onion was obtained after 3 weeks storage at 5°C in CA conditions of 0.1 to 0.2% O_2 with or without 7.5 to 9.0% CO_2 (Hong et al., 2000). Satisfactory results were also achieved with minimally processed green onion packed in moderate vacuum packaging (LDPE) and stored at 10°C for 3 weeks. The O_2 equilibrated at 1.5 to 3.0% and CO_2 at 4.1 to 6.6% (Hong and Kim, 2004).

If temperatures are not well controlled, as is common in air freight, then increase in temperature of the produce in MAP may result in anaerobic respiration and development of strong off-odors. Results of an experiment that simulated air freight conditions in which extreme temperature fluctuations may occur are depicted in Table 5.3. Trimmed green onion (20 cm long) in bulk packaging (2 kg) was stored for 1 day at 1°C followed by 4 days at 6°C and an additional day at 17°C.

The bunched onions were packed in macroperforated PE, microperforated PE, and microperforated Xtend (XF12). Packing in macroperforated PE resulted in severe decay, yellowing, and leaf elongation. Even though the perforation rate was the same in both microperforated PE and XF12, the latter packaging suppressed leaf elongation and decay more efficiently. The appearance index in XF12 was 2.8 (above marketability threshold), compared to 2.4 in microperforated PE (nonmarketable). The minimum for marketable produce was an appearance index of 2.5. It seems that the ability of XF packaging to preserve the quality of the bunched onions better than the other packaging was due to the higher level of CO_2 and the lower level of O_2 that were obtained. However, the reduced humidity in the XF packaging as expressed by a moderate weight loss (0.7%) and the almost complete lack of condensation (index of 1.3 compared with 3.5 in PE) may also contribute to

Table 5.3 Effect of Various Bulk Packages of Trimmed Green Onion on Water Condensation, CO_2 and O_2 Concentrations, and Quality Parameters of the Stored Onions[a,b]

Packaging	Condensation index[c] (1–5)	Weight loss (%)	Elongation (cm)	Yellowing index (1–5)[c]	Severe decay (%)	Appearance index[c] (1–5)	Gas Concentration (%)	
							O_2	CO_2
Perforated PE[d]	2.2 b	1.4 a	7.1 a	2.7 a	86.7 a	2.0 c	—	—
Microperforated PE[e]	3.5 a	0.2 c	4.4 b	1.9 b	16.7 b	2.4 b	3.3	11.2
Microperforated XF12[e]	1.3 c	0.7 b	1.8 c	1.8 b	10.0 c	2.8 a	2.5	20.1

[a] Bunched onions were trimmed to 20 cm length and roots were cut to about 0.5 cm below the leaf base. After washing in tap water and subsequent drying, bunches (2 kg) were packed vertically in plastic-lined cartons. Onions were stored for 1 day at 1°C followed by 4 days at 6°C and an additional day at 17°C.

[b] Data are presented as the mean of 3 cartons, with 10 bunches sampled from each. Mean values that are followed by the same letter are not significantly different at $p < 0.05$ according to Duncan's multiple range test.

[c] Indices: 1 = the lowest level; 5 = the highest level. Appearance indices of 2.5 are marketable.

[d] Total perforation area of macroperforated PE was 0.19%.

[e] Total perforation area of microperforated PE and XF12 was 0.00015%.

the reduction in leaf elongation and decay. Both fungal and bacterial decay, leaf elongation, and regrowth are known to be enhanced by humidity.

5.6.4 Retardation of leaf sprouting, decay, and discoloration in root vegetables

5.6.4.1 General background

The root vegetables, parsnips and radishes, which are temperate zone crops, were selected as models since they respond positively to MAP, and the use of hydrophilic packaging was found to improve MAP efficacy. Most root crops are mechanically harvested following topping of the foliage. The topped roots are transported in bulk for washing, recutting of roots and tops, hydrocooling, sizing, and packing. Therefore, mechanical damages to the roots occur very often during harvesting and subsequent handling, leading to development of decay in storage and shelf life (Brecht, 2003).

The storage temperature for both parsnips and radishes should be as close as possible to 0°C with the RH in the range of 95 to 100% (Hardenburg et al., 1986; Ryall and Lipton, 1979; van den Berg and Lentz, 1973). This low temperature minimizes textural changes, discoloration, and development of fungal and bacterial decay, and also inhibits regrowth of leaves and roots.

5.6.4.2 Parsnip

Undamaged and healthy roots can be stored 4 to 6 months at 0 to 1°C and 97 to 100% RH. The main problems in prolonged storage are decay, surface browning, and shriveling. The browning is caused by enzymatic oxidation of phenolic compounds (Chubey and Dorrell, 1972) and increases with storage duration (Kaldy et al., 1976). These problems are aggravated during shelf life at higher temperature, in addition to the development of leaves sprouting. A postharvest dip in a solution containing $CaCl_2$, ascorbic acid, and citric acid has been found to reduce considerably surface browning of parsnip roots in "Javelin" cultivar exhibiting an intermediate level of susceptibility to browning. However, the treatment was not effective enough in susceptible cultivars (Toivonen, 1992).

Packing parsnip roots in perforated plastic bags reduces water loss. However, little work has been done on the effect of CA or MAP on the keeping quality of the stored roots. Limited results suggest that there are no benefits to CA storage (Stoll and Weichmann, 1987). Geeson (1990) found that wrapping punnets of baby parsnips (400 g each) with nonperforated PVC retarded browning but resulted in an anaerobic atmosphere due to very high CO_2 and low O_2. The roots stained and developed a strong alcoholic odor and deteriorated. Packing the roots in microperforated PVC resulted in an atmosphere of 4.5 to 6.5% CO_2 and 12 to 13% O_2. In this packaging, discoloration and the problems of staining and breakdown were reduced to some extent.

We found in preliminary experiments (unpublished) that packaging parsnip roots (cv. Javelin) in macroperforated PE-lined cartons reduced water loss, increased decay, but did not reduce browning. Additional experiments revealed that high concentrations of CO_2 (12 to 18%) are required to effectively reduce surface browning, even though lower levels do have some effect. Higher levels of CO_2 achieved by using nonperforated PE packaging were found to be detrimental. The roots developed an alcoholic odor and severe sour smell due to the development of bacterial soft rot caused by *E. carotovora*.

We assumed that the combination of high CO_2 with low O_2, high humidity in the packaging, and condensation on the roots' surface was the reason for the detrimental effect. Therefore, experiments were performed in an attempt to control both humidity and atmosphere by using microperforated films with different permeabilities to water vapors. Results of a representative experiment (unpublished) are shown in Table 5.4 and Figure 5.4. The experiment comprised bulk microperforated packages (plastic-lined cartons) made of polyethylene and two types of Xtend films: XF10 and XF12. The three films had the same microperforation level. Each package contained 18.5 kg of parsnip roots.

The diffusion rates of O_2 and CO_2 through microperforated film are theoretically expected to be similar (Cameron et al., 1995; Emond et al., 1991; Fishman et al., 1996; Lee and Renault, 1998; Renault et al., 1994a, 1994b).

Table 5.4 Effect of Microperforated Polyethylene (PE) and Xtend (XF12, XF10) Packages on Condensation, Browning Discoloration, Decay, Leaf Sprouting, and Flavor in Parsnip Roots[a,b]

	Cold storage (1.0°C) in bulk packaging		Shelf life (20°C) in consumer microperforated PE bags[d]			
Bulk packaging	Condensation (1–5)[c]	Browning (1–5)[c]	Browning (1–5)[c]	Severe decay (%)	Sprouting (%)	Flavor
Microperforated PE[d]	5.0 a	2.5 a	3.5 a	20.0 a	10.0 a	Bland
Microperforated XF12[d]	3.5 b	2.0 b	2.5 b	15.0 ab	0.0 b	Sweet
Microperforated XF10[d]	1.5 c	1.5 c	2.2 bc	0.0 b	0.0 b	Sweeter

[a] Parsnip roots (cv. Javelin) packed in microperforated plastic-lined cartons (18.5 kg) were checked after 32 days at 1°C and after an additional 2 days' shelf life at 20°C in consumer microperforated PE bags (1 kg).

[b] Data for bulk packages are presented as the mean of 3 cartons, with 20 roots sampled from each. Data for consumer bags are presented as the mean of five bags per treatment. Mean values that are followed by the same letter are not significantly different at $p < 0.05$ according to Duncan's multiple range test.

[c] Indices: 1 = the lowest level; 5 = the highest level.

[d] Total perforation area of microperforated PE, XF10, and XF12 liners was 0.0004%, and of PE bags, 0.00035%.

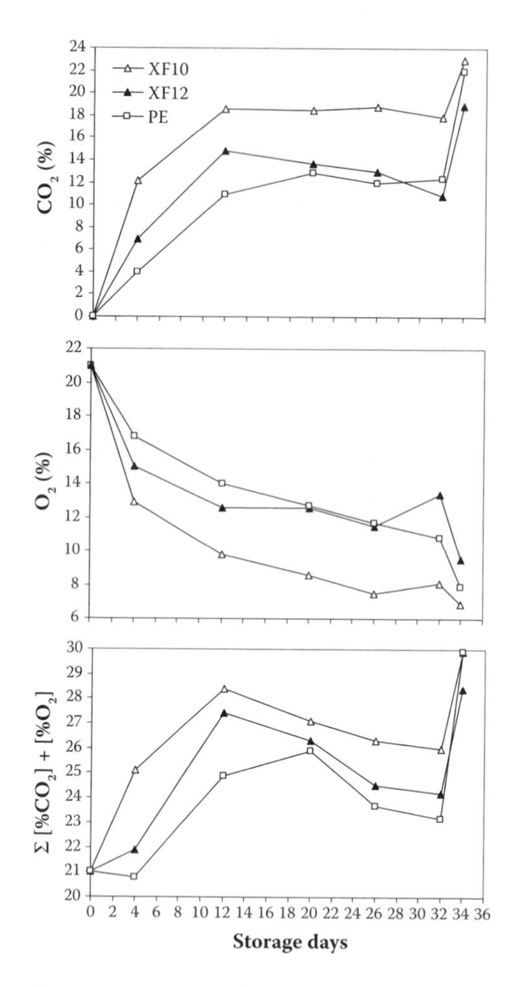

Figure 5.4 Changes in concentrations of CO_2, O_2, and their combined sum in bulk packaging containing parsnips (18.5 kg) during 32 days at 1°C followed by 2 days in consumer PE bags (1 kg) at 20°C. During cold storage roots were packed in microperforated liners of PE, XF12, and XF10, and during shelf life, in microperforated PE bags. Data are means of three gas samples, one sample from each carton. The percentage of perforation area of the films is given in Table 5.4.

Therefore, as discussed in Section 5.3, the sum of O_2 and CO_2 concentrations in microperforated packages under aerobic conditions at steady state is expected to be close to 21% (Mir and Beaudry, 2004; see also Figure 5.2). However, in contrast to this expectation, an increase of CO_2 in microperforated packages of parsnip roots during cold storage (1°C) and shelf life (20°C) resulted in an increase of the sum of CO_2 and O_2 in the range of 24 to 30% (Figure 5.4).

After 32 days at 1°C, concentrations (in %) of CO_2/O_2 in XF10, XF12, and PE packaging were 17.9/8.1, 10.8/13.4, and 12.4/10.8, respectively. After two additional days at 20°C in microperforated consumer PE bags, concentrations of CO_2/O_2 were 23.0/6.9, 18.9/9.5, and 22.0/7.9, respectively. Due to the safe levels of O_2, no anaerobic conditions occurred and no objectionable odor was perceived.

Similar results in which the sum value of CO_2 and O_2 in XF microperforated packaging is greater than 21% have also been found with other commodities containing high levels of starch or sugars. Sums of 23 to 25% were found for sweet corn (Aharoni and Richardson, 1997) and Charentais melon (Rodov et al., 2002). In these crops, the improved keeping quality was attributed to the high level of CO_2 (13 to 18%), accompanied by safe levels of O_2 (7 to 10%).

The high sum values of CO_2 and O_2 found in the parsnip packaging may suggest the existence of an additional respiration pathway — alternative oxidation (AOX) pathway. Both regular aerobic respiration and AOX require O_2. However, the alternative pathway can use organic acid as substrate, mostly pyruvate, resulting in the increase of respiratory quotient (RQ; the ratio of carbon dioxide produced to the oxygen consumed) above 1. Very low levels of pyruvate can activate the AOX (Millar et al., 1996; Millenaar and Lambers, 2003). Therefore, the increase of CO_2 in the parsnip packaging is not equal to the depletion of O_2. Due to the microperforation in the films, safe levels of O_2 (8 to 13%) allow both regular respiration and AOX.

The alternative pathway occurs under stress conditions such as in the presence of cyanide, nitric acid, sulfide, or limited phosphate supply, as well as low temperature and high CO_2. AOX (cyanide-resistant respiration) was increased in potato tubers that were stored at 5°C in the presence of 10 to 30% CO_2 (Laties, 1982; Perez-Trejo et al., 1981). It has been proposed that the physiological function of the AOX pathway is the prevention of the formation of oxygen free radicals that are responsible for plant disorders under severe environmental conditions (Millenaar and Lambers, 2003).

Differences in atmosphere compositions developed in the three package types in spite of the same perforation level indicate that gas exchange properties of the packages were not solely dictated by perforation area, but were also affected by other factors, such as polymer nature, water condensation, and water absorption in the packaging material. The results depicted in Table 5.4 show that packing parsnip roots in the highest water permeable film, XF10, resulted in the highest level of CO_2 and the highest-quality parameters. After the cold storage period, the roots remained dry without any sign of shriveling, almost no browning (index = 1.5), and smelled fresh. The good quality of the roots, including sweet flavor, was also maintained after 2 days' shelf life in microperforated PE bags.

In XF12 packaging, which is less permeable to water vapors than XF10, more condensation was observed (index = 3.5) and there was also more browning and decay. Even though XF12 film is slightly less permeable to O_2 than XF10, the humidity that accumulated in the packaging increased the permeability to gas diffusion (see properties of Xtend films in Section 5.6.1). Therefore, the reduced quality, compared to XF10, could be related to the higher water condensation and lower CO_2 levels in the packaging during both cold storage and shelf life.

Packing parsnip roots in PE liners resulted in very high condensation (index = 5) inside the packaging throughout the cold storage. Browning developed already during 32 days' storage at 1°C, but became more severe during 2 days of shelf life at 20°C in the consumer PE bags. During shelf life, decay and leaf sprouting were also observed, and the flavor was poor. It seems that the excess humidity on the roots packed in microperforated PE may have accelerated the microbial growth, as well as the activity of the oxidative enzymes that are responsible for the browning reaction.

Organoleptic tests performed at the end of the experiment revealed that the XF10-packed parsnips were sweeter than those from the other packaging (Table 5.4). In parsnip roots, as in potatoes, the starch is converted to sugars during cold storage. Parsnips stored for 2 weeks at 0 to 1°C developed better flavor (Boswell, 1923; Shattuck et al., 1989). It seems that the very high concentrations of CO_2 in the XF10 packages did not alter this process, and may even help to inhibit the decrease in sugar content during shelf life.

In other experiments (not published), we have found that XF10-packed parsnip roots stored at 1°C tolerated CO_2 concentrations above 20%, whereas when packed in PE packaging, the root surface of the parsnips was slimy and an offensive sour smell was perceived. Slimy roots also produced more ethylene that accumulated in the packaging. As for carrots, low concentrations of ethylene also induce bitterness in parsnips, probably due to accumulation of phenolic metabolites (Shattuck et al., 1988).

Studies with potatoes revealed that a layer of water on the surface of the tubers, which reduces O_2 permeation, increased incidence of bacterial soft rot, and drying of the tubers reduced the infection (Bartz and Kelman, 1986). Potato tubers are highly susceptible to infection by *E. carotovora* under anaerobic conditions (Burton and Wiggington, 1970). Therefore, it is plausible that in the wet parsnip roots packed in the PE package containing high CO_2 an anaerobic atmosphere is created inside the roots. In most cases, the decay in the parsnips was caused by *E. carotovora*, *B. cinerea*, and less commonly by *Penicilium* sp.

5.6.4.3 *Topped radishes*

Topped radishes can be stored for 3 to 4 weeks at 0°C and 95 to 100% RH (Ryall and Lipton, 1979; Hardenburg et al., 1986). Because of the high surface area of the small roots, they are very susceptible to water loss. Packing the roots in ventilated plastic bags helps to maintain their turgidity and freshness. The recommendation for CA storage of topped radishes is 1 to 2% O_2 and 2 to 3% CO_2 at 0 to 5°C, with the optimum at 0°C (Saltveit, 1997). The advantage of atmosphere modification is most notable when the storage temperature is increased above 2.5°C.

An oxygen concentration of 1% was optimal for retardation of softening and regrowth of both roots and leaves of radishes stored at either 5 or 10°C (Lipton, 1972). Polderdijk and van den Boogaard (1998) found that 1% O_2

and 15% CO_2 inhibited yellowing of the leaves and growth of new roots of bunched radishes held at 12°C. Elevated CO_2 concentrations were more efficient than decreased O_2 concentrations, although 15% CO_2 did slightly increase purplish color of the roots.

Little published data exist on the effect of MAP on topped radishes. Packaging of red radishes (cv. Cherry Bell) in microperforated PE-lined cartons (5 kg) markedly improved their storability (Aharoni et al., 1996a). After 2 weeks at 2°C, followed by 3 days at 20°C in the sealed packaging, there was a decrease in severe top regrowth from 29% in the control (macroperforated PE) to 3%, a decrease in internal sponginess from 60.3% to 40.5% and a decrease in hollowness from 13.8% to 5.2%. In this experiment, within 3 days at 2°C, CO_2 equilibrated at 5% and O_2 at 16%. After 3 days of holding at 20°C in the sealed packages CO_2 increased to 10% and O_2 decreased to 7%.

In further experiments (Aharoni et al., 1998) performed with topped radishes (cv. Sakata), it was found that packing 13.5-kg roots in microperforated XF12-lined cartons retarded leaf sprouting and reduced decay more efficiently than packing in a microperforated PE liner. Decay was mostly bacterial black spot caused by *Xanthomonas vesicotoria*. In these experiments levels of CO_2 and O_2 were in the range of 4 to 8% and 8 to 16%, respectively. Therefore, the beneficial effect of MAP in preserving the quality of the radishes was attributed to the elevated CO_2 rather than to the reduced O_2.

In-pack elevated CO_2 in cold storage caused purplish discoloration of the roots that was found to be reversible following removal from the packaging and ventilation (Aharoni et al., 1996a, 1998). In fact, at the end of the experiments, the percentage of purplish radishes was higher in macroperforated PE packaging. It was concluded that the purplish discoloration of red radishes in ventilated atmosphere is an irreversible process, part of senescence, whereas CO_2-induced purplish color is a reversible reaction. Therefore, when elevated CO_2 inhibits senescence, it also reduces purplish discoloration at the final stages of the postharvest chain.

5.6.5 *Retardation of ripening, decay, peel blemishes, and chilling injury in fruit vegetables*

5.6.5.1 *General background*

The vegetables selected in this section are snap beans, cucumbers, summer squash, and Charentais-type melons (cantaloupes) that are all warm-season crops susceptible to chilling injury. Their storage temperature is in the range of 6 to 12°C. These vegetables, except for melons, are classified as immature fruits since they are harvested when they reach the requested commercial size and color, before completing their natural ripening processes. They are highly perishable due to their high metabolic rates (Mohammed and Brecht, 2003). They exhibit medium (snap beans and summer squash) or high (cucumbers and Charentais melons) sensitivity to ethylene (Cantwell, 2001). Charentais-type melon is the only one among this group that is a climacteric fruit and produces a high rate of ethylene (Rodov et al., 2002). The storage life of these vegetables is relatively short, only 1 to 3 weeks, due to their high respiration rate and the inability to store them at lower temperatures. Chilling injury in these vegetables is manifested as peel blemishes, pitting, discoloration, shriveling, and decay.

Optimal RH for this group is in the range of 85 to 95%. Higher RH may result in development of fungi and bacteria and aggravation of physiological disorders. Therefore, MAP that results in a high RH is a problematic method for prolonging their storage life. Commercial application of MAP or CA in these crops is limited (Saltveit, 1997). However, CA experiments have shown a positive response to low O_2 and moderate to high CO_2. We concentrated on these crops since the use of hydrophilic packaging for humidity control in combination with microperforation for modifying in-pack atmosphere may lead to successful implementation on a commercial scale.

5.6.5.2 *Snap beans*

Snap beans can be stored, depending on cultivar, for only 7 to 12 days when held at the recommended temperatures of 5 to 7.5°C and 95% RH. However, at 5°C chilling injury may occur in sensitive cultivars after 7 to 8 days (Cantwell, 2004; Hardenburg et al., 1986). Typical symptoms of chilling injury in beans are rusty brown spots (russeting), which become more evident after transfer to higher temperatures. At temperatures below 5°C, discoloration, pitting, increased water loss, and increased susceptibility to decay are common. Storage of snap beans above 7.5°C results in undesirable seed development, toughening, yellowing, and water loss (Littmann, 1967; Watada and Morris, 1966). Plastic packaging can reduce water loss from the pods. However, condensed water on the pod surface may aggravate development of chilling-induced russeting, leading to severe decay (Hardenburg et al., 1986).

Beans benefit from CA conditions of 2 to 3% O_2 and 3 to 10% CO_2 because they retard yellowing and reduce discoloration and decay on mechanical damages (Cano et al., 1994; Hardenburg et al., 1986; Trail et al., 1992). Low O_2 and high CO_2 also reduce susceptibility to chilling injury (Costa et al., 1994). Groeschel et al. (1966) found that 2 to 5% O_2 suppresses respiration, but increasing the O_2 level to 10% was ineffective. However, a combination of 10% O_2 plus 5% CO_2 suppresses respiration more than 10% CO_2 alone. Beans produce low amounts of ethylene, but their exposure to very low concentrations of ethylene promotes chlorophyll degradation, increases browning, and reduces storage life by 30 to 50% (Wills and Kim, 1996).

It is well known that high RH in the environment may reduce susceptibility of plant organs to chilling injury (Wang, 1993). However, excess humidity on the produce surface may aggravate symptoms of chilling injury. Therefore, the use of MAP with polymeric films having very low permeability to water vapors is faced with serious problems. In experiments performed by several research teams at the Volcani Center, Israel, it was found that MAP using the hydrophilic film Xtend (XF) was very effective in reducing chilling injury and other low-temperature-related disorders in various fruits. This was found with different types of microperforated XF films for mango

(Pesis et al., 2000), stone fruits (Lurie and Aharoni, 1997), cucurbit vegetables (Rodov et al., 1998b, 2002, 2003), and citrus fruits (Porat et al., 2004).

Fallik et al. (2002) found in experiments with mechanically harvested snap beans that when Xtend (XF10) film was used with an appropriate microperforation (0.0012% perforations of total film area), a desirable atmospheric modification was created. During 7 storage days at 5°C, CO_2 gradually increased up to 9% and O_2 decreased to 12%. After 2 days of holding of the sealed packaging at 20°C, CO_2 increased to 13% and O_2 decreased to 7%. No off-odor was perceived upon opening of the packages.

The efficacy of the microperforated XF packaging in maintaining pod quality was compared to microperforated PE and control (unpacked beans). The XF-packed pods better retained their green color and had the lowest level of russet-like spots and decay. Percentages of total defects on the pods, including chilling injury, were 24, 18, and 5% for the control, PE, and XF, respectively. Percentages of weight loss in the above-mentioned treatments were 11.0, 1.2, and 4.6%, respectively. No significant difference in the level of pod firmness was found between the two types of film packaging tested. The very high condensation found in the PE-packed beans resulted in increased decay.

In another experiment (Fallik, unpublished) in which the beans (bulk packaging of 2.2 kg) were packed in microperforated XF10 and compared to microperforated PE, macroperforated PE, and nonpacked pods, a very high susceptibility to chilling injury was observed. After 6 days at 5°C followed by 2 days at 17°C, the percentages of rusty spots were 7, 18, 40, and 70%, respectively, and percentages of decay were 1, 4, 12.3, and 10%, respectively.

As found with broccoli, green onions, and parsnips, the use of XF films to prevent excess humidity in the packaging increased the tolerance of the beans to high CO_2, thereby allowing the use of effective MAP during both storage and marketing.

5.6.5.3 Cucumbers

Cucumbers, as with most immature fruit vegetables, are very susceptible to chilling injury. The recommended storage temperatures are 10 to 12°C with 95% RH or more (Hardenburg et al., 1986; Saltveit, 2004). Within this very narrow range of temperatures cucumbers can be kept for 10 to 14 days. Storage below 10°C results in development of chilling injury characterized by surface pitting, water soaking, yellowing, tissue deterioration, increased water loss, and increased susceptibility to pathogens, leading to decay. The symptoms of chilling injury are more prominent once the fruits are transferred to higher temperatures (Hardenburg et al., 1986; Leshuk and Saltveit, 1990; Mohammed and Brecht, 2003). Storage of cucumbers at temperatures above 10°C results in rapid senescence, decay, and, in some cultivars, development of warty fruit disorder.

Many studies have shown that high humidity can suppress chilling injury symptoms (Wang, 1993). High humidity can easily be achieved by film packaging. In this case, a combination of high humidity and modified atmosphere might be beneficial. However, Eaks (1956) reported that concentrations of CO_2 above 3% increased pitting and accelerated breakdown in cucumbers held at 5°C. Ryall and Lipton (1979) reported that CA storage of 5% CO_2 and 5% O_2 was beneficial, but at chilling temperatures, high CO_2 and low O_2 aggravated chilling injury.

Wang and Qi (1997) have shown that MAP is beneficial in reducing chilling injury in cucumbers. Packing of fruits in sealed low-density polyethylene (LDPE) bags (three fruits in each bag) and storage for 18 days at a chilling temperature of 5°C resulted in an increase of CO_2 up to 3% and a decrease of O_2 to 16%. The other treatments were macroperforated PE bags and nonpacked fruits. Evaluation of fruit quality after 24 hours at 20°C revealed that the fruits packed in the sealed bags had the lowest levels of chilling injury and weight loss. Evaluation also showed that fruits packed in the sealed bags had the highest level of the polyamine putrescine, which is known to increase tolerance of fruits and vegetables toward chilling temperatures and other stresses (Guye et al., 1986; Kramer and Wang, 1989). Wang and Qi (1997) suggested that besides the stress induced by chilling, additional stress imposed by elevated CO_2 within the sealed packages might have contributed to the increased level of putrescine.

While nonperforated consumer PE packaging allows the development of an appropriate modified atmosphere during cold storage and a further 24 hours at 20°C, it is likely to be a problem to reach desirable conditions in bulk packaging having a lower ratio between film surface and produce weight. In bulk packaging, undesirable concentrations of CO_2 and O_2 as well as accumulation of condensed water might occur.

Rodov et al. (1998b) have shown that a beneficial modified atmosphere has been created in closed bulk packaging of parthenocarpic Beit-Alpha-type cucumbers (4 kg of fruit per carton) by using microperforated Xtend film (XF10, total perforation area was 0.00075%). Preliminary experiments showed that MAP using microperforated PE resulted in excess humidity and decay. The fruits were stored for 14 days at optimal (10°C), suboptimal (7°C), and increased (12°C) temperatures followed by an additional 2 days at 20°C in folded liners.

At the suboptimal temperature of 7°C, XF10 packaging alleviated the external symptoms of chilling injury that in the control fruits manifested as surface pitting that was subsequently followed by increased softening and decay. However, in some experiments, internal yellowing, a symptom of chilling injury, was found in the XF10-packed cucumbers.

At 12°C the visual appearance of control fruits was impaired by development of warty disorder, exhibited as the emergence during storage of numerous irregular warts on the initially smooth fruit surface (Gubler, 1996). The disorder was greatly reduced by XF packaging. The alleviation of the warts was attributed to the modified atmosphere composition rather than to elevated RH inside the packaging. Macroperforated packaging that had high RH but without the modified atmosphere did not prevent the formation of the warts.

At the optimal storage temperature of 10°C, minimal levels of weight loss and decay, without any symptoms of chilling injury, were evident in the XF10-packed cucumbers. Levels of O_2 and CO_2 in XF10 packages held at 10°C leveled off at 13 to 15% and 5 to 7%, respectively. The packages were opened upon transfer to 20°C and subsequently kept folded. After 2 days at 20°C, the atmosphere composition in the folded liners was 15 to 17% O_2 and 4 to 6% CO_2. Levels of CO_2 above 10%, which developed when the rate of microperforation was too little, resulted in severe physiological damage and decay caused by *B. cinerea.*

In contrast to the literature recommendations based on other cucumber varieties, the parthenocarpic Beit-Alpha-type cucumbers responded favorably in the presence of up to 8 to 9% CO_2. Other experiments have shown that a combination of optimal atmosphere composition and humidity preserved the tender texture and turgidity of cucumbers.

In another experiment (Rodov et al., 1998b), the behavior of the hydrophilic XF10 and XF14 films was compared with that of high-density PE (HDPE). Beit-Alpha-type cucumbers (cv. Hassan) packed in HDPE and XF10 (3 kg of packaging) were stored for 3 days at 16°C and 92 to 95% RH and then transferred to the ambient conditions of 25 to 27°C and 60 to 65% RH, which simulates the extreme of retail marketing. Under these ambient conditions the RH levels inside the HDPE, XF10, and XF14 liners stabilized at ca. 100%, 97 to 98%, and 95 to 96%, respectively. These seemingly small differences in RH levels caused significant contrasts in terms of water vapor pressure deficit (VPD), a parameter that determines the actual rate of water evaporation (Ben-Yehoshua and Rodov, 2003). Values of VPD in the HDPE, XF10, and XF14 were 25, 69, and 140 Pa, respectively. As a result, in the XF10 and XF14 packages most of the condensed water disappeared within a short time but accumulated in the HDPE. It should be noted that at lower storage temperatures the amount of condensed water in PE packaging was much higher than that under the ambient conditions mentioned above.

5.6.5.4 Summer squash

Summer squash are quite perishable and highly susceptible to chilling injury. The skin is tender and easily damaged during handling. The recommended storage conditions for the various types of summer squash are in the range of 5 to 10°C and 95% RH. Under these conditions, the storage life is typically only 1 to 2 weeks. However, if storage exceeds more than 1 week followed by several days in distribution, then storage at 7 to 10°C is best (Hardenburg et al., 1986). Storage below 5°C for more than 3 to 4 days results in surface pitting and discoloration, followed by decay that becomes evident when the squash are transferred to higher temperatures (McCollum, 1989; Ryall and Lipton, 1979). For zucchini squash, 5°C was found to be optimal for 2 weeks of storage (Mencarelli et al., 1983). The tender skin makes the fruits very susceptible to water loss and shriveling. Senescence symptoms include color change, softening of the flesh, and increased susceptibility to decay, especially when there is mechanical damage or the produce has previously been stored at chilling temperatures. Little work has been done to test the effect of CA or MAP on the storability of summer squash.

Storage of zucchini squash at 5°C under low O_2 was found to be ineffective or of little value (Leshuk and Saltveit, 1990; Mencarelli et al., 1983), although a low level of O_2 (3 to 5%) delayed yellowing and development of decay by a few days. Zucchini was found to tolerate elevated CO_2 concentrations up to 10%, but the storage life was not greatly extended (Suslow and Cantwell, 1998b).

Rodov et al. (1998b) found beneficial effects of MAP when using the hydrophilic Xtend film. Summer squash fruits (cv. Erlica and Black Magic) were packed (4 kg) in microperforated XF10- and PE-lined cartons and stored for 16 days at 10°C followed by an additional 3 days at 20°C in folded liners. The commercial carton packages (fruit cover with paper) were used as control. Results of the experiment with cv. Erlica are depicted in Table 5.5.

Packaging in XF10 liners improved the general appearance of the fruits, inhibited yellowing, softening, and growth of superficial mold, and reduced by approximately two-fold the weight loss of the fruits, compared to the control. Similar positive effects were observed with light-green (cocozelle type) (Table 5.5) or dark-colored (zucchini-type) (Rodov et al., 1998b) summer squash varieties. The atmosphere composition in PE and XF10 liners was rather similar. Both inhibited yellowing of the fruits and reduced weight loss two-fold for XF10 and four-fold for PE, compared to control. However, the high condensation in PE liners favored the development of postharvest pathogens, mainly *Pseudomonas lachrymans.* It should be noted that the PE-packed squash still had better appearance than the control fruits due to

Table 5.5 Effect of Microperforated Packages on Keeping Quality of Summer Squash[a,b]

Packaging	Color Index (1–4)[c]	Decay (%)	Weight loss (%)	Appearance index (1–5)[d]	CO_2 steady-state range (%)
No liner	3.0	24.5	12.7	0.0	0.03–0.05
PE[e]	1.9	32.4	3.2	2.2	8–10.5
XF10[e]	1.8	5.4	6.6	3.1	7–9.5

[a] Squash fruits (cv. Erlica) packed in microperforated plastic-lined cartons (4 kg) were checked after 16 days at 10°C (in closed liners) followed by 3 days at 20°C (in folded liners).

[b] Data are presented as the means of 3 cartons, with 25 fruits sampled from each.

[c] Color index: 1 = light green; 4 = yellow.

[d] Appearance index: 1 = the lowest level; 5 = the highest level. Levels of 2.5 are marketable.

[e] Total perforation area of PE and XF10 was 0.00075%.

the absence of the most devastating pathogens (*Erwinia* sp., *B. cinerea, Rhizopus stolonifer*), which accounted for most of the decay in the control fruits. Packaging in Xtend MA liners accumulating 8 to 12% CO_2 markedly inhibited the ethylene production by summer squash fruit (not shown).

It seems that the combination of appropriate levels of CO_2 with the reduced humidity and condensation in the XF10 packaging was responsible for maintaining quality of the squashes for 19 days, which is beyond the maximum storage life reported so far (McCollum, 2004).

5.6.5.5 Charentais-type melons

Charentais are typically nonnetted melons that belong to the Cantalupensis group and are the most perishable variety of *Cucumi melo* L. due to a pronounced climacteric behavior. The climacteric rise of ethylene and respiration in melons harvested at commercial maturity (preripe) commences within days after harvesting, and soon they become overripe. The overripe stage is characterized by excessive softening, color change, flavor deterioration, reduction in sugar content, and increased susceptibility to pathogens (Rodov et al., 2002).

The recommended storage temperature for different varieties of cantaloupes is in the range of 2 to 7°C (Saltveit, 1997). However, storage at 10 to 11°C was recommended for Charentais melons due to their chilling sensitivity (Rodov et al., 1998c). Symptoms of chilling injury are surface discoloration, pitting, and enhanced decay.

CA storage has some potential for extending storage life of cantaloupe melons, and the recommended conditions are 3 to 5% O_2 and 10 to 20% CO_2 (Saltveit, 1997; Shellie and Lester, 2004). Yahia and Rivera (1992) have found that individual MA packaging of netted cantaloupes in LDPE reduced weight loss and preserved the quality of fruits held at 5°C. However, severe decay developed after transferring PE packaging to 20°C due to in-pack water condensation. Very little information is available regarding the effect of MAP for nonnetted Charentais fruits.

Rodov et al. (2002) have reported a beneficial effect of MAP for Charentais-type melon (cv. Luna) using the hydrophilic Xtend film (Table 5.6). This cultivar has relatively good storage potential in spite of its pronounced climacteric behavior combined with adequate flavor and aroma (Rodov et al., 1998c). Fruit were packed in microperforated XF14, which is the highest WVPR film of the Xtend series. Unpacked fruit served as control. In simulation of possible conditions for sea transportation from Israel to Europe and subsequent storage and display in a retail outlet, the fruits were stored for 12 days at 10 to 11°C or 6 to 7°C plus an additional 3 days at 20°C. The ambient humidity was 80 to 90% RH during cold storage and 60 to 70% RH during shelf life. Immediately after transfer to 20°C the liners were opened and kept folded for the first 2 days of shelf life. During the third day at 20°C the liners were completely opened.

The best keeping quality was obtained in fruits that were packed in the microperforated XF14 packages stored at 6 to 7°C. The beneficial effects were

Table 5.6 Effect of Xtend Packaging and Storage Temperature on Quality of Charentais-Type Melons[a,b]

Storage temperature (°C)	Packaging	Peel color index[c]	SSC[d] (%)	Firmness (N)[e]	Decay index (%)	Appearance index (1–5)[f]
Initial (before storage)		2.4 c	14.1 a	66.6 a	0.0 c	4.5 a
11	No liner	3.8 a	11.0 c	43.2 c	56.3 a	1.5 c
11	XF14[g]	3.6 ab	12.4 b	44.5 c	18.8 b	3.0 b
7	No liner	3.6 ab	11.0 c	45.0 c	43.8 a	2.0 c
7	XF14[g]	3.3 bc	12.8 b	58.2 b	0.0 c	4.0 a

[a] Melons (cv. Luna) were packed in XF14-lined cartons (5 kg per carton) and stored in closed liners for 12 days at 7 or 11°C. Thereafter, the liners were kept folded for 2 days at 20°C and kept open for an additional day at 20°C.

[b] Data are presented as the means of three cartons, with eight fruits sampled from each. Mean values that are followed by the same letter are not significantly different at $p < 0.05$ according to Duncan's multiple range test.

[c] Peel color index: 1 = grayish green; 2 = black; 3 = yellowish beige; 4 = yellow-orange.

[d] SSC = soluble solids concentration.

[e] N = firmness values in newtons determined by Chatillion penetrometer.

[f] Appearance index: 1 = the lowest level; 5 = the highest level. Levels of 2.5 are marketable.

[g] Total perforation area of XF14 was 0.00025%.

Source: Adapted from Rodov, V. et al., *HortScience*, 37, 950–953, 2002.

retardation of the overripening symptoms, including excessive softening, color change of the peel, decrease of soluble solids, and development of postharvest pathogens (Table 5.6).

Atmospheric composition in the closed (at 7°C) and folded (at 20°C) liners was 13 to 14% CO_2 and 7 to 10% O_2. This MA composition was consistent with general CA/MA recommendations for cantaloupes: CO_2 concentration no higher than 15 to 20% and O_2 no lower than 3 to 5% (Saltveit, 1997). The composition is also similar to the beneficial concentrations for CA storage of "Galia" melons, which are 10% CO_2 and 10% O_2 (Aharoni et al., 1993b). It should be noted that the delay of overripening by MAP was not disturbed by the high concentration of ethylene, which reached a peak of 120 ppm on the second day of storage at 7°C. It may be concluded that the MA conditions and in particular high CO_2 drastically reduced the fruit's ethylene sensitivity.

Yahia and Rivera (1992) showed that humidity control with NaCl-containing bags reduced water condensation and prevented decay in individual LDPE packaging of muskmelons kept for 10 days at 5°C. However, this type of modified humidity packaging did not prevent the increase of decay up to at least 15% during a subsequent 2 days of shelf life at 20°C. In contrast, in the XF14 packaging, no water condensation and no decay were observed during both cold storage and shelf life at 20°C. Therefore, packaging Charentais melons in the microperforated high-WVTR XF14 film facilitates their long-distance transportation due to the optimal MA achieved together with the reduced humidity.

5.7 Conclusions

MAP based on microperforated films is a suitable method for extending postharvest life of fruits and vegetables, especially for highly respiring produce that respond positively to elevated concentrations of CO_2. Microperforation technology allows the use of different polymeric films regardless of their permeability to O_2 and CO_2. Adequate microperforation results in safe levels of O_2 in the packages during prolonged cold storage and even at abused temperatures, thereby preventing anaerobic respiration. Most of the commercially available plastic films used for MAP have very low permeability to water vapors. The high in-pack humidity and condensed water lead to increased decay, physiological disorders, and objectionable odors. The use of microperforated hydrophilic Xtend packaging was found to modify effectively both atmosphere composition and humidity inside the packages of various fruits and vegetables. The combination of reduced in-pack humidity with a safe level of O_2 allows the accumulation of high noninjurious levels of CO_2 within the fungistatic range (>10%), and thereby a reduction of microbial contamination. Additionally, the elevated concentrations of CO_2 in Xtend packages were found to effectively retard ripening and senescence processes as well as physiological disorders. The hydrophilic Xtend packaging is a suitable method for commodities that respond positively to moderate–high CO_2 concentrations, and especially for those that are sensitive to excessive humidity and water condensation in the packaging. Most of the technologies based on the Xtend packaging reviewed in this chapter have been successfully implemented in postharvest practice in Israel, the U.S., and other countries, particularly for long-distance shipments.

Acknowledgments

The authors warmly thank Dr. Susan Lurie of the ARO, The Volcani Center, and Dr. Gary Ward of StePac L.A. company, for their helpful advice and constructive review of the manuscript.

References

Adamicki, F. 2004. Onion. In *The Commercial Storage of Fruits, Vegetables, and Florist and Nursery Crops*, Agricultural Handbook 66. Gross, K.C., Wang, C.Y., and Saltveit, M., Eds. USDA, www.ba.ars.usda.gov/hb66/contents.html.

Aharoni, N. 1994. Postharvest physiology and technology of fresh culinary herbs. *Israel Agresearch*, 7:35–59 (Hebrew with English summary).

Aharoni, N., Afek, U., Dvir, O., Chalupowicz, D., and Aharon, Z. 1996a. Modified atmosphere packaging extends the storability of radishes. *Gan Sadeh VaMeshek*, 6:77–83 (in Hebrew).

Aharoni, N., Afek, U., Dvir, O., Chalupowicz, D., and Aharon, Z. 1998. Retardation of senescence processes by modified atmosphere packaging in radishes for export. *Gan Sadeh VaMeshek*, 9:64–66 (in Hebrew).

Aharoni, N., Aharoni, Y., Fallik, E., Dvir, O., Chalupowicz, D., Aharon, Z., Copel, A., Grinberg, S., Alkalai, S., Nir, M.M., and Ben-Tzur, I. 1997. Use of newly-developed films for modified atmosphere and modified humidity packaging of fresh vegetables. In *International Congress for Plastics in Agriculture (CIPA)*, Tel-Aviv, Israel, Congress Handbook p. 99.

Aharoni, N., Dvir, O., Chalupowicz, D., and Aharon, Z. 1993a. Coping with postharvest physiology of fresh culinary herbs. *Acta Hort.*, 344:69–78.

Aharoni, N., Dvir, O., Chalupowicz, D., and Aharon, Z. 1996b. Modified atmosphere packaging of vegetables and fresh herbs. In *International Conference on Postharvest Science*, Taupo, New Zealand, Conference Handbook, p. 108.

Aharoni, N., Lieberman, M., and Sisler, H.D. 1979. Patterns of ethylene production in senescing leaves. *Plant Physiol.*, 64:796–800.

Aharoni, N., Philosoph-Hadas, S., and Barkai-Golan, R. 1986. Modified atmospheres to delay senescence and decay of broccoli. In *Controlled Atmospheres for Storage and Transport of Perishable Agricultural Commodities*, Blankenship, S.M., Ed. 4th Controlled Atmosphere Research Conference, Raleigh, NC, pp. 169–177.

Aharoni, N., Reuveni, A., and Dvir, O. 1989. Modified atmospheres in film packages delay senescence and decay of green vegetables and herbs. *Acta Hort.*, 258:255–262

Aharoni, Y., Copel, A., and Fallik, E. 1993b. Storing "Galia" melons in controlled atmosphere with ethylene absorbent. *HortScience*, 28:725–726.

Aharoni, Y. and Richardson, D.G. 1997. New, higher water permeable films for modified atmosphere packaging of fruits and vegetables prolonged MA storage of sweet corn. In *Proceedings of the 7th International Controlled Atmosphere Research Conference: Vegetables and Ornamentals*, Davis, CA, pp. 73–77.

Ahvenainen, R. 1996. New approaches in improving the shelf life of minimally processed fruits and vegetables. *Trends Food Sci. Technol.*, 7:179–187.

Anelli, G., Mencarelli, F., and Guaraldi, F. 1984. Short storage of *Brassica oleracea* L. and *Brassica campestris* L. in different types of modified atmospheres. *Acta Hort.*, 157:177–184.

Ballantyne, A., Stark, R., and Selman, J.D. 1988. Modified atmosphere packaging of broccoli florets. *Int. J. Food Sci. Technol.*, 23:353–360.

Banks, N.H., Cleland, D.J., Cameron, A.C., Beaudry, R.M., and Kader, A.A. 1995. Proposal for a rationalized system of units for postharvest research in gas exchange. *HortScience* 30:1129–1131.

Barkai-Golan, R. 1990. Postharvest diseases suppression by atmospheric modification. In *Food Preservation by Modified Atmospheres*, Calderon, M. and Barkai-Golan, R., Eds. CRC Press, Boca Raton, FL, pp. 237–264.

Barmor, C.R. 1987. Packaging technology for fresh and minimally processed fruits and vegetables. *J. Food Qual.*, 10:207–217.

Barron, C., Varoquaux, P., Guilbert, S., Gontard, N., and Gouble, B. 2002. Modified atmosphere packaging of cultivated mushrooms (*Agaricus bisporus* L.) with hydrophilic films. *J. Food Sci.*, 67:251–255.

Barth, M. and Zhuang, H. 1996. Packaging design affects antioxidant vitamin retention and quality of broccoli florets during postharvest storage. *Postharvest Biol. Technol.*, 9:141–150.

Bartz, J.A. and Kelman, A. 1986. Reducing the potential for bacterial soft rot in potato tubers by chemical treatments and drying. *Am. Potato J.*, 63:481–493.

Beaudry, R.M. 1999. Effect of O_2 and CO_2 partial pressure on selected phenomena affecting fruit and vegetable quality. *Postharvest Biol. Technol.*, 15:293–303.

Beaudry, R.M. 2000. Responses of horticultural commodities to low oxygen: limits to the expanded use of modified atmosphere packaging. *HortTechnology*, 10:491–500.

Ben-Yehoshua, S. and Rodov, V. 2003. Transpiration and water stress. In *Postharvest Physiology and Pathology of Vegetables*, Bartz, J.A. and Brecht, J.K., Eds. Marcel Dekker, New York, pp. 111–159.

Ben-Yehoshua, S., Rodov, V., Fishman, S., and Peretz, J. 1996. Modified-atmosphere packaging of fruits and vegetables: reducing condensation of water in bell peppers and mangoes. *Acta Hort.*, 464:387–392.

Boswell, V.R. 1923. Changes in quality and chemical composition of parsnips under various storage conditions. *Md. Agric. Exp. Stn. Bull.*, 258:61–86.

Brackett, R.E. 1994. Microbiological spoilage and pathogens in minimally processed refrigerated fruits and vegetables. In *Minimally Processed Refrigerated Fruits and Vegetables*, Wiley, R.C., Ed. Chapman & Hall, New York, pp. 269–312.

Brecht, J.K. 2003. Underground storage organs. In *Postharvest Physiology and Pathology of Vegetables*, Bartz, J.A. and Brecht, J.K., Eds. Marcel Dekker, New York, pp. 625–647.

Brown, W. 1922. On the germination and growth of fungi at various temperatures and in various concentrations of oxygen and carbon dioxide. *Ann. Bot.*, 36:257–283.

Burton, W.G. and Wiggington, M.J. 1970. The effect of film of water upon the oxygen status of a potato tuber. *Potato Res.*, 13:180–186.

Cameron, A.C., Talasila, P.C., and Joles, D.J. 1995. Predicting the film permeability needs for modified atmosphere packaging of lightly processed fruits and vegetables. *HortScience*, 30:25–34.

Cano, M.P., Monreal, M., de Ancos, B., and Alique, R. 1997. Controlled atmosphere effects on chlorophylls and carotenoids changes in green beans (*Phaseolus vulgaris* L., cv. Perona). In *Proceedings of the 7th International Controlled Atmosphere Research Conference*, Davis, CA, pp. 46–52.

Cantwell, M. 2001. Properties and Recommended Conditions for Storage of Fresh Fruits and Vegetables. www.postharvest.ucdavis.edu/produce/storage/prop.

Cantwell, M. 2004. Beans. In *The Commercial Storage of Fruits, Vegetables, and Florist and Nursery Crops*, Agricultural Handbook 66, Gross, K.C., Wang, C.Y., and Saltveit, M., Eds. USDA. www.ba.ars.usda.gov/hb66/contents.html.

Cantwell, M. and Reid, M.S. 1993. Postharvest physiology and handling of fresh culinary herbs. *J. Herbs Spices Medicinal Plants*, 1:93–127.

Cantwell, M. and Suslow, T. 1999. Broccoli: Recommendations for Maintaining Postharvest Quality. postharvest.ucdavis.edu/produce/producefacts/veg/broccoli.html.

Chubey, B.B. and Dorrell, D.G. 1972. *J. Am. Soc. Hort. Sci.*, 97:107–109.

Church, I.J. and Parsons, A.L. 1995. Modified atmosphere packaging technology: a review. *J. Sci. Food Agric.*, 67:143–152.

Costa, M.A.C., Brecht, J.K., Sargent, S.A., and Huber, D.J. 1994. Tolerance of snap beans to elevated CO_2 levels. *Proc. Fla. State Hort. Soc.*, 107:271–273.

Day, B.P.F. 1993. Fruit and vegetables. In *Principles and Application of Modified Atmosphere Packaging of Food*, Parry, R.T., Ed. Blackie, London, pp. 115–133.

DeEll, J.R. and Toivonen, P.M.A. 2000. Chlorophyll fluorescence as a nondestructive indicator of broccoli quality during storage in modified atmosphere packaging. *HortScience*, 35:256–259.

DeEll, J.R., Toivonen, P.M.A., Cornut, F., Roger, C., and Vigneault, C. 2006. Addition of sorbitol with $KMNO_4$ improves broccoli quality retention in modified atmosphere packages. *J. Food Qual.*, 29:65–75.

Eaks, L.L. 1956. Effect of modified atmospheres on cucumbers at chilling and nonchilling temperatures. *Proc. Am. Soc. Hort. Sci.*, 67:473–478.

El-Goorani, M.A. and Sommer, N.F. 1981. Effects of modified atmospheres on postharvest pathogens of fruits and vegetables. *Hort. Rev.*, 3:412–461.

Emond, J.P., Castaigne, F., Toupin, C.J., and Desilets, D. 1991. Mathematical modeling of gas exchange in modified atmosphere packaging. *Trans. Am. Soc. Agric. Eng.*, 34:239–245.

Exama, A., Arul, J., Lencki, R.W., Lee, L.Z., and Toupin, C. 1993. Suitability of plastic films for modified atmosphere packaging of fruits and vegetables. *J. Food Sci.*, 58:1365–1370.

Fallik, E., Chalupowicz, D., Aharon, Z., and Aharoni, N. 2002. Modified atmosphere in a water permeable film maintains snap bean quality after harvest. *Folia Hort.*, 14:85–94.

Fishman, S., Rodov, V., and Ben-Yehoshua, S. 1996. Mathematical model for perforation effect on oxygen and water vapor dynamics in modified-atmosphere packages. *J. Food Sci.*, 61:956–961.

Forney, C.F., Mattheis, J.P., and Austin, R.K. 1991. Volatile compounds produced by broccoli under anaerobic conditions. *J. Agric. Food Chem.*, 39:2257–2259.

Forney, C.F., Rij, R.E., and Ross, S.R. 1989. Measurements of broccoli respiration rate in film-wrapped packages. *HortScience*, 24:111–113.

Geeson, J.D. 1989. Modified atmosphere packaging of fruits and vegetables. *Acta Hort.*, 258:143–150.

Geeson, J.D. 1990. Micro-perforated films for fruit and vegetable packaging. *Prof. Hort.*, 4:32–35.

Geeson, J.D., Everson, H.P., and Browne, K.M. 1988. Microperforated films for fresh produce. *Grower*, 109:31–34.

Gillies, S.L., Cliff, M.A., Toivonen, P.M.E., and King, M.C. 1997. Effect of atmosphere on broccoli sensory attributes in commercial MAP and microperforated packages. *J. Food Qual.*, 20:105–115.

Groeschel, E.C., Nelson, A.I., and Steinberg, M.P. 1966. Changes in color and other characteristics of green beans stored in controlled refrigerated atmospheres. *J. Food Sci.*, 31:488–496.

Gubler, W.D. 1996. Measles. In *Compendium of Cucurbit Diseases*, Zitter, T.A., Hopkins, D.L., and Thomas, C.E., Eds. American Phytopathological Society Press, St. Paul, 87 pp.

Guye, M.G., Vigh, L., and Wilson, J.M. 1986. Polyamine titre in relation to chill-sensitivity in *Phaseolus* sp. *J. Exp. Bot.*, 37:1036–1043.

Hansen, M., Buttery, R.G., Stern, D.J., Cantwell, M.I., and Ling, L.C. 1992. Broccoli storage under low oxygen atmosphere: identification of higher boiling volatiles. *J. Agric. Food Chem.*, 40:850–852.

Hao, Y.Y., Brackett., R.E., Beuchat, L.R., and Doyle, M.P. 1999. Microbiological quality and production of botulinal toxin in film-packaged broccoli, carrots and green beans. *J. Food Prot.*, 62:499–508.

Hardenburg, R.F. 1971. Effect of in-package environment on quality of fruits and vegetables. *HortScience*, 6:190–201.

Hardenburg, R.E., Watada, A.E., and Wang, C.Y. 1986. *The Commercial Storage of Fruits, Vegetables, and Florist and Nursery Stocks*, USDA Handbook 66. Washington, DC.

Hernandez, R. 1994. Effect of water vapor on the transport properties of oxygen through polyamide packaging materials. *J. Food Eng.*, 22:495–507.

Herner, R.C. 1987. High CO_2 effects on plant organs. In *Postharvest Physiology of Vegetables*, Weichman, J., Ed. Marcel Dekker, New York, pp. 239–253.

Hintlian, C.B. and Hotchkiss, J.H. 1986. The safety of modified atmosphere packaging: a review. *Food Technol.*, 40:70–76.

Hong, G., Peiser, G., and Cantwell, M.I. 2000. Use of controlled atmospheres and heat treatment to maintain quality of intact and minimally processed green onions. *Postharvest Biol. Technol.*, 20:53–61.

Hong, S.I. and Kim, D.M. 2004. The effect of packaging treatment on the storage quality of minimally processed bunched onions. *Int. J. Food Sci. Technol.*, 39:1033–1041.

Hruschka, H.W. 1974. *Storage and Shelf Life of Packaged Green Onion*, Marketing Research Report. USDA, Washington, DC, p. 1015.

Hruschka, H.W. and Wang, C.W. 1979. *Storage and Shelf Life of Packaged Watercress, Parsley, and Mint*, Marketing Research Report. USDA, Washington, DC, p. 1102.

Hyodo, H., Morozumi, S., Kato, C., Tanaka, K., and Terai, H. 1995. Ethylene production and ACC oxidase activity in broccoli flower buds and the effect of endogenous ethylene on their senescence. *Acta Hort.*, 394:191–198.

Isenberg, F.M.R. 1979. Controlled atmosphere storage of vegetables. *Hort. Rev.*, 1:337–394.

Izumi, H., Watada, A., and Douglas, W. 1996. Optimum O_2 and CO_2 atmosphere for storing broccoli florets at various temperatures. *J. Am. Soc. Hort. Sci.*, 121:127 131.

Joyce, D.C. 1988. Evaluation of a ceramic-impregnated plastic film as a postharvest wrap. *HortScience*, 23:1188.

Kader, A.A. 1986. Biochemical and physiological basis for effects of controlled and modified atmospheres on fruits and vegetables. *Food Technol.*, 40:99–100, 102–104.

Kader, A.A. and Saltveit, A.E. 2003. Atmosphere modification. In *Postharvest Physiology and Pathology of Vegetables*, Bartz, J.A. and Brecht, J.K., Eds. Marcel Dekker, New York, pp. 229–246.

Kader, A.A., Zagory, D., and Kerbel, E.L. 1989. Modified atmosphere packaging of fruits or vegetables. *CRC Crit. Rev. Food Sci. Nutr.*, 28:1–30.

Kaldy, M.S., Dormaar, J.F., Molnar, S.A., and Ragan, P. 1976. Browning of parsnips as related to varieties, soil conditions and length of storage. In *55th Annual Report of the Canadian Horticulture Council*, pp. 101–102.

Kasmire, R.F., Kader, A.A., and Klaustermeyer, J.A. 1974. Influence of aeration rate and atmosphere composition during simulated transit on visual quality and off-odor production by broccoli. *HortScience*, 9:228–229.

Kays, S.J. 1997. *Postharvest Physiology of Perishable Plant Products*. 2nd ed. AVI Book, Van Nostrand Reinhold, New York.

Kenigsbuch, D., Chalupowicz, D., Aharon, Z., Maurer, D., and Aharoni, N. 2002. The effect of CO_2 and 1-methylcyclopropene on the regulation of postharvest senescence of mint, *Mentha longifolia* L. *Postharvest Biol. Technol.*, 43:165–173.

King, G.A. and Morris, S.C. 1992. Physiological changes of broccoli during early postharvest senescence and through the preharvest-postharvest continuum. *J. Am. Soc. Hort. Sci.*, 119:270–275.

Kramer, G.F. and Wang, C.Y. 1989. Correlation of reduced chilling injury with increased spermine and spermidine levels in zucchini squash. *Plant Physiol.*, 76:479–484.

Laties, G.G. 1982. The cyanide-resistant, alternative path in higher plants respiration. *Ann. Rev. Plant Physiol.*, 33:519–555.

Lebermann, K.W., Nelson, A.I., and Steinberg, M.P. 1968. Post-harvest changes of broccoli stored in modified atmosphere. 1. Respiration of shoot and color of flower heads. *Food Technol.*, 22:143–146.

Lee, D.S. and Renault, P. 1998. Using pinholes as tool to attain optimum modified atmospheres in packages of fresh produce. *Packaging Technol. Sci.*, 11:119–130.

Leshuk, J.A. and Saltveit, M.E. 1990. Controlled atmosphere storage requirements and recommendations for vegetables. In *Food Preservation by Modified Atmospheres*, Calderon, M. and Barkai-Golan, R., Eds. CRC Press, Boca Raton, FL, pp. 315–352.

Lieberman, M. and Hardenburg, R.E. 1954. Effect of modified atmosphere on respiration and yellowing of broccoli at 75 degrees F. *Proc. Am. Soc. Hort. Sci.*, 63:409–414.

Lipton, W.J. 1972. Market quality of radishes stored in low-O_2 atmospheres. *J. Am. Hort. Sci.*, 97:164–167.

Lipton, W.J. and Harris, C.M. 1974. Controlled atmosphere effects on the market quality of stored broccoli (*Brassica oleracea* L., Italica group). *J. Am. Soc. Hort. Sci.*, 99:200–205.

Littmann, M.D. 1967. Effect of temperature on the post-harvest deterioration in quality of beans. *Queensland J. Agric. Anim. Sci.*, 24:271–278.

Loaiza, J. and Cantwell, M. 1997. Postharvest physiology and quality of cilantro (*Coriandrum sativum* L.). *HortScience*, 32:104–107.

Lurie, S. and Aharoni, N. 1997. Modified atmosphere storage of stone fruits. In *Proceedings of International Congress for Plastics in Agriculture (CIPA)*, Tel-Aviv, Israel, pp. 536–541.

Makhlouf, J., Castaigne, F., Arul, J., Willemot, C., and Gosselin, A. 1989a. Long-term storage of broccoli under controlled atmosphere. *HortScience*, 24:637–639.

Makhlouf, J., Willemot, C., Arul, J., Castaigne, F., and Emond, J. 1989b. Regulation of ethylene biosynthesis in broccoli flower buds in controlled atmospheres. *J. Am. Soc. Hort. Sci.*, 114:955–958.

Mannapperuma, J.D. and Singh, R.P. 1994. Design of perforated polymeric packages for the modified atmosphere storage of broccoli. In *Minimal Processing of Foods and Process Optimization*, Singh, R.P. and Oliveira, F.A.R., Eds. CRC Press, Boca Raton, FL, pp. 784–786.

Mattheis, J.P. and Fellman, J.P. 2000. Impact of modified atmosphere packaging and controlled atmosphere on aroma, flavor and quality of horticultural produce. *HortTechnology*, 10:507–510.

McCollum, T.G. 2004. Squash. In *The Commercial Storage of Fruits, Vegetables, and Florist and Nursery Crops*, Agricultural Handbook 66, Gross, K.C., Wang, C.Y., and Saltveit, M., Eds. USDA. www.ba.ars.usda.gov/hb66/contents.html.

McCollum, T.G. 1989. Physiological changes in yellow Summer squash at chilling and nonchilling temperatures. *HortScience*, 24:633–635.

Meir, S., Philosoph-Hadas, S., and Aharoni, N. 1992. Ethylene-increased accumulation of fluorescent lipid-peroxidation products detected during senescence of parsley by a newly developed method. *J. Am. Soc. Hort. Sci.*, 117:128–132.

Mencarelli, F., Lipton, W.J., and Peterson, S.J. 1983. Response of "zucchini" squash to storage in low-O_2 atmospheres at chilling and non-chilling temperatures. *J. Am. Soc. Hort. Sci.*, 108:884–890.

Millar, A.H., Marcel, H., Hoefnagel, N., Day, D.A., and Wiskich, J.T. 1996. Specifically of the organic acid activation of alternative oxidase in plant mitochondria. *Plant Physiol.*, 111:613–618.

Millenaar, F.F. and Lambers, H. 2003. The alternative oxidase: *in vivo* regulation and function. *Plant Biol.*, 5:2–15 (review).

Mir, N. and Beaudry, R.M. 2004. Modified atmosphere packaging. In *The Commercial Storage of Fruits, Vegetables, and Florist and Nursery Crops*, Agricultural Handbook 66, Gross, K.C., Wang, C.Y., and Saltveit, M., Eds. USDA, Washington, DC, www.ba.ars.usda.gov/hb66/contents.html.

Mohammed, M. and Brecht, J.K. 2003. Immature fruit vegetables. In *Postharvest Physiology and Pathology of Vegetables*, Bartz, J.A. and Brecht, J.K., Eds. Marcel Dekker, New York, pp. 671–690

Obenland, D.M., Aung, L.H., and Rij, R.E. 1994. Timing and control of methanethiol emission from broccoli florets induced by atmospheric modification. *J. Hort. Sci.*, 69:1061–1065.

Oliveira, F.A.R., Fonseca, S.C., Oliveira, J.C., Brecht, J.K., and Chau, K.V. 1998. Development of perforation-mediated atmosphere packaging to preserve fresh fruit and vegetable quality after harvest. *Food Sci. Technol. Int.*, 4:339–352.

Patterson, B.D., Jobling, J.J., and Moradi, S. 1993. Water relations after harvest: new technology helps translate theory into practice. In *Proceedings of the Australasian Postharvest Conference*, Gatton, Queensland, Australia, pp. 99–102.

Perez Trejo, M.S., Janes, H.W., and Frenkel, C. 1981. Mobilization of respiratory metabolism in potato tubers by carbon dioxide. *Plant Physiol.*, 67:514–517.

Pesis, E., Aharoni, D., Aharon, Z., Ben-Arie, R., Aharoni, N., and Fuchs, Y. 2000. Modified atmosphere and modified humidity packaging alleviates chilling injury symptoms in mango fruit. *Postharvest Biol. Technol.*, 19:93–101.

Phillips, C.A. 1996. Modified atmosphere packaging and its effects on the microbiological quality and safety of produce. *Int. J. Food Sci. Technol.*, 31:463–479 (Review).

Philosoph-Hadas, S., Jacob, D., Meir, S., and Aharoni, N. 1993. Mode of action of CO_2 in delaying senescence of chervil leaves. *Acta Hort.*, 343:117–122.

Philosoph-Hadas, S., Pesis, E., Meir, S., Reuveni, A., and Aharoni, N. 1989. Ethylene-enhanced senescence of leafy vegetables and green herbs. *Acta Hort.* 258:37–45.

Polderdijk, J.J. and van den Boogaard, G.J.P.M. 1998. Effect of reduced levels of O_2 and elevated levels of CO_2 on the quality of bunched radishes. *Gartenbauwissenschaft*, 63:250–253.

Porat, R., Weiss, B., Cohen, L., Daus, A., and Aharoni, N. 2004. Reduction of postharvest rind disorders in citrus fruits by modified atmosphere packaging. *Postharvest Biol. Technol.*, 33:35–43.

Renault, P., Houal, L., Jacquemin, G., and Chambroy, Y. 1994a. Gas exchange in modified atmosphere packaging. 2. Experimental results with strawberries. *Int. J. Food Sci. Technol.*, 29:379–394.

Renault, P., Souty, M., and Chambroy, Y. 1994b. Gas exchange in modified atmosphere packaging. 1. A new theoretical approach for micro-perforated packs. *J. Food Sci. Technol.*, 29:365–378.

Rodov, V., Ben-Yehoshua, S., Fierman, T., and Fang, D. 1995. Modified humidity packaging reduces decay of red bell pepper fruit. *HortScience*, 30:299–302.

Rodov, V., Ben-Yehoshua, S., Fishman, S., Peretz, J., and De la Asuncion, R. 1998a. Modified-humidity and modified-atmosphere packaging of bell pepper: mathematical modeling and experimental trials. In *The Post-Harvest Treatment of Fruit and Vegetables: Current Status and Future Prospects*, Woltering, E. et al., Eds. COST94, Oosterbeek, The Netherlands, pp. 61–66.

Rodov, V., Copel, A., Aharoni, Y., Aharoni, N., Nir, M., Shapira, A., and Gur, G. 1998b. Modified-atmosphere packaging of cucurbit vegetables. In *Proceedings of the COST915 Conference on Physiological and Technological Aspects of Gaseous and Thermal Treatments of Fresh Fruit and Vegetables*, Madrid, p. 26.

Rodov, V., Copel, A., Aharoni, N., Aharoni, Y., Wiseblum, A., Horev, B., and Vinokur, Y. 2000. Nested modified-atmosphere packaging maintains quality of trimmed sweet corn during cold storage and the shelf life period. *Postharvest Biol. Technol.*, 18:259–266.

Rodov, V., Copel, A., Aharoni, Y., Omer, S., Yehezkel, H., and Cohen, S. 1998c. Testing the quality of melon varieties of Charentais type. *Gan Sade VeMeshek* 7:59–65 (in Hebrew).

Rodov, V., Horev, B., Vinokur, Y., Copel, A., Aharoni, Y., and Aharoni, N. 2002. Modified-atmosphere packaging improves keeping quality of Charentais-type melons. *HortScience*, 37:950–953.

Rodov, V., Horev, B., Vinokur, Y., Goldman, G., and Aharoni, N. 2003. Modified-atmosphere and humidity packaging of whole and lightly processed cucurbit commodities: melons, cucumbers, squash. In *Postharvest Horticulture Conference*, Queensland, Australia, p. 39.

Rooney, M.L. 1995. Overview of active food packaging. In *Active Food Packaging*, Rooney, M.L., Ed. Blackie Academic & Professional, London, pp. 1–37.

Roy, S., Anantheswaran, R.C., and Beelman, R.B. 1996. Modified atmosphere and modified humidity packaging of fresh mushrooms. *J. Food Sci.*, 61:391–397.

Ryall, A.L. and Lipton, W.J. 1979. *Handling, Transportation and Storage of Fruits and Vegetables*, Vol. 1, *Vegetables and Melons*, 2nd ed. AVI Publ., Westport, CT.

Saltveit, M.E. 1997. A summary of CA and MA requirements and recommendations for harvested vegetables. In *Proceedings of the 7th International Controlled Atmosphere Research Conference*, Davis, CA, pp. 98–117.

Saltveit, M.E. 2004. Cucumbers. In *The Commercial Storage of Fruits, Vegetables, and Florist and Nursery Crops*, Agricultural Handbook 66, Gross, K.C., Wang, C.Y., and Saltveit, M., Eds. USDA, Washington, DC. www.ba.ars.usda.gov/hb66/contents.html.

Serrano, M., Martinez-Romero, D., Guillen, F., Castilo, S., and Valero, D. 2006. Maintenance of broccoli quality and functional properties during cold storage as affected by modified atmosphere packaging. *Postharvest Biol. Technol.*, 39:61–68.

Sfirakis, A. and Rogers, C.E. 1980. Effects of sorption modes on the transport and physical properties of nylon 6. *Polym. Eng. Sci.*, 20:294–299.

Shattuck, V.I., Kakuda, Y., and Yada, R. 1989. Sweetening of parsnip roots during short-term cold storage. *Can. Inst. Food Sci. Technol. J.*, 22:378–382.

Shattuck, V.I., Yada, R., and Lougheed, E.C. 1988. Ethylene-induced bitterness in stored parsnips, *HortScience*, 23:912.

Shellie, K.C. and Lester, G. 2004. Netted melons. In *The Commercial Storage of Fruits, Vegetables, and Florist and Nursery Crops*, Agricultural Handbook 66, Gross, K.C., Wang, C.Y., and Saltveit, M., Eds. USDA, Washington, DC. www.ba.ars.usda.gov/hb66/contents.html.

Shirazi, A. and Cameron, A.C. 1992. Controlling relative humidity in modified-atmosphere packages of tomato fruit. *HortScience*, 27:336–339.

Stoll, K. and Weichmann, J. 1987. Root vegetables. In *Postharvest Physiology of Vegetables*, Weichmann, J., Ed. Marcel Dekker, New York, pp. 541–553.

Suslow, T.V. and Cantwell, M. 1998a. Dry onions. Fresh Produce Facts, www.postharvest.ucdavis.edu.

Suslow, T.V. and Cantwell, M. 1998b. Squash (soft rind). Fresh Produce Facts, www.postharvest.ucdavis.edu.

Tian, M.S., Downs, C.G., Lill, R.E., and King, G.A. 1994. A role for ethylene in the yellowing of broccoli after harvest. *J. Am. Soc. Hort. Sci.*, 119:276–281.

Toivonen, P.M.A. 1992. The reduction of browning in parsnips. *J. Hort. Sci.*, 67:547–551.

Toivonen, P.M.A. 1997a. Quality changes in packaged, diced onion (*Allium cepa* L.) containing two different absorbent materials. In *Proceedings of the 7th International Controlled Atmosphere Research Conference*, Davis, CA, pp. 1–6.

Toivonen, P.M.A. 1997b. Non-ethylene, non-respiratory volatiles in harvested fruits and vegetables: their occurrence, biological activity and control. *Postharvest Biol. Technol.*, 12:109–125.

Trail, M.A., Wahem, I.A., and Bizri, J.N. 1992. Snap bean quality changed minimally when stored in low density polyolefin film package. *J. Food Sci.*, 57:977–979.

van den Berg, L. and Lentz, C.P. 1973. High humidity storage of carrots, parsnips, rutabagas and cabbage. *J. Am. Soc. Hort. Sci.*, 98:129–132.

Wang, C.Y. 1993. Approaches to reduce chilling injury of fruits and vegetables. *Hort. Rev.*, 15:63–95.

Wang, C.Y. 1977. Effect of aminoethoxy analog of rhizobitoxine and sodium benzoate on senescence of broccoli. *HortScience*, 12:54–56.

Wang, C.Y. 1979. Effect of short-term high CO_2 treatment on the market quality of stored broccoli. *J. Food Sci.*, 44:1478–1482.

Wang, C.Y. and Hruschka, H.W. 1977. Quality Maintenance and Anaerobic Off-Odor and Off-Flavor Production and Dissipation in Consumer Polyethylene-Packaged Broccoli, Marketing Research Report. USDA, Washington, DC, p. 1085.

Wang, C.Y. and Qi, L. 1997. Modified atmosphere packaging alleviates chilling injury in cucumbers. *Postharvest Biol. Technol.*, 10:195–200.

Watada, A.E. and Morris, L.L. 1966. Effect of chilling and non-chilling temperatures on snap bean fruits. *Proc. Am. Soc. Hort. Sci.*, 89:368–374.

Werner, B.G. and Hotchkiss, J.H. 2006. Modified atmosphere packaging. In *Microbiology of Fruits and Vegetables*, Sapers, G.M., Gorny, J.R., and Yousef, A.E., Eds. CRC Press, Boca Raton, FL, pp. 437–459.

Wills, R.B.H. 1968. Influence of water loss on the loss of volatiles by apples. *J. Sci. Food Agric.*, 19:354–356.

Wills, R.B.H. and Kim, G.H. 1996. Effect of ethylene on postharvest quality of green beans. *Aust. J. Exp. Agric.* 36:335–337.

Wills, R.B.H. and McGlasson, W.B. 1970. Loss of volatiles by apples in cool storage: a differential response to increased water loss. *J. Hort. Sci.*, 45:283–286.

Yahia, E.M. and Rivera, M. 1992. Modified atmosphere packaging of muskmelon. *Lebensm. Wiss. Technol.*, 25:38–42.

Yamauchi, N. and Watada, A.E. 1998. Chlorophyll and xanthophyll changes in broccoli florets stored under elevated CO_2 or ethylene containing atmosphere. *HortScience*, 33:114–117.

Zagory, D. and Kader, A.A. 1988. Modified atmosphere packaging of fresh produce. *Food Technol.*, 42:70–74, 76–77.

Zhuang, H., Barth, M., and Hilderbrand, D.F. 1994. Packaging influenced total chlorophyll, soluble protein, fatty acid composition and lipoxygenase activity in broccoli florets. *J. Food Sci.*, 59:1171–1174.

chapter six

MA packaging combined with other preserving factors

Jan Thomas Rosnes, Morten Sivertsvik, and Torstein Skåra

Contents

6.1 Introduction

Modified atmosphere (MA) packaging has been used for decades and the effects on different raw materials are well understood. The fact that gas modifications only are bacteriostatic has encouraged producers to use additional preservation technology to enhance or support safety and retard spoilage. By combining different preservation factors or techniques, reliable and mild multitargeted preservation can be achieved that facilitates improvements in food safety, quality, and economical aspects. Minimally processing and MA have been used for novel foods, but during the last years traditional foods have also been developed with a modern and fresh appearance. This chapter describes the general benefits of combining MA with other preservative methods, without special focus on nonrespiring foods. The main aim is to make a general reflection on the possibilities and limitations of the combined use of MA and other preservatives.

6.1.1 Background

The examination of gas atmosphere on the effect on food shelf life has a long history, and the first pioneers started experiments more than 100 years ago (Fränkel, 1889; Kolbe, 1882). Since the early development of MA it has been shown that MA can, on its own, inhibit the growth of microorganisms. High levels of CO_2 have a bacteriostatic effect on microorganisms, and properly designed MA packaging can double a product's shelf life (Davies, 1995). Still, only during the last two decades has MA packaging become a widely commercially used technology for the storage and distribution of foods. This trend is mainly driven by the modern consumer's demand for preprocessed products that have a fresh appearance and are convenient and easy to prepare, combined with the industry's desire for longer shelf life. Product safety is a major concern associated with the use of MA packaging. The desired suppression of spoilage microorganisms extends the shelf life, compared to food products stored in a normal air environment, which may create opportunities for slower-growing pathogenic bacteria. In particular, the growth of psychrotrophic pathogens in refrigerated ready-to-eat food may create a health risk before the product is overtly spoiled (Farber, 1991).

Some preservation procedures (e.g., chemical additives) used in food products act by inhibiting growth and not by inactivating microorganisms.

Thus, their contribution may be most beneficial when used against pathogens that form toxins in foods or those that need to reach high numbers to cause foodborne illnesses. In order to protect consumers at risk from foodborne illnesses or against microbes with low infectious doses, there is a need for the complete inactivation of pathogens and avoidance of recontamination of foods during processing, distribution, and preparation for consumption. For each specific MA-packaged product this must be done either prior to packaging or later by adjusting to correct preservation intensity in the product.

In food environments, the conditions for bacterial growth are seldom optimal since most nutrients are in the form of complex substrates, and hence generally stressful for bacteria. Also, free moisture may be restricted, acids or other chemicals may be at stressful levels, and competition from other microorganisms present may limit growth. Replacing the normal atmosphere with a modified atmosphere, i.e., altered concentrations of O_2, CO_2, N_2, or even other gases, will add additional stress to microorganisms and change the composition of the initial microbial flora.

6.1.2 *Preservation, hurdles, and homeostasis*

The basis of hurdle technology (Leistner, 1995) is the combination of traditional and innovative preservation techniques in small doses with the aim of establishing a series of factors that interact cumulative or synergically to control the microbial population in food. The hurdle concept is widely accepted as a food preservation strategy, but its potential, when using MA, has yet to be fully realized. The intelligent selection of hurdles in terms of the number required, the intensity of each, and the sequence of applications to achieve a specified outcome is supposed to have significant potential for the future (McMeekin et al., 2002).

In this way, the same inhibitory effect is reached with a lower-intensity treatment, inhibiting or delaying the multiplication of the microorganisms surviving the treatment. This fact contributes to a better preservation of the sensory properties after treatment. The effectiveness of the different preservation methods can be studied through different responses, such as microbial growth, sensory analysis, changes in lipids or proteins, volatile compounds, etc.

The most commonly used preservation methods, which are used at different intensities, are high temperature (heat treatment), low temperature (refrigerated or frozen storage), water activity (a_w), acidity (pH), redox potential (Eh), preservatives, and a competitive flora (Leistner, 1992). These may be combined at intensities adjusted to the product and required shelf life. Some of these combinations are still in the exploratory and developmental stages, whereas other methods have obtained regulatory approval and been introduced in the food industry.

Homeostatsis is the tendency toward uniformity and stability in the internal status of living organisms. For instance, the maintenance of a defined pH value within narrow limits is a prerequisite and feature of all living cells, and this applies to higher organisms as well as microorganisms. Homeostasis of microorganisms is a key phenomenon in foods because if it is disturbed by the preservation methods, the microorganisms will not multiply, i.e., they will remain in the lag phase or may even die before their homeostasis is reestablished. Hence, in actual fact, the preservation of food is achieved by disturbing, temporarily or permanently, the homeostasis of microorganisms in the food (Leistner, 1995). In most foods, microorganisms are able to operate homeostatically in order to react to the environmental stresses imposed by the applied preservation procedures. Applying additional preservation will inhibit the repair of disturbed homeostasis, and this requires extra energy from the microorganisms concerned. In MA products energy depletion increases as the intensity or concentration of preservation is increased, and the restriction of the energy supply will inhibit the repair mechanisms of the microbial cells and lead to growth inhibition or death.

6.2 *Targeting specific groups of microorganisms*

To release the potential of the emerging preservation technologies, combined with MA, systematic kinetic data are required describing their efficiency against key target microorganisms. The number and types of microorganisms in the raw material directly affect the efficiency of MA in inhibiting both spoilage organisms and pathogens. The use of additional preservation to packaged food requires insight regarding which part of the bacterial population is inhibited and which is not. It is important to stress that the shelf life extension obtained with MA does not always give the same extension in safety. Pathogenic bacteria may benefit when the competing flora is inhibited, e.g., *Listeria monocytogenes* increased in numbers on raw chicken in a 72.5:22.5:5 (CO_2:N_2:O_2) atmosphere at 4°C, irrespective of a decrease in the aerobic spoilage flora (Wimpfheimer et al., 1990).

Many MA-packaged products of meat, vegetable, and seafood origin have common key target organisms. For chilled products, psychrotrophic pathogens are targeted, while in heat-treated ready-meals the spore-forming *Clostridium* and *Bacillus* species are the target organisms.

Five foodborne pathogenic bacteria are known to grow below 5°C: *Bacillus cereus*, nonproteolytic *Clostridium botulinum* (types B, E, and F), *L. monocytogenes*, *Yersinia enterocolitica*, and *Aeromonas hydrophila*. Hence, the ability of MA to inhibit the growth of these organisms in foods under refrigerated storage is of vital importance, and additional preservation factors have therefore been combined with MA (Table 6.1). The main cause of concern, however, is the possible growth of nonproteolytic *C. botulinum*, because it is both anaerobic and low temperature tolerant. Of particular concern is the fact that it may grow and produce toxins in the product even before spoilage is detectable to the consumer.

Few nonthermal treatments can currently be relied upon to inactivate bacterial spores. Hence, low-temperature storage must be combined with an additional preservation hurdle, such as acidic formulation or salt, to prevent spore outgrowth.

Most food spoilage mold species have an absolute requirement for oxygen and appear to be sensitive to high levels of CO_2. Consequently, foods with low a_w values, such as bakery products, susceptible to spoilage by molds, can have their shelf life extended by MA. A number of yeasts are capable of growing in the complete absence of oxygen, and most are comparatively resistant to CO_2. Although MA can inhibit the growth of bacterial and fungal spoilage microorganisms, its effect on the survival of enteric viruses, including hepatitis A viruses (HAVs), has not been well investigated. Due to contact with contaminated water, both mussels and lettuce packaged in MA may be vehicles in the transmission of HAV (Cliver, 1997). Experiments by Bidawid et al. (2001) indicated that MA does not influence HAV survival when present on the surface of fresh produce with high CO_2 levels. This has been attributed to the inhibition of spoilage-causing enzymatic activities in the lettuce, which have reduced the exposure of viruses to potential toxic by-products.

6.3 Additional preservation methods

Products are generally preserved by preventing the access of microorganisms to foods, or by inactivating the microorganisms in case they have gained access, or by preventing or slowing down their growth in case they have gained access but not been inactivated. These approaches can be modified and further developed in order to reduce the severity of the more extreme techniques. New combinations of currently existing techniques are often tested in order to inhibit microbial growth. Approaches where lower intensity or lower concentration of the preservation techniques is used cause bacterial inactivation and bacterial growth inhibition to overlap. It is the safety level, the quality level, or the extent of inactivation or growth inhibition of the target organisms that determines the final use of the chosen preservation method.

6.3.1 Prepackaging

Depending on the product and the desired shelf life, a number of measures and processing steps may be included prior to packaging. Although hygienic production is not a preservation method, ingredients or raw material used in MA packaging should always be of superior quality, i.e., low bacterial numbers and preferably without pathogenic bacteria.

6.3.1.1 Hygiene

Good production hygiene is a prerequisite for obtaining fresh products with increased shelf life, and preservation should not be used to compensate for inadequate hygiene or poor raw material quality. A strategy for the control of pathogens and, to a large extent, spoilage organisms is to reduce or

Table 6.1 Preservations Used to Inhibit Specific Pathogens in Combination with MA

Organism	Relevant food	Preservation	Reference
Clostridium perfringens	Beef	Ozone or heat treatment	Novak and Yuan, 2004
Nonproteolytic *Clostridium botulinum*	Ready-to eat food	Irradiation	Lambert et al., 1991
	Steamed rice	Heat process	Kasai et al., 2005
	Raw, cooked, and sterilized surimi nuggets	Microbial inhibition by *Bacillus* species	Lyver et al., 1998
	Hard-boiled eggs	Heat process	Claire et al., 2004
	White asparagus	Heat process	Valero et al., 2006a
	Cold and hot smoked trout	NaCl, smoke	Dufresne et al., 2000
	Laboratory medium	NaCl	Gibson et al., 2000
Listeria monocytogenes	Fish, meat, vegetables, fresh produce	Competitive microbial flora	Bennik et al., 1999; Francis and O'Beirne, 1998; Liserre et al., 2002; Wimpfheimer et al., 1990
	Cooked pork	Nisin	Fang and Lin, 1994; Szabo and Cahill, 1998
	Cooked meat	Na-lactate	Devlieghere et al., 2001; Pothuri et al., 1996; Zeitoun and Debevere, 1991
	Cooked ham	Protective culture	Vermeiren et al., 2006
	Turkey meat	Irradiation	Thayer and Boyd, 1999, 2000; Lafortune et al., 2005
	Ready-to-eat carrots, antimicrobial compounds	Irradiation	Thayer and Boyd, 1999, 2000; Caillet et al., 2006a, 2006b; Lafortune et al., 2005
	Vegetables	pH	Francis and O'Beirne, 2001
	Meat	Oregano essential oils	Tsigarida et al., 2000
	Fresh-cut produce	High O_2 level	Jacxsens et al., 2001
Yersinia enterocolitica	Pork	Lactate	Barakat and Harris, 1999
	Poultry	Lactic acid	Grau, 1981
	Minced meat	Background flora	Kleinlein and Untermann, 1990
	Chicken meats	Heat process	Wei et al., 2001
	Beef	Low temperature	Gill and Reichel, 1989

Aeromonas hydrophila	Fish, shellfish, mussels	Heat	Devlieghere et al., 2000b
	Lamb meat	pH	Doherty et al., 1996
Salmonella	Ground beef	Irradiation	Chiasson et al., 2004, 2005
	Beef fillets	Oregano essential oils	Skandamis et al., 2002
	Cherry tomatoes	Gaseous ozone	Das et al., 2006
	Chicken portions	Gaseous ozone	Al Haddad et al., 2005
	Poultry	Sorbate	Elliott and Gray, 1981

Source: Adapted from Rosnes, J.T. et al., in *Novel Food Packaging Techniques*, Ahvenainen, R., Ed., Woodhead Publishing Ltd., Cambridge, England, 2003, pp. 287–311.

eliminate the initial microbial load and to minimize further contamination. Since MA-packaged products are hermetically sealed, recontamination is eliminated and the hygienic prepackaging conditions are the most important steps. An appropriate design and construction of the prepackaging premises is therefore necessary to limit the entry and spread of microorganisms.

6.3.1.2 *Low temperature (freezing, partial freezing, superchilling)*

Since both enzymatic and microbiological activities are substantially influenced by temperature, low and stable temperature is a general prerequisite for fresh MA products in particular. At temperatures below 10°C, many bacteria are unable to grow, and even psychrotrophic organisms grow slowly and sometimes with extended lag phases as temperatures approach 0°C.

Freeze-chilling involves freezing and frozen storage followed by thawing and chilled retail display (Gormley et al., 2003). It has an advantage over chilling as it allows bulk preparation of frozen products followed by controlled batch release of thawed products into the chill chain. It has the logistic advantage of enabling chilled products to reach foreign markets more easily. Lasagna has been found suitable for freeze-chilling (Redmond et al., 2005). Lasagna stored frozen (up to 12 months) followed by storage at 4°C for 6 days was still found acceptable by an experienced taste panel. The MA packaging had little impact on the quality of freeze-chilled storage.

Temperature can be used to achieve special effects in MA products. Guldager et al. (1998) and Bøknæs et al. (2001) have found that frozen (–20°C) and thawed cod fillets in MA had longer shelf life than raw cod in MA. This shelf life extension was most likely due to the inactivation of the spoilage bacterium *Photobacterium phosphoreum* during frozen storage. Frozen fillets as a raw material provide a more stable MA product and allow increased flexibility for production and distribution. A similar effect was found when frozen and thawed salmon was packaged in MA; the freezing eliminated *P. phosphoreum* and extended the shelf life of MA salmon at 2°C by 1 to 2 weeks (Emborg et al., 2002).

Studies with whole gutted salmon have shown that MA can be combined with superchilling to further extend the shelf life and safety of fresh fish (Rosnes et al., 1998, 2001; Sivertsvik et al., 1999a). In this technique the temperature of the fish is reduced to between 1 and 2°C below the initial freezing point and some ice is formed inside the product (Gould and Peters, 1971). Under normal conditions, the gas atmosphere will insulate the product, thus prolonging the required time for satisfactory chilling. Superchilling can decrease the temperature before packaging and increase stored refrigeration capacity during storage, thereby significantly decreasing microbial growth at temperatures of 2 to 6°C, which is often found in chilled retail counters. Sikorski and Sun (1994) found that superchilling can store enough refrigeration capacity to keep a core temperature of <0°C during the first 3 weeks of chilled storage. A shelf life extension of 7 days has been obtained for superchilled fish when compared to traditionally ice stored (Leblanc and

Leblanc, 1992). Commercially prepared smoked blue cod (*Parapercis colias*), packaged aerobically, under vacuum and in carbon dioxide-controlled atmosphere packs, was stored at 3 and –1.5°C (Penney et al., 1994). Product stored aerobically on overwrapped polystyrene trays spoiled by 14 and 28 days, respectively, and that in vacuum packs spoiled by 14 and 35 days, respectively, when held at 3 and –1.5°C. In contrast, product in CO_2 packs remained acceptable until the 3 and –1.5°C storage trials ended after 49 and 113 days, respectively. Superchilled salmon steaks in MA had an acceptable microbiological quality after 22 days at 0°C (but were rejected due to off-odor after 17 days), whereas salmon steaks in air had an acceptable microbiological quality for only 8 days at 0°C (Rosnes et al., 2001). Superchilled MA-packaged salmon fillets (–2°C) had a shelf life of more than 24 days based on sensory and microbiological analysis (Sivertsvik et al., 2003), while MA-packaged wolf fish portions at –1°C only had a maximum shelf life of 15 days compared to 8 to 10 days in air at the same temperature (Rosnes et al., 2006).

6.3.1.3 Alteration of pH

The product pH affects enzyme activity and growth of microorganisms, thus influencing spoilage (Ashie et al., 1996). The CO_2/bicarbonate ion has an observed effect on the permeability of the cell membranes (Daniels et al., 1985), and CO_2 is able to produce rapid acidification of the internal pH of the microbial cell, with possible ramifications relating to metabolic activities. Fey and Regenstein (1982) noted that CO_2 did not lower the pH of the fish. In most fish products, with the isoelectric point of fish proteins approximately 5.5, lowered pH could lead to reduced water-holding capacity as well as textural changes.

The effect of pH on the solubility of CO_2 has been modeled, revealing that higher amounts of CO_2 can be dissolved in aqueous foods with high pH levels (Devlieghere et al., 1998). In spite of the fact that higher concentrations of CO_2 are dissolved at high pH levels, the preservative effect seems to be greater at low pH levels when in combination with a modified atmosphere. Beef with pH 6.3 in 100% N_2 at 5°C supported the growth of *Enterobacter cloacae* but not at pH 5.4 (Grau, 1981). For example, growth of *Serratia liquefaciens* was inhibited in beef products with pH 5.4 in 100% N_2 at 5°C but grew to levels of 10^8 CFU/g in 8 days on meat with pH 6.3. *Y. enterocolitica* failed to grow on beef ranging in pH from 5.4 to 5.9 under 100% N_2 but grew at pH 6.0 to 6.2. Under aerobic conditions pH had little effect on the growth of *Y. enterocolitica*.

In muscle foods, the initial decline in pH is reversed during later stages of postmortem changes as a result of the decomposition of nitrogenous compounds (Ashie et al., 1996).

6.3.1.4 Salt

Salt (NaCl) binds water, reduces a_w, and consequently inhibits bacterial growth. Focus on the adverse health effects of a high sodium intake on blood pressure has led to a sharp decrease in salt consumption. Hence, salt is used less for the preservation of food and more for taste purposes. MA-packaged foods with high water content have been reported to cause a higher drip loss in CO_2-rich atmospheres (Rosnes et al., 2006). Bøknæs et al. (2001) studied the effect of trimethylamine oxide (TMAO) and NaCl content of thawed cod fillets on drip loss and shelf life. They suggest that varying NaCl (from the environment) and TMAO content in fish from different fishing grounds may explain different shelf lifes of thawed MA-packaged cod.

A farmed Japanese fish species (*seriola aurevittata*) obtained extended shelf life after being subjected to a brine solution (5% NaCl) and then packaged in different gas mixtures (Mitsuda et al., 1980). Pastoriza et al. (1998) examined the effects of an optimum gas mixture on hake slices when combined with a NaCl dip. A delay in chemical, microbiological, and sensorial alteration was found, and total volatile bases (TVBs) and microbiological levels were significantly lower when MA-stored samples had been previously dipped in NaCl solution. Additional effects, which are important for MA-packaged fish, were reduced exudation, higher water-binding capacity, and increased time before MA-stored samples were rejected due to off-odors. The same effect of reduced drip loss after NaCl dipping was observed for cod (Bjerkeng et al., 1995). The antimicrobial contribution of NaCl in a food system may also be influenced by the presence of other preservatives, e.g., benzoate, sorbate, phosphates, antioxidants, spices, and liquid smoke.

6.3.1.5 Polyphosphates

Polyphosphates are often used in muscle foods to improve water binding and texture (increased firmness). They may enhance the preservative effect by (1) acting as metal ion chelators, (2) acting as pH buffers, (3) interacting with proteins to promote hydration and water-binding capacity, and thus (4) preventing lipid oxidation and microbial growth (Ellinger, 1972). Fresh hake slices packed in a modified atmosphere with a cryoprotectant agent (sodium tripolyphosphate) and under chilling conditions showed a reduction of exudate production. However, no improvement was achieved in microbiological and chemical parameters used as quality control factors compared with those packed only in a MA (Alvarez et al., 1996). Masniyom et al. (2005) studied the combination effect of three phosphate compounds and MA on quality and shelf life of seabass slices and found that although there was little effect on microbial growth, sodium pyrophosphate was most effective in reducing the amount of exudates and keeping the sensory properties. Effects were examined on beef steaks injected with potassium lactate, sodium chloride, sodium tripolyphosphate, and sodium acetate (Knock et al., 2006). Injection enhancement of beef will increase brown-roasted and beef flavors while limiting rancid flavor development. These effects are noticeable just 2 days postinjection and are more pronounced several days postinjection. Effects of injections containing phosphate have also been examined on beef round muscles (Seyfert et al., 2005) and beef and bison steaks (Pietrasik et al., 2006).

6.3.1.6 Organic acids

Lipophilic organic acids, e.g., acetic, lactic, citric, and malic acids, all have antimicrobial properties that are utilized by the food industry for food preservation. The undissociated molecule of the organic acid or ester, claimed to be responsible for the antimicrobial activity, can penetrate the cell membrane and accumulate in the cytoplasm. The antimicrobial activity of lactate has also been attributed to its lowering of the water activity of the food product, but this can only partly explain its antimicrobial effects on meat products (Debevere, 1989; Houtsma et al., 1993). The ability of the acid to enter the cell in the undissociated form, dissociate within the cell, and acidify its interior has been proposed as a possible means of action. Salts of organic acids, such as sodium and potassium lactate, are fully dissociated in aqueous solutions, and at the pH of an unfermented meat product, which is typically 6.0 to 6.5, the concentration of the undissociated form of the added lactate is low. The increased permeability of cellular membranes for lactic acid at higher pH values may be an important factor in understanding the antimicrobial activity of Na-lactate, as observed in neutral food media and food products. Cooked meat products packaged in oxygen-free atmospheres will spoil due to psychrotrophic lactic acid bacteria (Borch et al., 1996), but with addition of Na-lactate the shelf life will be prolonged (Debevere, 1989). Devlieghere et al. (2000a) examined the shelf life of MA-packaged cooked meat products after the addition of Na-lactate and found that a significant shelf life extension was obtained through the use of Na-lactate, and this was more pronounced at low temperatures. A synergistic effect was reported between Na-lactate and CO_2, which could partly be explained by the pH-lowering effect of CO_2. Also, beef steaks treated with a low level of lactic acid (1.5%) alone or supplemented with antioxidants and packaged in 40% CO_2 had an extended shelf life (Djenane et al., 2003). The use of buffered lactic acid systems on poultry enhanced the decontaminating effect and increased the shelf life of poultry (Zeitoun and Debevere, 1992). Buffered lactic acid treatment and MA have also shown an inhibitory effect on *L. monocytogenes* and increased the shelf life of this product (Zeitoun and Debevere, 1991). Jimenez et al. (1997) first characterized the spoilage flora in fresh chicken breast at 4°C in MA. Thereafter, they combined the use of acetic acid treatment (1%) and modified atmosphere (70% CO_2 and 30% N_2) packaging of chicken breast portions (Jimenez et al., 1999). Acetic acid treatment produced decreases in counts in all genera studied, and the combined action of acid and MA extended the samples' shelf life.

In crawfish tail meat homogenate a combination of 200 $\mu g\ g^{-1}$ monolaurin, 0.5% lactic acid, and MA had the greatest potential to inhibit growth of this bacterium (Oh and Marshall, 1995). Recent studies of fresh pork ham portions sprayed with 1, 2, 4, and 6% lactic acid solution at an amount of 2 ml/100 g and packaged in a modified atmosphere of 45% O_2 and 20% CO_2 show that the bacterial load was reduced, but the better effects could be observed only at 4 and 6% lactic acid concentration (1 log reduction in bacterial counts) (Shrestha and Min, 2006). There was, however, a significant loss of color and overall acceptability scores along with a significant increment in the thiobarbituric acid (TBA) values. From this finding, it was suggested that a suitable color and lipid stabilizer should be used to further enhance the effect of lactic acid application in fresh MA-packaged meat.

6.3.1.7 Essential oils

Essential oils (EOs) are aromatic oily liquids obtained from plant material (flowers, buds, seeds, leaves, twigs, herbs, fruits, and roots). They can be obtained by expression, fermentation, enfleurage, or extraction. EOs are well-known inhibitors of microorganisms (Burt, 2004), and they are regarded as natural alternatives to chemical preservatives. In a recent work, antimicrobial activity of cinnamon and clove oils under modified atmosphere conditions was systematically examined in a laboratory (Matan et al., 2006). The volatile gas phase of combinations of cinnamon oil and clove oil showed promising potential to inhibit growth of spoilage fungi, yeast, and bacteria normally found on intermediate-moisture foods when combined with a modified atmosphere comprising a high concentration of CO_2 (40%) and low concentration of O_2 (<0.05%). In general, EOs' practical application is limited because of flavor considerations, and their effectiveness is moderate due to their interaction with food ingredients and structures. The results obtained by Skandamis et al. (2002) showed that volatile compounds of oregano essential oil affect both the growth and metabolic activity of the microbial association of meat stored at modified atmospheres. This inhibition was not as strong as that found in the contact of pure essential oil with microorganisms, when added directly on the surface of meat (Skandamis and Nychas, 2001). These authors conclude that the volatile compounds of oregano essential oils can expand their application to extend the shelf life of meat by (1) delaying growth of specific spoilage organisms, (2) inhibiting or restricting metabolic activities that cause spoilage through the production of spoilage microbial metabolites, and (3) minimizing the flavor concentration.

In fish, as in meat products, a high fat content appears to reduce the effectiveness of antibacterial EOs. For example, oregano oil at 0.5 $\mu l\ g^{-1}$ is more effective against the spoilage organism *P. phosphoreum* on MA-packaged cod fillets than on salmon (Mejlholm and Dalgaard, 2002).

Eugenol, menthol, or thymol in combination with MA have been found to maintain quality and safety by reducing microbial spoilage in sweet cherry (Serrano et al., 2005) and 'Crimson' table grapes (Valverde et al., 2005; Valero et al., 2006b). The combination of MA and EOs has also been examined for foods like bread (Nielsen and Rios, 2000) and banana (Ranasinghe et al., 2005).

6.3.1.8 $Na_2CaEDTA$

Low levels (25 to 500 ppm) of the chelating agent $Na_2CaEDTA$ have been approved for use in some foods. Although this compound rarely affects the microorganisms found in seafood (Dalgaard et al., 1998), it has been found to inhibit *P. phosphoreum* in MA-packaged cod. In naturally contaminated MA-packaged cod fillets, 500 ppm $Na_2CaEDTA$ reduced the growth rate of *P. phosphoreum* by 40% and shelf life was increased proportionally by 40%. In aerobic stored cod fillets, however, other microorganisms were responsible for spoilage and $Na_2CaEDTA$ had no influence on shelf life. The shelf life of poultry has been improved by a treatment of EDTA and nisine under modified atmosphere packaging (Cosby et al., 1999). EDTA has been included as an agent in other combined packaging systems, e.g., to prevent discoloration in MA-packaged pork bones (Nicolalde et al., 2006), for postharvest treatment of MA-packaged litchi (Sivakumar and Korsten, 2006b), and for preservation of MA-packaged blue whiting (Cabo et al., 2005).

6.3.1.9 Smoke

Smoked products were originally well preserved by reduction of water activity through salting and drying, as well as the potential bacteriostatic effect of smoke compounds (Sunen, 1998). These preservation factors are less in use today, to accommodate consumer preferences for less salt. Thus, additional preservation may be required to achieve the required shelf life. Penney et al. (1994) studied the bacterial growth in smoked blue cod (*P. colias*), packaged aerobically, under vacuum and in carbon dioxide-controlled atmosphere packs chilled at +3°C and superchilled at –1.5°C. They found that the two-fold extension of shelf life by the use of carbon dioxide-controlled atmosphere packaging was attributable to a significant extension of the lag phase before spoilage microflora proliferation commenced and to the selection of low-spoilage-potential lactic-acid-bacteria-dominated flora. Bell et al. (1995) inoculated samples of smoked blue cod with a two-strain cocktail of one of the psychrotrophic pathogens *A. hydrophila*, *L. monocytogenes*, or *Y. enterocolitica* and stored the samples either vacuum or carbon dioxide packed, at 3 or –1.5°C. In vacuum packs all three psychrotrophic pathogens were able to grow during storage at 3°C. Reduction of the storage temperature to –1.5°C retarded but did not prevent pathogen proliferation. Under carbon dioxide, only *A. hydrophila* was able to grow at 3°C and then only after a 21-day lag period. None of the psychrotrophic pathogens were able to grow under carbon dioxide at –1.5°C.

Swordfish fillets, dry-salted and treated with liquid smoke, were sliced and packaged in vacuum conditions and under MA (5% O_2, 45% CO_2, and 50% N_2) and stored at 4°C to determine the effect of different packaging on shelf life (Muratore and Licciardello, 2005). For this product the time to sensory rejection (at least ≤5 in a scale from 1 to 9) was shorter for MA-packaged samples (12 days) than for vacuum-packaged ones (42 days).

6.3.1.10 High temperature (thermal processing)

MA can be applied to foods that have been heated, chilled, and portioned or sliced prior to packaging. Products that are heat-treated to reduce their original microbiological population are also called refrigerated and processed foods of extended durability (REPFEDs), or simply chilled ready-to-eat or ready-to-cook products.

Several guidelines and codes of practice are made with respect to safe production of REPFEDs (ECFF, 1996; ACMSF, 1992, 1995; Betts, 1996). Most of these are targeted at preventing growth and toxin production by nonproteolytic *C. botulinum*. In these guidelines several product extremes are recommended, e.g., storage temperature, heat treatment, and shelf life. According to guidelines from the European Chilled Food Federation (ECFF, 1996), there are three main designs for cook–chill products:

1. Products designed to be free from vegetative pathogens such as *L. monocytogenes* should get a pasteurization of 2 minutes at 70°C.
2. Products designed to be free from vegetative pathogens such as *L. monocytogenes* and with control of psychrotrophic (nonproteolytic) *C. botulinum* should obtain a pasteurization of 2 minutes at 70°C (and additional hurdles).
3. Products designed to be free from vegetative pathogens such as *L. monocytogenes* and from the spores of psychrotrophic (nonproteolytic) *C. botulinum* should obtain a pasteurization of 10 minutes at 90°C.

Due to the possibility of recontamination of the cooked products before assembly and packaging, hygienic production and controlled heating and chilling must be performed in a high-risk area. Specific requirements are needed for the packaging materials, seal integrity, and gas composition. These MA-packaged ready-to-eat meals have a shelf life of 5 to 10 days, depending on the amount of heat used and the hygienic conditions prior to sealing of the packages. Guidelines with recommendations for the MA packaging of cook–chill products are available (Day, 1992). New and improved heating technologies, in combination with other hurdles, can contribute to the continued safe development of REPFEDs.

According to Phillips (1996), the increase in shelf life of MA-packaged products may provide sufficient time for human pathogens to multiply to levels that render the seafood unsafe while still edible. Microbial safety is highly important since MA-packaged cook–chill food is commonly used in many nursing homes, hospitals, and canteens. The popularity of such products is increasing since they effectively can be produced at a central kitchen, and still give a minimal loss of sensory quality and preserve much of the nutrients.

Thus, due to the extended shelf life of heat-treated MA-packaged foods, much effort has been focused on safety issues and microbial risk assessments. Challenge studies were performed in MA-packaged steamed rice, using a mixture of proteolytic *C. botulinum* strains (Kasai et al., 2005). No neurotoxins were detected, and organoleptically acceptable conditions persisted for 24 weeks at 15% oxygen conditions. Under 0% oxygen conditions, neurotoxin was detected after 1 week, and an atmosphere containing 10% oxygen to prevent activity of obligate anaerobes was therefore recommended. Hintlian and Hotchkiss (1987) studied the growth of *Pseudomonas fragi*, *Salmonella typhimurium*, *Staphylococcus aureus*, and *Clostridium perfringens* on inoculated roast beef under modified atmospheres (75% CO_2, 15% N_2, and 10% O_2) at two abuse temperature regimes. They found that the MA inhibited the growth of *S. aureus*, and that atmospheres with 5 to 10% O_2 inhibited the outgrowth of *C. perfringens*. Heat-shocked or non-heat-shocked (control) *Y. enterocolotica* was added to ground pork in MA (Shenoy and Murano, 1996). All samples were heat-treated at 55°C for 15 minutes and then stored at 25 or 4°C. The results showed that there was no significant difference in growth rates between heat-shocked and control samples under different storage atmospheres, and *Y. enterocolotica* grew rapidly and the survivors remained pathogenic. Pressurized CO_2 exerts a lethal effect on bacteria, yeasts, and molds. In a study where heat and high pressure with CO_2 were combined, Ballestra and Cuq (1998) showed that this combination increases the inactivation rate of *Bacillus subtilis* spores and *Bacillus fulva* ascospores (at temperatures above 80°C) and *A. niger* conidia (above 50°C).

Traditional dishes with ready-to-eat food packaged in MA were examined by Murcia et al. (2003). The study included foods like lentil soup, meat stew, and meat, legume, and vegetable soup stored for 7 and 29 days at 3°C in MA. The modified atmosphere packaging was effective for prolonging the shelf life of the studied products up to 29 days with minimal changes in the proximate composition.

The shelf life of major commercial cooked meat products (i.e., bologna sausage, Italian-type cooked sausage, and cooked ham) was tested at chilled conditions and different combinations of headspace CO_2 concentrations (Szalai et al., 2004). A CO_2 level of 60% had beneficial effects on both microbiological and sensory properties of sliced sausages and cooked ham.

Cooked shrimp has a short shelf life at chilled temperatures. Brined shrimp containing NaCl, reduced pH, and preservatives (benzoate, sorbate), however, has a shelf life of 9 weeks at chilled temperatures. Spoilage and safety of cooked, brined, and MA-packed shrimp were studied at 0, 5, 8, 15, and 25°C. Shrimp from two sources, cold and warm waters, was brined in a sodium chloride brine containing benzoic, citric, and sorbic acids. Shelf life was above 7 months at 0°C but only 4 to 6 days at 25°C (Dalgaard and Jørgensen, 2000). A safety evaluation on MA-packaged shrimp was carried out with challenge tests, including *L. monocytogenes* or *C. botulinum* (Mejlholm et al., 2005). Ready-to-eat shrimp can also obtain an extended shelf life without preservatives if soluble gas stabilization (SGS) is used in combination with MA and chilled storage (Sivertsvik and Birkeland, 2006). Both cooked crab legs and cooked shrimp obtain a substantial increased shelf life in MA, even without any supplementary preservatives, as shown in Figure 6.1 and Figure 6.2.

Postharvest heating is a noncontaminating physical treatment that delays the ripening process, reduces chilling injury, and controls activity of pathogens. Due to these beneficial effects, heat treatments are currently used commercially for quality control of fresh products (Ferguson et al., 2000). In

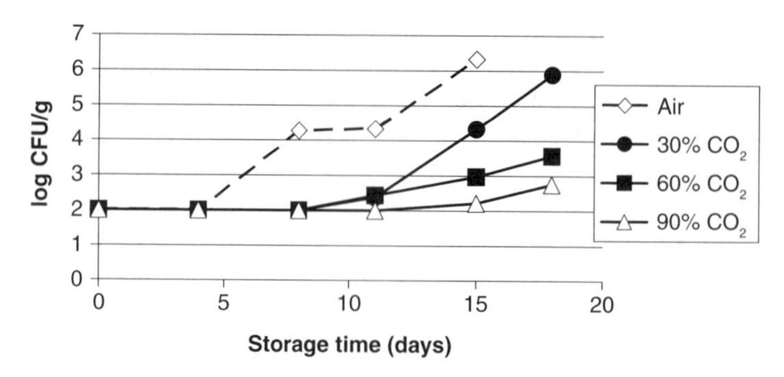

Figure 6.1 Total viable counts (log CFU/g) in cooked crab legs packaged in different mixtures of CO_2 and stored at 4°C.

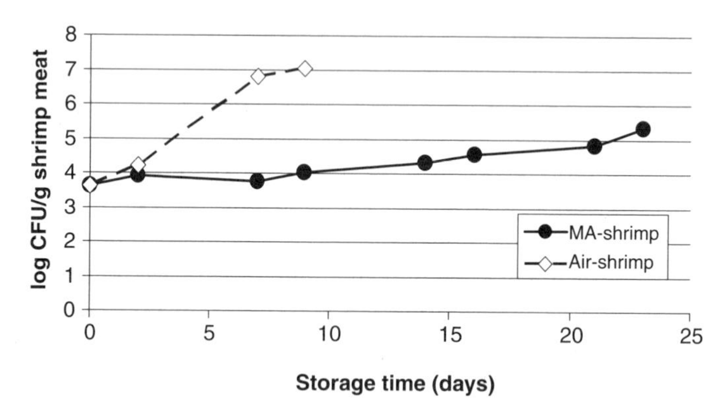

Figure 6.2 Total viable counts (log CFU/g) in cooked shrimp packaged in MA (60% CO_2:40% N_2) and stored at 3°C.

recent years, the use of combined techniques in the postharvest handling of fresh products has increased and numerous authors have obtained promising results using a combined treatment. Mature green cherry tomato fruit was dipped in hot water (39°C for 90 minutes) and subsequently stored in plastic films with various O_2 and CO_2 permeabilities at 15°C. The combination of hot water treatment and lower O_2 had a substantial effect in delaying color development on mature green cherry tomatoes (Ali et al., 2004). Mild heat treatment is also evaluated for use on litchi (Sivakumar and Korsten, 2006a) and on fresh-cut peach (Steiner et al., 2006) packaged in MA.

6.3.1.11 *New emerging and alternative techniques*

Light pulses have been used successfully as a new technique for the inactivation of bacteria and fungi on the surface of food products when the major composition of the emitted spectrum is UV (Marquenie et al., 2002). High UV doses can cause damage to the treated tissue, e.g., of fruits and berries. The possibility of decreasing the treatment intensity by combining two or more treatments to preserve the food quality without decreasing the inactivation properties therefore appears promising. Microbial reductions of up to 2.04 log units have been reported by combination of UV and MA. The shelf life is, however, not always extended (Gomez-Lopez et al., 2005), and Lopez-Rubira et al. (2005) found that microbial counts were not systematically reduced throughout the shelf life of processed pomegranate arils in MA. An appropriate dose should therefore be used that can reduce microbial loads without adversely affecting sensorial quality of fresh produce (Allende and Artes, 2003a, 2003b).

High-pressure (HP) processing at low temperatures combined with modified atmosphere packaging has been used for the preservation of salmon (Amanatidou et al., 2000). When salmon had been subjected to HP treatment in the presence of 50% O_2, the threshold value for microbial spoilage of salmon (7.0 to 7.2 log CFU/g) was not reached for at least 18 days at 5°C. Spoilage microorganisms (lactic acid bacteria, *Shewanella putrefaciens*) as well as pathogens (*L. monocytogenes* Scott A, *S. typhimurium*) spiked on salmon prior to the treatment were susceptible to HP in the presence of 50% O_2 and 50% CO_2.

Chitosan has recently gained more interest due to its applications in food and pharmaceutics. In a study where chitosan was used as a coating on fruits and vegetables, Devlieghere et al. (2004b) suggest that chitosan as an antimicrobial preservative for food will be limited to food products with low protein and NaCl content. Fruit and vegetables belong to this category, and they found a clear antimicrobial activity of the treatment of lettuce and strawberries. The applicability for lettuce can be hampered due to a pronounced bitter taste developed after treatment.

Combinations of MA and postpackaging thermal treatments are rarely seen. This is due to the fact that headspace gas in thermally processed products has an insulating effect in classic heat transfer. With noncontact dielectric heating this is not an issue, but with dielectric heating, e.g., microwave or radio frequency heating, a rapid noncontact heating is possible. Still, one has to deal with steam that is generated inside the package with the risk that such a package may explode. Counterpressure or venting is therefore necessary to prevent overpressure. A number of new valves are now available that release steam during thermal processing, and during chilling the condensation of the vapor will generate a vacuum in the processed package. Although not yet commercially available, technologies may also be developed to reintroduce preserving gas mixtures after processing, if found commercially viable.

6.3.2 *Before and during the packaging process*

In most MA packages with nonrespiring products an anoxic gas mixture of CO_2 and N_2 is used, although high levels of O_2 could be used to increase color stability of red meat (Sivertsvik et al., 2002) or reduce trimethylamine production in lean fish (Sivertsvik, 2007). During the packaging process the air surrounding the product is evacuated and replaced with the desired gas mixture before hermetic sealing of the package. Due to the high solubility of CO_2 in wet foods, the gas mixture cannot contain CO_2 alone, but must be balanced with N_2 or O_2 and must be combined with a relatively high gas-to-product volume ratio to prevent package collapse. After closure MA packages are passive systems, only subjected to internal equilibration, but can be combined with many of the active packaging technologies described in detail in other chapters, e.g., oxygen scavengers and packaging materials with antimicrobial or antioxidative properties. Technologies that must be combined with MA are the use of in-package CO_2-producing or -emitting systems and prepackaging dissolvement of CO_2.

6.3.2.1 CO_2-emitting systems

The amount of dissolved CO_2 in a product is proportional with the partial pressure of CO_2 gas above the product according to Henry's law (Sivertsvik et al., 2004a). CO_2 emitters must therefore be used together with a CO_2-enriched MA to avoid package inflation due to the release of CO_2 gas. CO_2 could be produced inside the packages after packaging by allowing the exudates from the product to react with a mixture of sodium carbonate and citric acid inside the drip pad, an approach used successfully for cod fillets (Bjerkeng et al., 1995), increasing shelf life, as compared to traditional MA packaging. A similar system, the Verifrais package manufactured by Codimer, consists of a standard MA tray but has a perforated false bottom under which a porous sachet containing sodium bicarbonate/ascorbate is positioned. The main benefit of a CO_2 emitter is package size reduction and the possibility to use high partial pressures of CO_2 without packaging collapse. Solid CO_2 (dry ice) could also be used as a CO_2 producer inside a closed package; however, this requires a very precise ratio of dry ice to product amount or the use of one-way valve systems releasing excessive CO_2 (Sivertsvik et al., 1999b).

6.3.2.2 Soluble gas stabilization

The same benefits can be obtained by dissolving the CO_2 gas into the product immediately prior to packaging. The effectiveness of CO_2 on inhibiting bacterial growth is proportional to the amount of CO_2 dissolved (Devlieghere and Debevere, 2000). The solubility of CO_2 increases at lower temperatures and at higher partial and total pressures (Sivertsvik et al., 2004a); thus, a sufficient amount of CO_2 can be dissolved into the product in pure CO_2 prior to retail packaging. This method is called soluble gas stabilization (SGS) (Sivertsvik, 2000), and in order to withhold the dissolved CO_2 in the products, SGS must be combined with a MA. SGS has the potential to prevent package collapse even at low gas-to-product volume ratios (Sivertsvik et al., 2004a, 2004b; Sivertsvik and Jensen, 2005) without compromising the quality of the packaged food. The application of SGS has shown promising results on raw fillets of Atlantic salmon (*Salmo salar*) (Sivertsvik, 2000), on sliced meat products (cold cuts) (Jensen, 2004), for dairy products (Chen and Hotchkiss, 1991; Loss and Hotchkiss, 2002), on chicken breast fillets (Rotabakk et al., 2006), and for ready-to-eat shrimp (Sivertsvik and Birkeland, 2006).

6.3.2.3 Preservatives

Benzoic and sorbic acids are most inhibitory at acidic pH, and their use is rarely recommended if the pH exceeds 6. These acids are usually added as their more water soluble salts, sodium benzoate and potassium sorbate. These molecules inhibit the outgrowth of both bacterial and fungal cells; sorbic acid is also reported to inhibit the germination and outgrowth of bacterial spores.

The effect of adding potassium sorbate to ice, used in the cooling of red hake and salmon packaged in a modified atmosphere, was studied by Fey and Regenstein (1982). They found that a 60% CO_2, 20% N_2, and 20% O_2 atmosphere combined with 1% potassium sorbate ice was most satisfactory.

It is well known that specific spoilage organisms, like *P. phosphoreum*, *S. putrefaciens*, *Brocotrix thermospacta*, and lactic acid bacteria, develop on fish products with elevated levels of CO_2. In order to inhibit the spoilage flora, Drosinos et al. (1997) concluded that the use of sorbate plus gluconate was more effective than sorbate alone on fresh MA-packaged seabream (*Sparus aurata*). Potassium sorbate and citric acid have also been incorporated as additives to MA-packaged desalted cod to obtain increased shelf life (Fernandez-Segovia et al., 2006). Guynot et al. (2004) examined a possibility to use MA to reduce the amount of potassium sorbate added for preservation of bakery goods. At 0.2% potassium sorbate in 30% N_2 and 70% CO_2, MA packages prevented fungal spoilage of neutral cake analogues at all a_w levels examined, compared to visible growth after 6 days in air-packaged controls.

Also, other studies, regarding the use of sorbates in fish and fish products, suggest that sorbates in combination with other compounds or techniques can be used as an effective preservative tool for extending the shelf life of fish products (Thakur and Patel, 1994).

Elliott and Gray (1981) discovered growth inhibition of *Salmonella enteritidis* following exposure to a combination treatment of potassium sorbate (0.5, 1.5, or 2.5%) and modified atmospheres of 20, 60, and 100% CO_2 at pH 6.5, 6.0, or 5.5 at 10°C. At pH 6 and the same temperatures and atmospheres, Gray et al. (1984) found that a combination of CO_2 and sorbate dipping (0 to 5%) was more efficient in inhibiting microbial growth than either treatment alone. Dalgaard et al. (1998) found that potassium sorbate was effective in reducing the growth of the specific spoilage organisms *P. phosphoreum* in model substrates. This may have a practical use in extending the shelf life of MA-packaged seafood. Cooked and brined shrimp, including benzoic, citric, and sorbic acids, packaged in modified atmosphere were stored at 0, 5, 8, and 25°C (Dalgaard and Jørgensen, 2000). The shrimp had a shelf life of >7 months at 0°C, but spoiled in 4 to 6 days at 25°C. This pronounced effect of temperature was explained by changes in spoilage pattern at different storage temperatures.

6.3.2.4 *Biopreservation: protective microbes and their bacteriocins*

In biopreservation, storage life is extended or safety of food products is enhanced by using natural or controlled microflora, mainly lactic acid bacteria or their antibacterial products, such as lactic acid, bacteriocins, and others (Hugas, 1998). Bacteriocins are defined as antibacterial proteins that kill or inhibit closely related bacteria (Tagg et al., 1976). This definition has been broadened somewhat in recent years, in that it has become evident that some bacteriocins may have inhibitory spectra that include unrelated genera. In cases where the mode of action is known, the cell membrane is usually the mode of action (Montville and Bruno, 1994). Bacteriocins with potential use in foods are produced by strains of *Carnobacterium, Lactobacillus, Lactococcus, Leuconostoc, Pediococcus,* and *Propionibacterium*. These inhibitory proteins in combination with traditional methods of preservation and proper hygienic processing can be effective in controlling spoilage and pathogenic bacteria. Although naturally produced bacteriocins are found in fermented and nonfermented foods, nisin is currently the only bacteriocin widely used as a food preservative. Nisin is approved for use in over 40 countries and has been in use as a food preservative for over 50 years (Cleveland et al., 2001). Since bacteriocins are isolated from foods such as meat and dairy products, which normally contain lactic acid bacteria, they have unknowingly been consumed for centuries. Today there are many examples of the effective use of nisin in food systems without MA, e.g., cottage cheese (Ferreira and Lund, 1996), ricotta cheese (Davies et al., 1997), skimmed milk (Wandling et al., 1999), lean beef (Cutter and Siragusa, 1998), Kimchi (Choi and Parrish, 2000), and pasteurized mashed potato (Thomas et al., 2002). In principle, there are two common ways to use bacteriocins: by the addition of a starter culture that produces a bacteriocin that has the necessary inhibitory spectrum (Stiles, 1996), or the bacteriocin itself may be added as an ingredient at an early stage of the production process. A third way is by immobilizing the bacteriocins on the packaging materials (Scannell et al., 2000).

Fang and Lin (1994) found that the numbers of *P. fragi* on cooked tenderloin pork were reduced by MA storage but were unaffected by nisin. In contrast to this, the growth of *L. monocytogenes* was prevented when samples were treated with 1×10^4 nisin IU/ml. In addition, the MA (100% CO_2, 80% CO_2 + 20% air)/nisin (10^3, 10^4 IU/ml) combination system used in the same study decreased the growth of both organisms, and the inhibition was more pronounced at 4°C than at 20°C. In a cocktail of seven *L. monocytogenes* isolates of food, human, and environmental origin, Szabo and Cahill (1998) found an increase in lag phase in all atmospheres when nisin was used. Increasing the concentration of nisin to 1250 IU/ml inhibited the growth of *L. monocytogenes* in all atmosphere combinations at 4 and 12°C. The addition of nisin or a CO_2 atmosphere increased the shelf life of cold smoked salmon from 4 weeks (5°C) to 5 or 6 weeks (Paludan-Muller et al., 1998).

Scannell et al. (2000) developed bioactive food packaging materials using immobilized bacteriocins lacticin 3147 and nisaplin. They found antimicrobial activity against the indicator strains *Lactococcus lactis, Listeria innocua*, and *S. aureus*. Adsorption of lacticin 3147 into plastic film was unsuccessful, but nisin bound well and the resulting film maintained its activity for 3 months, both at room temperature and under refrigeration.

Although several studies have been published about bacteriocin production lactic acid bacteria being successfully used to control *L. monocytogenes*, there are also some reporting antagonistic interactions between non-bacteriocin-producing lactic acid bacteria (Bredholt et al., 1999; Buchanan and Bagi, 1997). Vermeiren et al. (2006) found that the non-bacteriocinogenic *Lactobacillus sakei* 10 A, isolated from cooked turkey fillet, offered opportunities as a biopreservative, as this protective culture can improve the safety of cooked meat products by inhibiting the growth of contaminating *L. monocytogenes*. The combination of the biopreservative *L. sakei* 10 A and a modified atmosphere containing 50% of CO_2 fully preserved growth of *L. monocytogenes*.

6.3.3 *Postpackaging*

Preservation methods that can be utilized postpackaging benefit from reduced risk of recontamination. Thus, it can be useful, especially for products that are consumed without further heating, like meat cuts, sliced sausages, and smoked fish.

6.3.3.1 *Irradiation*

Since modified atmospheres are not lethal to spoilage organisms and pathogens, irradiation below the threshold dose, i.e., the level at which spoilage organisms and pathogens are killed, and below the level where undesirable sensorial changes are introduced, can be used to enhance the attractiveness of MA. The effects of MA/irradiation on sensory properties, and its effect upon the depletion of vitamin content during storage, compared to untreated items, have been examined in detail. Studies on the effects of MA/irradiation methods on nutritional quality showed that the deleterious effects of irradiation on vitamins can be removed by modifying storage atmospheres (Robins, 1991). For a radiation dose of 0.25 kGy and in an air atmosphere, 60% of the thiamine content was lost over the storage period, compared to a minimal loss in the nonirradiated control sample over the same period. The loss of α-tocopherol, after exposure to 1 kGy irradiation, was some 50% over this period, compared to a minimal loss in the nonirradiated control sample. In both cases there were considerable reduced loss rates in N_2 atmospheres, which demonstrated that the effects of irradiation on these vitamins could be removed by modifying storage atmospheres.

The growth rate of the surviving microorganisms was measured as a function of atmospheric composition for the irradiated and nonirradiated food samples, and the optimum lethal atmospheres were found to range from 25% CO_2:75% N_2 to 50% CO_2:50% N_2. Tests at 10°C showed a similar trend, although the effectiveness of high concentrations of CO_2 was reduced.

The major surviving organisms even in the irradiated packages were lactobacilli, in accordance with general expectations due to their resistance to radiation.

A series of studies on MA/irradiation combinations have been carried out for different food categories. Combinations used for poultry and pork products, aiming to optimize sensory quality, have shown that each particular food item requires careful evaluation and that generalization can lead to incorrect and inappropriate specifications for optimum storage.

Thayer and Boyd (2000) showed that irradiation of ground turkey did not decrease its safety when it was contaminated following processing with *L. monocytogenes*. Irradiation treatments were significantly more lethal in the presence of air packaging than in vacuum packaging or MA packaging. In samples that received >1.0 kGy, there was a concentration-dependent CO_2 inhibition of *L. monocytogenes* multiplication or recovery (Thayer and Boyd, 1999). Radiosensitivities of *L. monocytogenes* have been determined in the presence of different antimicrobial compounds like traps-cinnamaldehyde, Spanish oregano, winter savory, and Chinese cinnamon on peeled minicarrots, packaged under MA (Caillet et al., 2006b). Results indicated that the bacterium was more resistant to irradiation under air in the absence of the active compound. The most active compound was traps-cinnamaldehyde, where a mean 3.8-fold increase in relative radiation sensitivity was observed for the modified atmosphere (60% O_2, 30% CO_2, and 10% N_2) compared with the control. Experimental batches of a stuffed pasta product, tortellini, and slightly prefried breaded reconstituted turkey steaks with cheese and ham filling, Cordon Bleu, were prepared according to commercial recipes, then inoculated with 10^4 CFU/g of *S. aureus* (in case of tortellini) and with 10^6 CFU/g of *L. monocytogenes* (in case of Cordon Bleu) prior to packing in plastic bags under a gas atmosphere of 20% CO_2 and 80% N_2. The inoculated packages were irradiated at 3 kGy (tortellini) and 2 kGy (Cordon Bleu) with a Co-60 radiation source (Farkas et al., 2005). The limiting factor of the shelf life of the unirradiated poultry products was the growth of lactic acid bacteria at 9°C, whereas enhanced lipid oxidation was an unwanted side effect of radiation treatment. From these studies it can be concluded that the potential risk posed by the investigated non-spore-forming pathogenic bacteria could be considerably reduced by gamma irradiation; however, storage temperature remains a crucial factor of safety and methods should be developed to counteract the lipid-oxidative effect of the radiation processing.

Several studies have been carried out on the use of MA/irradiation treatments in fish products, e.g., low-dose irradiation extended the shelf life of haddock and cod fillets (Licciardello et al., 1984) more than either process on its own. Przybylski et al. (1989) examined fresh catfish fillets, processed with low-dose irradiation in combination with MA, and demonstrated that irradiation treatments with or without elevated CO_2 MA packaging significantly reduced the bacterial load and extended shelf life from 5 to 7 days to 20 to 30 days. For cod fillets packaged in MA (80% CO_2:20% N_2) and under vacuum prior to irradiation with 2.2 kGy, and subsequent storage at 4°C, irradiation showed a large inhibitory effect on microorganisms (Rosnes et al., 2003). The most promising results were observed when combining irradiation with MA packaging. The sensory shelf life of irradiated MA cod and irradiated vacuum-packaged cod was >24 days and 24 days, respectively. For non-irradiated MA cod the shelf life was <14 days, and for vacuum-packaged cod, <9 days. This indicates a large potential for seafood product shelf life extensions through the use of MA combined with low-dose irradiation.

The combination of MA/irradiation has also been suggested to preserve sensory quality and improve the shelf life of fruits and vegetables. Lacroix and Lafortune (2004) showed that the alternative combination of MA and irradiation treatment can be used to maintain the quality of fresh minimally processed carrots, with a significant decrease in the inoculated *Escherichia coli* population. Promising results on sensory scores and microbiological quality have been found in combination of irradiation and MA-packaged products like fresh-cut cantaloupe (Boynton et al., 2006), ready-to-use carrots (Caillet et al., 2006a, 2006b), endive (Niemira et al., 2005), and Chinese cabbage (Ahn et al., 2005).

Before this method is widely accepted, several issues need to be resolved, such as legislative, scientific (food safety), and consumer attitudes toward irradiated foods. Irradiation treatments have been a matter of debate for a long time. Despite the advantages of irradiation for the processor, retailer, and consumer, irradiation is not widely used because of uncertainty regarding consumer acceptance, due to the requirement to label irradiated food products in most countries. Research on consumer attitudes and marked responses to irradiated foods has shown that the public's knowledge is sparse and that public acceptance of food in the fresh product category is limited (Bruhn, 1995; Lusk et al., 1999).

6.4 Consumer acceptance

Both the number of food types involved in carrying foodborne illnesses and the number of pathogenic microorganisms documented as being transmitted through food have increased in recent years. Hence, the approach to food preservation and pathogen control must be reconsidered in order to meet these new challenges and enhance food safety. MA packaging is perceived as a mild preservation method by most consumers, inducing minor changes in the inherent raw material qualities. A development toward using more preservatives in combination with MA packaging, e.g., additives or preservatives, and in some cases technologies with less recognized effects (e.g., irradiation), may lead to a lowering of consumer acceptance for MA-packaged foods. Consumer demands for fresh and safe food provide the producer with a dilemma: Should he produce a product with a modest shelf life or use preservatives to enhance product safety?

Most legislative authorities in Europe and the U.S. aim to give the consumer complete information about the processes and packaging conditions. Thus, the producer must clearly state on the label which additives, preservatives, or methods have been used. This can lead to special choices to satisfy the consumer. Some of the preservatives examined for use in combination with MA packaging in products for daily use in households may therefore be met with skepticism. Furthermore, the food additives benzoate and sorbate are often associated with a negative image. The food control authorities are also concerned, because some preservation techniques may mask poor and improper raw material quality. For example, irradiation and, to a certain extent, preservatives used at harvest may decrease or delay microbial growth onset without delaying biochemical reactions. In the end, this may provide for a long shelf life in MA products, measured by microbiological analysis, without offering improvements in eating quality.

6.5 Future aspects

In a recent market analysis report (Datamonitor Europe, 2005), the prepared meal consumption in Europe and the U.S. is expected to double in 10 years, to exceed U.S.$40 billion by 2009, and strong opportunities are predicted in the crossover trend between health and convenience. For many products this can be met by maintaining the nutritional value through minimal processing. Minimally processed modified atmosphere-packaged food products are known to be of very high quality and are increasingly being used in gastronomy and food service. Their taste and texture in many cases compare to that of freshly prepared foods. The convenience aspect they offer has also led to an increasing demand by quality-conscious consumers in retail markets. These products, however, have a limited shelf life and require chilled distribution.

Further developments can be foreseen in the optimization of gases (combinations and SGS), the use of low temperatures, and the utilization of protective microbes and their bacteriocins.

Novel processes are now under development for microbial control of foods, many of which aim to prevent microorganisms access to foods (Devlieghere et al., 2004a). Further developments in heat processing, like infrared heating, electric volume heating, electric resistance/ohmic heating, high-frequency (HF) or radio frequency heating, microwave heating, and inductive electric heating, aim to maintain food safety while reducing the required thermal load. There are also nonthermal methods, such as high pressure, pulsed electric fields, pulsed white light, ultrasound, and ultraviolet radiation. These are promising as pretreatments to MA packaging and have been proposed for use as part of combinations in multiple hurdle systems. However, many of them are still at an experimental stage, with expensive and ineffective batch production.

The purpose or need for using additional preservations in MA products is an important question producers should address. Increased safety, as previously described in this chapter, is both obvious and sensible. By using preservations together with MA, it is possible to get a long shelf life where target pathogens are under control. An increasing number of predictive

Table 6.2 Examples of Literature Where Combined Effects of Preserving Factors in Addition to MA Have Been Modeled

Food system	Main factors in the model	Reference
Cooked meat products	Na-lactate	Devlieghere et al., 2000a
Sliced pork	Temperature (–2 to 10°C)	Liu et al., 2006
Prickley pear cactus cladodes	Temperature, relative humidity	Guevara-Arauza et al., 2006
Cold smoked salmon	Effect of temperature, water-phase salt, and phenolic content on *Listeria monocytogenes*	Cornu et al., 2006
Nutrient agar	Effects of O_2 (>20%) and CO_2 on growth of *Listeria innocua*	Geysen et al., 2005
Ready-to-eat cooked meat products	Effect of salt, moisture, potassium lactate, and sodium diacetate on growth of *Listeria monocytogenes*	Legan et al., 2004
Cured ready-to-eat processed meat products	Effects of sodium chloride, sodium diacetate, potassium lactate, and moisture on growth of *Listeria monocytogenes*	Seman et al., 2002
Fruits and vegetables	Packaging design and membranes	Paul and Clarke, 2002
Fruits and vegetables	Respiration rates	Fonseca et al., 2002
Homogenized mushroom	*Pseudomonas fluorescens* and *Candida sake*	Masson et al., 2002
Laboratory medium	Effect of CO_2, pH, temperature, and NaCl on nonproteolytic *Clostridium botulinum*	Fernandez et al., 2001
Laboratory medium	Effects of CO_2, pH, temperature, and NaCl on *Listeria monocytogenes*	Fernandez et al., 1997

models have been developed during the last few years, and modeling will most likely be a tool for quality aspects as well as for risk assessment. These models include different parts of the food chain, e.g., packaging material, respiration of fruits and vegetables, raw materials, and heated ready-to-eat food (Table 6.2).

The definition of shelf life, however, is not obvious. Most chilled raw or partly processed food products packaged in MA will have a limited period of superior quality, then chemical and biochemical processes together with microbiological spoilage will decrease the sensory quality. After the period of superior quality, a period with regular or even poor quality may follow, without introducing safety hazards to the consumer. Future use of preservation, next to safeguarding safety, should focus on prolonging the superior-quality life span of MA products. For heat-treated products, new methods that allow a faster and more even heat penetration may improve eating quality and the survival of nutrients. Most processes or preservations used together with MA packaging do not prolong the high-quality period. An exception to this is low temperature and superchilling treatments, which may inhibit both microbial spoilage and biochemical reactions. Time is the most endangered commodity of modern life. On-the-go consumers are forever searching for meal solutions that are quick to fix, nutritious, and tasty. Products such as so-called dinner kits that are quick and easy to use but encourage consumers to prepare at least part of the meal are beginning to take off. This gives an indication that the consumer wants a minimum of control over the end result.

References

ACMSF. (1992). *Report on Vacuum Packaging and Associated Processes.* HMSO, London.

ACMSF. (1995). Annual report. HMSO, London.

Ahn, H.J., Kim, J.H., Kim, J.K., Kim, D.H., Yook, H.S., and Byun, M.W. (2005). Combined effects of irradiation and modified atmosphere packaging on minimally processed Chinese cabbage (*Brassica rapa* L.). *Food Chem* 89, 589–597.

Al Haddad, K.S.H., Al Qassemi, R.A.S., and Robinson, R.K. (2005). The use of gaseous ozone and gas packaging to control populations of *Salmonella infantis* and *Pseudomonas aeruginosa* on the skin of chicken portions. *Food Control* 16, 405–410.

Ali, M.S., Nakano, K., and Maezawa, S. (2004). Combined effect of heat treatment and modified atmosphere packaging on the color development of cherry tomato. *Postharvest Biol Technol* 34, 113–116.

Allende, A. and Artes, F. (2003a). Combined ultraviolet-C and modified atmosphere packaging treatments for reducing microbial growth of fresh processed lettuce. *Lebensmittel-Wissenschaft Technol* 36, 779–786.

Allende, A. and Artes, F. (2003b). UV-C radiation as a novel technique for keeping quality of fresh processed 'Lollo Rosso' lettuce. *Food Res Int* 36, 739–746.

Alvarez, J.A., Pozo, R., and Pastoriza, L. (1996). Note. Effect of a cryoprotectant agent (sodium tripolyphosphate) on hake slices preserved in modified atmosphere packaging. *Food Sci Technol Int* 2, 177–181.

Amanatidou, A., Schluter, O., Lemkau, K., Gorris, L.G.M., Smid, E.J., and Knorr, D. (2000). Effect of combined application of high pressure treatment and modified atmospheres on the shelf life of fresh Atlantic salmon. *Innov Food Sci Emerging Technol* 1, 87–98.

Ashie, I.N., Smith, J.P., and Simpson, B.K. (1996). Spoilage and shelf-life extension of fresh fish and shellfish. *Crit Rev Food Sci Nutr* 36, 87–121.

Ballestra, P. and Cuq, J.L. (1998). Influence of pressurized carbon dioxide on the thermal inactivation of bacterial and fungal spores. *Lebensmittel-Wissenschaft Technol* 31, 84–88.

Barakat, R.K. and Harris, L.J. (1999). Growth of *Listeria monocytogenes* and *Yersinia enterocolitica* on cooked modified-atmosphere-packaged poultry in the presence and absence of a naturally occurring microbiota. *Appl Environ Microbiol* 65, 342–345.

Bell, R.G., Penney, N., and Moorhead, S.M. (1995). Growth of the psychrotrophic pathogens *Aeromonas hydrophila, Listeria monocytogenes* and *Yersinia enterocolitica* on smoked blue cod (*Parapercis colias*) packed under vacuum or carbon dioxide. *Int J Food Sci Technol* 30, 515–521.

Bennik, M.H., van Overbeek, W., Smid, E.J., and Gorris, L.G. (1999). Biopreservation in modified atmosphere stored mungbean sprouts: the use of vegetable-associated bacteriocinogenic lactic acid bacteria to control the growth of *Listeria monocytogenes. Lett Appl Microbiol* 28, 226–232.

Betts, G.D. (1996). *Code of Practice for the Manufacture of Vacuum and Modified Atmosphere Packaged Chilled Foods with Particular Regard to the Risk of Botulism.* Campden and Chorleywood Food Research Association, Chipping Campden, U.K.

Bidawid, S., Farber, J.M., and Sattar, S.A. (2001). Survival of hepatitis A virus on modified atmosphere-packaged (MAP) lettuce. *Food Microbiol* 18, 95–102.

Bjerkeng, B., Sivertsvik, M., Rosnes, J.T., and Bergslien, H. (1995). Reducing package deformation and increasing filling degree in packages of cod fillets in CO_2-enriched atmospheres by adding sodium carbonate and citric acid to an exudate absorber. In *Foods and Packaging Materials: Chemical Interactions*, Ackermann, P., Jägerstad, M., and Ohlsson, T., Eds. Royal Society of Chemistry, Cambridge, U.K., pp. 222–227.

Borch, E., Kant-Meuermans, M.L., and Blixt, Y. (1996). Bacterial spoilage of meat and cured meat products. *Int J Food Microbiol* 33, 103–120.

Bøknæs, N., Osterberg, C., Nielsen, J., and Dalgaard, P. (2000). Influence of freshness and frozen storage temperature on quality of thawed cod fillets stored in modified atmosphere packaging. *Lebensmittel-Wissenschaft und Technologie–Food Science and Technology* 33, 244–248.

Bøknæs, N., Osterberg, C., Sorensen, R., Nielsen, J., and Dalgaard, P. (2001). Effects of technological parameters and fishing ground on quality attributes of thawed, chilled cod fillets stored in modified atmosphere packaging. *Lebensmittel Wissenschaft Technol* 34, 513–520.

Boynton, B.B., Welt, B.A., Sims, C.A., Balaban, M.O., Brecht, J.K., and Marshall, M.R. (2006). Effects of low-dose electron beam irradiation on respiration, microbiology, texture, color, and sensory characteristics of fresh-cut cantaloupe stored in modified-atmosphere packages. *J Food Sci* 71, S149–S155.

Bredholt, S., Nesbakken, T., and Holck, A. (1999). Protective cultures inhibit growth of *Listeria monocytogenes* and *Escherichia coli* O157:H7 in cooked, sliced, vacuum- and gas-packaged meat. *Int J Food Microbiol* 53, 43–52.

Bruhn, C.M. (1995). Consumer attitudes and market response to irradiated food. *J Food Prot* 58, 175.

Buchanan, R.L. and Bagi, L.K. (1997). Microbial competition: effect of culture conditions on the suppression of *Listeria monocytogenes* Scott A by *Carnobacterium piscicola. J Food Prot* 60, 254–261.

Burt, S. (2004). Essential oils: their antibacterial properties and potential applications in foods: a review. *Int J Food Microbiol* 94, 223–253.

Cabo, M.L., Herrera, J.J.R., Sampedro, G., and Pastoriza, L. (2005). Application of nisin, CO_2 and a permeabilizing agent in the preservation of refrigerated blue whiting (*Micromesistius poutassou*). *J Sci Food Agric* 85, 1733–1740.

Caillet, S., Millette, M., Salmieri, S., and Lacroix, M. (2006a). Combined effects of antimicrobial coating, modified atmosphere packaging, and gamma irradiation on *Listeria innocua* present in ready-to-use carrots (*Daucus carota*). *J Food Prot* 69, 80–85.

Caillet, S., Millette, M., Turgis, M., Salmieri, S., and Lacroix, M. (2006b). Influence of antimicrobial compounds and modified atmosphere packaging on radiation sensitivity of *Listeria monocytogenes* present in ready-to-use carrots (*Daucus carota*). *J Food Prot* 69, 221–227.

Chen, J.H. and Hotchkiss, J.H. (1991). Effect of dissolved carbon dioxide on the growth of psychrotrophic organisms in cottage cheese. *J Dairy Sci* 74, 2941–2945.

Chiasson, F., Borsa, J., and Lacroix, M. (2005). Combined effect of carvacrol and packaging conditions on radiosensitivity of *Escherichia coli* and *Salmonella typhi* in ground beef. *J Food Prot* 68, 2567–2570.

Chiasson, F., Borsa, J., Ouattara, B., and Lacroix, M. (2004). Radiosensitization of *Escherichia coli* and *Salmonella typhi* in ground beef. *J Food Prot* 67, 1157–1162.

Choi, Y.M. and Parrish, F.C. (2000). Selective control of lactobacilli in kimchi with nisin. *Lett Appl Microbiol* 30, 173–177.

Claire, B., Smith, J.P., El Khoury, W., Cayouette, B., Ngadi, M., Blanchfield, B., and Austin, J.W. (2004). Challenge studies with *Listeria monocytogenes* and proteolytic *Clostridium botulinum* in hard-boiled eggs packaged under modified atmospheres. *Food Microbiol* 21, 131–141.

Cleveland, J., Montville, T.J., Nes, I.F., and Chikindas, M.L. (2001). Bacteriocins: safe, natural antimicrobials for food preservation. *Int J Food Microbiol* 71, 1–20.

Cliver, D.O. (1997). Virus transmission via food. *Food Technol* 51, 71–78.

Cornu, M., Beaufort, A., Rudelle, S., Laloux, L., Bergis, H., Miconnet, N., Serot, T., and Delignette-Muller, M.L. (2006). Effect of temperature, water-phase salt and phenolic contents on *Listeria monocytogenes* growth rates on cold-smoked salmon and evaluation of secondary models. *Int J Food Microbiol* 106, 159–168.

Cosby, D.E., Harrison, M.A., Toledo, R.T., and Craven, S.E. (1999). Vacuum or modified atmosphere packaging and EDTA-nisin treatment to increase poultry product shelf life. *J Appl Poultry Res* 8, 185–190.

Cutter, C.N. and Siragusa, G.R. (1998). Incorporation of nisin into a meat binding system to inhibit bacteria on beef surfaces. *Lett Appl Microbiol* 27, 19–23.

Dalgaard, P., Garcia, M.L., and Mejlholm, O. (1998). Specific inhibition of *Photobacterium phosphoreum* extends the shelf life of modified-atmosphere-packed cod fillets. *J Food Prot* 61, 1191–1194.

Dalgaard, P. and Jørgensen, L.V. (2000). Cooked and brined shrimps packed in a modified atmosphere have a shelf-life of >7 months at 0°C, but spoil in 4–days at 25°C. *Int J Food Sci Technol* 35, 431–442.

Daniels, J.A., Krishnamurthi, R., and Rizvi, S.S.H. (1985). A review of effects of carbon dioxide on microbial growth and food quality. *J Food Prot* 48, 532–537.

Das, E., Gurakan, G.C., and Bayindirli, A. (2006). Effect of controlled atmosphere storage, modified atmosphere packaging and gaseous ozone treatment on the survival of *Salmonella enteritidis* on cherry tomatoes. *Food Microbiol* 23, 430–438.

Datamonitor Europe. (2005). *Evolution of Global Consumer Trends*, DMCM2367. London.

Davies, A.R. (1995). Advances in modified atmosphere packaging. In *New Methods of Food Preservation*, Gould, G.W., Ed. Blackie Academic & Professional, Glasgow, pp. 304–320.

Davies, E.A., Bevis, H.E., and Delves-Broughton, J. (1997). The use of the bacteriocin, nisin, as a preservative in ricotta-type cheeses to control the food-borne pathogen *Listeria monocytogenes*. *Lett Appl Microbiol* 24, 343–346.

Day, B.P.F. (1992). *Guidelines for the Manufacturing and Handling of Modified Atmosphere Packaged Food Products*. Campden and Chorleywood Food Research Association, Chipping Campden, UK.

Debevere, J.M. (1989). The effect of sodium lactate on shelf life of vacuum-packaged coarce liver pâté. *Fleischwirtschaft* 69, 223–224.

Devlieghere, F. and Debevere, J. (2000). Influence of dissolved carbon dioxide on the growth of spoilage bacteria. *Lebensmittel Wissenschaft Technol* 33, 531–537.

Devlieghere, F., Debevere, J., and Van Impe, J. (1998). Concentration of carbon dioxide in the water-phase as a parameter to model the effect of a modified atmosphere on microorganisms. *Int J Food Microbiol* 43, 105–113.

Devlieghere, F., Geeraerd, A.H., Versyck, K.J., Bernaert, H., Van Impe, J.F., and Debevere, J. (2000a). Shelf life of modified atmosphere packed cooked meat products: addition of Na-lactate as a fourth shelf life determinative factor in a model and product validation. *Int J Food Microbiol* 58, 93–106.

Devlieghere, F., Geeraerd, A.H., Versyck, K.J., Vandewaetere, B., Van Impe, J., and Debevere, J. (2001). Growth of *Listeria monocytogenes* in modified atmosphere packed cooked meat products: a predictive model. *Food Microbiol* 18, 53–66.

Devlieghere, F., Lefevere, I., Magnin, A., and Debevere, J. (2000b). Growth of *Aeromonas hydrophila* in modified-atmosphere-packed cooked meat products. *Food Microbiol* 17, 185–196.

Devlieghere, F., Vermeiren, L., and Debevere, J. (2004a). New preservation technologies: possibilities and limitations. *Int Dairy J* 14, 273–285.

Devlieghere, F., Vermeulen, A., and Debevere, J. (2004b). Chitosan: antimicrobial activity, interactions with food components and applicability as a coating on fruit and vegetables. *Food Microbiol* 21, 703–714.

Djenane, D., Sanchez-Escalante, A., Beltran, J.A., and Roncales, P. (2003). The shelf-life of beef steaks treated with lactic acid and antioxidants and stored under modified atmospheres. *Food Microbiol* 20, 1–7.

Doherty, A., Sheridan, J.J., Allen, P., McDowell, D.A., Blair, I.S., and Harrington, D. (1996). Survival and growth of *Aeromonas hydrophila* on modified atmosphere packaged normal and high pH lamb. *Int J Food Microbiol* 28, 379–392.

Drosinos, E.H., Lambropoulou, K., Mitre, E., and Nychas, G.J.E. (1997). Attributes of fresh gilt-head seabream (*Sparus aurata*) fillets treated with potassium sorbate, sodium gluconate and stored under a modified atmosphere at 0 ± 1 degrees C. *J Appl Microbiol* 83, 569–575.

Dufresne, I., Smith, J.P., Liu, J.N., Tarte, I., Blanchfield, B., and Austin, J.W. (2000). Effect of films of different oxygen transmission rate on toxin production by *Clostridium botulinum* type E in vacuum packaged cold and hot smoked trout fillets. *J Food Safety* 20, 251–268.

ECFF. (1996). Guideline for the Hygienic Manufacture of Chilled Foods. The European Chilled Food Federation, Helsinki, Finland.

Ellinger, R.H. (1972), Phosphates in food processing. In *Handbook of Food Additives*, Furia, T., Ed. CRC Press, Boca Raton, FL, pp. 617–780.

Elliott, P. and Gray, R.J.H. (1981). *Salmonella* sensitivity in a sorbate/modified atmosphere combination system. *J Food Prot* 44, 903–908.

Emborg, J., Laursen, B.G., Rathjen, T., and Dalgaard, P. (2002). Microbial spoilage and formation of biogenic amines in fresh and thawed modified atmosphere-packed salmon (*Salmo salar*) at 2°C. *J Appl Microbiol* 92, 790–799.

Fang, T.J. and Lin, L.-W. (1994). Growth of *Listeria monocytogenes* and *Pseudomonas fragi* on cooked pork in a modified atmosphere packaging/nisin combination. *J Food Prot* 57, 485.

Farber, J.M. (1991). Microbiological aspects of modified-atmosphere packaging technology: a review. *J Food Prot* 54, 58–70.

Farkas, J., Andrassy, E., and Polyak-Feher, K. (2005). Improvement of the microbiological safety of two chilled semi-prepared meals by gamma irradiation. *Food Technol Biotechnol* 43, 263–269.

Ferguson, I.B., Ben Yehoshua, S., Mitcham, E.J., McDonald, R.E., and Lurie, S. (2000). Postharvest heat treatments: introduction and workshop summary. *Postharvest Biol Technol* 21, 1–6.

Fernandez, P.S., Baranyi, J., and Peck, M.W. (2001). A predictive model of growth from spores of non-proteolytic *Clostridium botulinum* in the presence of different CO_2 concentrations as influenced by chill temperature, pH and NaCl. *Food Microbiol* 18, 453–461.

Fernandez, P.S., George, S.M., Sills, C.C., and Peck, M.W. (1997). Predictive model of the effect of CO_2, pH, temperature and NaCl on the growth of *Listeria monocytogenes*. *Int J Food Microbiol* 37, 37–45.

Fernandez-Segovia, I., Escriche, I., Gomez-Sintes, M., Fuentes, A., and Serra, J.A. (2006). Influence of different preservation treatments on the volatile fraction of desalted cod. *Food Chem* 98, 473–482.

Ferreira, M.A. and Lund, B.M. (1996). The effect of nisin on *Listeria monocytogenes* in culture medium and long-life cottage cheese. *Lett Appl Microbiol* 22, 433–438.

Fey, M.S. and Regenstein, J.M. (1982). Extending shelf-life of fresh wet red hake and salmon using CO_2-O_2 modified atmosphere and potassium sorbate ice at 1°C. *J Food Sci* 47, 1048–1054.

Fonseca, S.C., Oliveira, F.A.R., and Brecht, J.K. (2002). Modelling respiration rate of fresh fruits and vegetables for modified atmosphere packages: a review. *J Food Eng* 52, 99–119.

Francis, G.A. and O'Beirne, D. (1998). Effects of storage atmosphere on *Listeria monocytogenes* and competing microflora using a surface model system. *Int J Food Sci Technol* 33, 465–476.

Francis, G.A. and O'Beirne, D. (2001). Effects of acid adaptation on the survival of *Listeria monocytogenes* on modified atmosphere packaged vegetables. *Int J Food Sci Technol* 36, 477–487.

Fränkel, C. (1889). Die Einwirkung der Kohlensäure auf die Lebensthätigkeit der Mikroorganismen. *Zeitschrift Hygiene* 5, 332–362.

Geysen, S., Geeraerd, A.H., Verlinden, B.E., Michiels, C.W., Van Impe, J.F., and Nicolai, B.M. (2005). Predictive modelling and validation of *Pseudomonas fluorescens* growth at superatmospheric oxygen and carbon dioxide concentrations. *Food Microbiol* 22, 149–158.

Gibson, A.M., Ellis-Brownlee, R.C.L., Cahill, M.E., Szabo, E.A., Fletcher, G.C., and Bremer, P.J. (2000). The effect of 100% CO_2 on the growth of nonproteolytic *Clostridium botulinum* at chill temperatures. *Int J Food Microbiol* 54, 39–48.

Gill, C.O. and Reichel, M.P. (1989). Growth of the cold-tolerant pathogens *Yersinia enterocolitica, Aeromonas hydrophila* and *Listeria monocytogenes* on high-pH beef packaged under vacuum or carbon dioxide. *Food Microbiol* 6, 223–230.

Gomez-Lopez, V.M., Devlieghere, F., Bonduelle, V., and Debevere, J. (2005). Intense light pulses decontamination of minimally processed vegetables and their shelf-life. *Int J Food Microbiol* 103, 79–89.

Gormley, T.R., Redmond, G.A., and Fagen, J. (2003). Freeze-chill applications in the food industry. *New Food* 2, 65–67.

Gould, E. and Peters, J.A. (1971). On testing the freshness of frozen fish. In *Fishing News,* Books Ltd., London, pp. 45–47.

Grau, R.D. (1981). Role of pH, lactate, and anaerobiosis in controlling the growth of some fermentative, gram-negative bacteria on beef. *Appl Environ Microbiol* 42, 1043–1050.

Gray, R.J.H., Elliott, P.H., and Tomlins, R.I. (1984). Control of 2 major pathogens on fresh poultry using a combination potassium sorbate carbon-dioxide packaging treatment. *J Food Sci* 49, 142–145.

Guevara-Arauza, J.C., Yahia, E.M., Cedeno, L., and Tijskens, L.M.M. (2006). Modeling the effects of temperature and relative humidity on gas exchange of prickly pear cactus (*Opuntia* spp.) stems. *Lebensmittel-Wissenschaft Technol* 39, 796–805.

Guldager, H.S., Bøknæs, N., Østerberg, C., Nielsen, J., and Dalgaard, P. (1998) Thawed cod fillets spoil less rapidly than unfrozen fillets when stored under modified atmosphere at 2°C. *J Food Prot* 61, 1129–1136.

Guynot, M.E., Marin, S., Sanchis, V., and Ramos, A.J. (2004). An attempt to minimize potassium sorbate concentration in sponge cakes by modified atmosphere packaging combination to prevent fungal spoilage. *Food Microbiol* 21, 449–457.

Hintlian, C.B. and Hotchkiss, J.H. (1987). Comparative growth of spoilage and pathogenic organisms on modified atmosphere-packaged cooked beef. *J Food Prot* 50, 218–223.

Houtsma, P.C., de Wit, J.C., and Rombouts, F.M. (1993). Minimum inhibitory concentration (MIC) of sodium lactate for pathogens and spoilage organisms occurring in meat products. *Int J Food Microbiol* 20, 247–257.

Hugas, M. (1998). Bacteriocinogenic lactic acid bacteria for the biopreservation of meat and meat products. *Meat Sci* 49, S139–S150.

Jacxsens, L., Devlieghere, F., Van der Steen, C., and Debevere, J. (2001). Effect of high oxygen modified atmosphere packaging on microbial growth and sensorial qualities of fresh-cut produce. *Int J Food Microbiol* 71, 197–210.

Jensen, J.S. (2004). Sliced cold cuts. *DMRI Newsletter,* June 14.

Jimenez, S.M., Salsi, M.S., Tiburzi, M.C., Rafaghelli, R.C., and Pirovani, M.E. (1999). Combined use of acetic acid treatment and modified atmosphere packaging for extending the shelf-life of chilled chicken breast portions. *J Appl Microbiol* 87, 339–344.

Jimenez, S.M., Salsi, M.S., Tiburzi, M.C., Rafaghelli, R.C., Tessi, M.A., and Coutaz, V.R. (1997). Spoilage microflora in fresh chicken breast stored at 4 degrees C: influence of packaging methods. *J Appl Microbiol* 83, 613–618.

Kasai, Y., Kimura, B., Kawasaki, S., Fukaya, T., Sakuma, K., and Fujii, T. (2005). Growth and toxin production by *Clostridium botulinum* in steamed rice aseptically packed under modified atmosphere. *J Food Prot* 68, 1005–1011.

Kleinlein, N. and Untermann, F. (1990). Growth of pathogenic *Yersinia enterocolitica* strains in minced meat with and without protective gas with consideration of the competitive background flora. *Int J Food Microbiol* 10, 65–71.

Knock, R.C., Seyfert, M., Hunt, M.C., Dikeman, M.E., Mancini, R.A., Unruh, J.A., Higgins, J.J., and Monderen, R.A. (2006). Effects of potassium lactate, sodium chloride, sodium tripolyphosphate, and sodium acetate on colour, colour stability, and oxidative properties of injection-enhanced beef rib steaks. *Meat Sci* 74, 312–318.

Kolbe, H. (1882). Antiseptische Eigenschaften der Kohlensäure. *J Prakt Chem* 26, 249–255.

Lacroix, M. and Lafortune, R. (2004). Combined effects of gamma irradiation and modified atmosphere packaging on bacterial resistance in grated carrots (*Daucus carota*). *Radiation Physics Chem* 71, 79–82.

Lafortune, R., Caillet, S., and Lacroix, M. (2005). Combined effects of coating, modified atmosphere packaging, and gamma irradiation on quality maintenance of ready-to-use carrots (*Daucus carota*). *J Food Prot* 68, 353–359.

Lambert, A.D., Smith, J.P., and Dodds, K.L. (1991). Combined effect of modified atmosphere packaging and low-dose irradiation on toxin production by *Clostridium botulinum* in fresh pork. *J Food Prot* 54, 94–101.

Leblanc, E.L. and Leblanc, R.J. (1992). Determination of hydrophobicity and reactive groups in proteins of cod (*Gadus morhua*) muscle during frozen storage. *Food Chem.* 43, 3–11.

Legan, J.D., Seman, D.L., Milkowski, A.L., Hirschey, J.A., and Vandeven, M.H. (2004). Modeling the growth boundary of *Listeria monocytogenes* in ready-to-eat cooked meat products as a function of the product salt, moisture, potassium lactate, and sodium diacetate concentrations. *J Food Prot* 67, 2195–2204.

Leistner, L. (1992). Food preservation by combined methods. *Food Res Int* 25, 151–158.

Leistner, L. (1995). Principles and applications of hurdle technology. In *New Methods of Food Preservation*, Gould, G.W., Ed. Blackie Academic & Professional, Glasgow, pp. 1–21.

Licciardello, J., Ravesi, E., Tuhkunen, B., and Racicot, L. (1984). Effects of some potentially synergistic treatments in combination with 100 krad irradiation on the shelf life of cod fillets. *J Food Technol* 49, 1341.

Liserre, A.M., Landgraf, M., Destro, M.T., and Franco, B.D.G.M. (2002). Inhibition of *Listeria monocytogenes* by a bacteriocinogenic *Lactobacillus sake* strain in modified atmosphere-packaged Brazilian sausage. *Meat Sci* 61, 449–455.

Liu, F., Yang, R.Q., and Li, Y.F. (2006). Correlations between growth parameters of spoilage micro-organisms and shelf-life of pork stored under air and modified atmosphere at –2, 4 and 10°C. *Food Microbiol* 23, 578–583.

Lopez-Rubira, V., Conesa, A., Allende, A., and Artes, F. (2005). Shelf life and overall quality of minimally processed pomegranate arils modified atmosphere packaged and treated with UV-C. *Postharvest Biol Technol* 37, 174–185.

Loss, C.R. and Hotchkiss, J.H. (2002). Effect of dissolved carbon dioxide on thermal inactivation of microorganisms in milk. *J Food Prot* 65, 1924–1929.

Lusk, J.L., Fox, J.A., and McIlvain, C.L. (1999). Consumer acceptance of irradiated meat. *Food Technol* 53, 56–59.

Lyver, A., Smith, J.P., Austin, J., and Blanchfield, B. (1998). Competitive inhibition of *Clostridium botulinum* type E by *Bacillus* species in a value-added seafood product packaged under a modified atmosphere. *Food Res Int* 31, 311–319.

Marquenie, D., Michiels, C.W., Geeraerd, A.H., Schenk, A., Soontjens, C., Van Impe, J.F., and Nicolai, B.M. (2002). Using survival analysis to investigate the effect of UV-C and heat treatment on storage rot of strawberry and sweet cherry. *Int J Food Microbiol* 73, 187–196.

Masniyom, P., Benjakul, S., and Visessanguan, W. (2005). Combination effect of phosphate and modified atmosphere on quality and shelf-life extension of refrigerated seabass slices. *Lebensmittel-Wissenschaft Technol* 38, 745–756.

Masson, Y., Ainsworth, P., Fuller, D., Bozkurt, H., and Ibanoglu, S. (2002). Growth of *Pseudomonas fluorescens* and *Candida sake* in homogenized mushrooms under modified atmosphere. *J Food Eng* 54, 125–131.

Matan, N., Rimkeeree, H., Mawson, A.J., Chompreeda, P., Haruthaithanasan, V., and Parker, M. (2006). Antimicrobial activity of cinnamon and clove oils under modified atmosphere conditions. *Int J Food Microbiol* 107, 180–185.

McMeekin, T.A., Olley, J., Ratkowsky, D.A., and Ross, T. (2002). Predictive microbiology: towards the interface and beyond. *Int J Food Microbiol* 73, 395–407.

Mejlholm, O., Boknaes, N., and Dalgaard, P. (2005). Shelf life and safety aspects of chilled cooked and peeled shrimps (*Pandalus borealis*) in modified atmosphere packaging. *J Appl Microbiol* 99, 66–76.

Mejlholm, O. and Dalgaard, P. (2002). Antimicrobial effect of essential oils on the seafood spoilage micro-organism *Photobacterium phosphoreum* in liquid media and fish products. *Lett Appl Microbiol* 34, 27–31.

Mitsuda, H., Nakajima, K., Mizuno, H., and Kawai, F. (1980). Use of sodium chloride solution and carbon dioxide for extending shelf-life of fish fillets. *J Food Sci* 45, 661–666.

Montville, T.J. and Bruno, M.E. (1994). Evidence that dissipation of proton motive force is a common mechanism of action for bacteriocins and other antimicrobial proteins. *Int J Food Microbiol* 24, 53–74.

Muratore, G. and Licciardello, F. (2005). Effect of vacuum and modified atmosphere packaging on the shelf-life of liquid-smoked swordfish (*Xiphias gladius*) slices. *J Food Sci* 70, 359–363.

Murcia, M.A., Martinez-Tome, M., Nicolas, M.C., and Vera, A.M. (2003). Extending the shelf-life and proximate composition stability of ready to eat foods in vacuum or modified atmosphere packaging. *Food Microbiol* 20, 671–679.

Nicolalde, C., Stetzer, A.J., Tucker, E.M., McKeith, F.K., and Brewer, M.S. (2006). Antioxidant and modified atmosphere packaging prevention of discoloration in pork bones during retail display. *Meat Sci* 72, 713–718.

Nielsen, P.V. and Rios, R. (2000). Inhibition of fungal growth on bread by volatile components from spices and herbs, and the possible application in active packaging, with special emphasis on mustard essential oil. *Int J Food Microbiol* 60, 219–229.

Niemira, B.A., Fan, X., and Sokorai, K.J.B. (2005). Irradiation and modified atmosphere packaging of endive influences survival and regrowth of *Listeria monocytogenes* and product sensory qualities. *Radiation Physics Chem* 72, 41–48.

Novak, J.S. and Yuan, J.T.C. (2004). The fate of *Clostridium perfringens* spores exposed to ozone and/or mild heat pretreatment on beef surfaces followed by modified atmosphere packaging. *Food Microbiol* 21, 667–673.

Oh, D.H. and Marshall, D.L. (1995). Influence of packaging method, lactic acid and monolaurin on *Listeria monocytogenes* in crawfish tail meat homogenate. *Food Microbiol* 12, 159–163.

Paludan-Muller, C., Dalgaard, P., Huss, H.H., and Gram, L. (1998). Evaluation of the role of *Carnobacterium piscicola* in spoilage of vacuum- and modified-atmosphere-packed cold-smoked salmon stored at 5°C. *Int J Food Microbiol* 39, 155–166.

Pastoriza, L., Sampedro, G., Herrera, J.J., and Cabo, M.L. (1998). Influence of sodium chloride and modified atmosphere packaging on microbiological, chemical and sensorial properties in ice storage of slices of hake (*Merluccius merluccius*). *Food Chem* 61, 23–28.

Paul, D.R. and Clarke, R. (2002). Modelling of modified atmosphere packaging based on designs with a membrane and perforations. *J Membrane Sci* 208, 269–283.

Penney, N., Bell, R.G., and Cummings, T.L. (1994). Extension of the chilled storage life of smoked blue cod (*Parapercis colias*) by carbon-dioxide packaging. *Int J Food Sci Technol* 29, 167–178.

Phillips, C.A. (1996). Review: modified atmosphere packaging and its effects on the microbiological quality and safety of produce. *Int J Food Sci Technol* 31, 463–479.

Pietrasik, Z., Dhanda, J.S., Shand, P.J., and Pegg, R.B. (2006). Influence of injection, packaging, and storage conditions on the quality of beef and bison steaks. *J Food Sci* 71, S110–S118.

Pothuri, P., Marshall, D.L., and McMillin, K.W. (1996). Combined effects of packaging atmosphere and lactic acid on growth and survival of *Listeria monocytogenes* in crayfish tail meat at 4°C. *J Food Prot* 59, 253–256.

Przybylski, L.A., Finerty, M.W., Grodner, R.M., and Gerdes, D.L. (1989). Extension of shelf-life of iced fresh channel catfish fillets using modified atmosphere packaging and low dose irradiation. *J Food Sci* 54, 269–273.

Ranasinghe, L., Jayawardena, B., and Abeywickrama, K. (2005). An integrated strategy to control post-harvest decay of Embul banana by combining essential oils with modified atmosphere packaging. *Int J Food Sci Technol* 40, 97–103.

Redmond, G.A., Gormley, T.R., and Butler, F. (2005). Effect of short- and long-term frozen storage with MAP on the quality of freeze-chilled lasagne. *Lebensmittel Wissenschaft Technol* 38, 81–87.

Robins, D. (1991). Combination treatments with food irradiation. In *The Preservation of Food by Irradiation. A Factual Guide to the Process and Its Effect on Food.* Dotesios Ltd., Trowbridge, Wiltshire, pp. 53–61.

Rosnes, J.T., Kleiberg, G.H., and Folkvord, L. (2001). Increased shelf life of salmon (*Salmo salar*) steaks using partial freezing (liquid nitrogen) and modified atmosphere packaging (MAP). *Ann Soc Sci Færoensis Suppl.* XXVIII, 83–91.

Rosnes, J.T., Kleiberg, G.H., Sivertsvik, M., Lunestad, B.T., and Lorentzen, G. (2006). Effect of modified atmosphere packaging and superchilled storage on the shelf-life of farmed ready-to-cook spotted wolf-fish (*Anarhichas minor*). *Packaging Technol Sci* 19, available online May 24.

Rosnes, J.T., Sivertsvik, M., and Skåra, T. (2003). Combining MAP with other preservation techniques. In *Novel Food Packaging Techniques*, Ahvenainen, R., Ed. Woodhead Publishing Ltd., Cambridge, England, pp. 287–311.

Rosnes, J.T., Sivertsvik, M., Skipnes, D., Nordtvedt, T.S., Corneliussem, C., and Jakobsen, Ø. (1998). *Transport of Superchilled Salmon in Modified Atmosphere.* International Insitute of Refrigeration–IIF/IIR, Nantes, France.

Rotabakk, B.T., Birkeland, S., Jeksrud, W.K., and Sivertsvik, M. (2006). Effect of modified atmosphere packaging and soluble gas stabilization on the shelf life of skinless chicken breast fillets. *J Food Sci* 71, S124–S131.

Scannell, A.G., Hill, C., Ross, R.P., Marx, S., Hartmeier, W., and Arendt, E. K. (2000). Development of bioactive food packaging materials using immobilised bacteriocins lacticin 3147 and nisaplin. *Int J Food Microbiol* 60, 241–249.

Seman, D.L., Borger, A.C., Meyer, J.D., Hall, P.A., and Milkowski, A.L. (2002). Modelling the growth of *Listeria monocytogenes* in cured ready-to-eat processed meat products by manipulation of sodium chloride, sodium diacetate, potassium lactate, and product moisture content. *J Food Prot* 65, 651–658.

Serrano, M., Martinez-Romero, D., Castillo, S., Guillen, F., and Valero, D. (2005). The use of natural antifungal compounds improves the beneficial effect of MAP in sweet cherry storage. *Innov Food Sci Emerging Technol* 6, 115–123.

Seyfert, M., Hunt, M.C., Mancini, R.A., Hachmeister, K.A., Kropf, D.H., Unruh, J.A., and Loughin, T.M. (2005). Beef quadriceps hot boning and modified-atmosphere packaging influence properties of injection-enhanced beef round muscles. *J Anim Sci* 83, 686–693.

Shenoy, K. and Murano, E.A. (1996). Effect of storage conditions on growth of heat-stressed *Yersinia enterocolitica* in ground pork. *J Food Prot* 59, 365–369.

Shrestha, S. and Min, Z. (2006). Effect of lactic acid pretreatment on the quality of fresh pork packed in modified atmosphere. *J Food Eng* 72, 254–260.

Sikorski, Z.E. and Sun, P. (1994). Preservation of seafood quality. In *Seafoods: Chemistry, Processing, Technology and Quality*, Shahidi, F. and Botta, J.R., Eds. Blackie Academic & Professionals, Glasgow, p. 168.

Sivakumar, D. and Korsten, L. (2006a). Evaluation of the integrated application of two types of modified atmosphere packaging and hot water treatments on quality retention in the litchi cultivar 'McLean's Red'. *J Hort Sci Biotechnol* 81, 639–644.

Sivakumar, D. and Korsten, L. (2006b). Influence of modified atmosphere packaging and postharvest treatments on quality retention of litchi cv. Mauritius. *Postharvest Biol Technol* 41, 135–142.

Sivertsvik, M. (2000). Use of soluble gas stabilisation to extend shelf-life of fish. In *Proceedings of 29th WEFTA Meeting*, Leptocarya, Pieria, October 10–14, 1999, pp. 79–91.

Sivertsvik, M. (2007). The optimized modified atmosphere for packaging of pre-rigor filleted farmed cod (*Gadus morhua*) is 63 ml/100 ml oxygen and 37 ml/100 ml carbon dioxide. *LWT, Food Sci Technol* 40, 430–438.

Sivertsvik, M. and Birkeland, S. (2006). Effects of soluble gas stabilisation, modified atmosphere, gas to product volume ratio and storage on the microbiological and sensory characteristics of ready-to-eat shrimp (*Panadalus borealis*). *Food Sci Technol Int* 12, 445–454.

Sivertsvik, M., Jeksrud, W., Vågane, Aa., and Rosnes, J.T. (2004a). Solubility and absorption rate of carbon dioxide into non-respiring foods. 1. Development and validation of experimental apparatus using a manometric method. *J Food Eng* 61, 449–458.

Sivertsvik, M. and Jensen, J.S. (2005). Solubility and absorption rate of carbon dioxide into non-respiring foods. 3. Meat products. *J Food Eng* 70, 499–505.

Sivertsvik, M., Nordtvedt, T.S., Aune, E.J., and Rosnes, J.T. (1999a). Storage Quality of Superchilled and Modified Atmosphere Packaged Whole Salmon. Paper presented at the 20th International Congress of Refrigeration, International Institute of Refrigeration, Sydney, Australia.

Sivertsvik, M., Rosnes, J.T., and Bergslien, H. (2002). Modified atmosphere packaging (MAP). In *Minimal Processing Technologies in The Food Industry*, Ohlsson, T. and Bengtsson, N., Eds. Woodhead Publishing Ltd., Cambridge, U.K., pp. 61–86.

Sivertsvik, M., Rosnes, J.T., and Jeksrud, W. (2004b). Solubility and absorption rate of carbon dioxide into non-respiring foods. 2. Raw fish fillets. *J Food Eng* 63, 451–458.

Sivertsvik, M., Rosnes, J.T., and Kleiberg, G.H. (2003). Effect of modified atmosphere packaging and superchilled storage on the microbial and sensory quality of Atlantic salmon (*Salmo salar*) fillets. *J Food Sci* 68, 1467–1472.

Sivertsvik, M., Rosnes, J.T., Vorre, A., Randell, K., Ahvenainen, R., and Bergslien, H. (1999b). Quality of whole gutted salmon in various bulk packages. *J Food Qual* 22, 387–401.

Skandamis, P., Tsigarida, E., and Nychas, G.J.E. (2002). The effect of oregano essential oil on survival/death of *Salmonella typhimurium* in meat stored at 5°C under aerobic, VP/MAP conditions. *Food Microbiol* 19, 97–103.

Skandamis, P.N. and Nychas, G.J.E. (2001). Effect of oregano essential oil on microbiological and physico-chemical attributes of minced meat stored in air and modified atmospheres. *J Appl Microbiol* 91, 1011–1022.

Steiner, A., Abreu, M., Correia, L., Beirao-da-Costa, S., Leitao, E., Beirao-da-Costa, M., Empis, J., and Moldao-Martins, M. (2006). Metabolic response to combined mild heat pre-treatments and modified atmosphere packaging on fresh-cut peach. *Eur Food Res Technol* 222, 217–222.

Stiles, M.E. (1996). Biopreservation by lactic acid bacteria. *Antonie Van Leeuwenhoek* 70, 331–345.

Sunen, E. (1998). Minimum inhibitory concentration of smoke wood extracts against spoilage and pathogenic micro-organisms associated with foods. *Lett Appl Microbiol* 27, 45–48.

Szabo, E.A. and Cahill, M.E. (1998). The combined effects of modified atmosphere, temperature, nisin and ALTA 2341 on the growth of *Listeria monocytogenes*. *Int J Food Microbiol* 43, 21–31.

Szalai, M., Szigeti, J., Farkas, L., Varga, L., Reti, A., and Zukal, E. (2004). Effect of headspace CO_2 concentration on shelf-life of cooked meat products. *Acta Alimentaria* 33, 141–155.

Tagg, J.R., Dajani, A.S., and Wannamaker, L.W. (1976). Bacteriocins of gram positive bacteria. *Bacteriol Rev* 40, 722–756.

Thakur, B.R. and Patel, T.R. (1994). Sorbates in fish and fish products: a review. *Food Rev Int* 10, 93–107.

Thayer, D.W. and Boyd, G. (1999). Irradiation and modified atmosphere packaging for the control of *Listeria monocytogenes* on turkey meat. *J Food Prot* 62, 1136–1142.

Thayer, D.W. and Boyd, G. (2000). Reduction of normal flora by irradiation and its effect on the ability of *Listeria monocytogenes* to multiply on ground turkey stored at PC when packaged under a modified atmosphere. *J Food Prot* 63, 1702–1706.

Thomas, L.V., Ingram, R.E., Bevis, H.E., Davies, E.A., Milne, C.F., and Delves-Broughton, J. (2002). Effective use of nisin to control *Bacillus* and *Clostridium* spoilage of a pasteurized mashed potato product. *J Food Prot* 65, 1580–1585.

Tsigarida, E., Skandamis, P., and Nychas, G.J.E. (2000). Behaviour of *Listeria monocytogenes* and autochthonous flora on meat stored under aerobic, vacuum and modified atmosphere packaging conditions with or without the presence of oregano essential oil at 5°C. *J Appl Microbiol* 89, 901–909.

Valero, A., Carrasco, E., Perez-Rodriguez, F., Garcia-Gimeno, R.M., Blanco, C., and Zurera, G. (2006a). Monitoring the sensorial and microbiological quality of pasteurized white asparagus at different storage temperatures. *J Sci Food Agric* 86, 1281–1288.

Valero, D., Valverde, J.M., Martinez-Romero, A., Guillen, F., Castillo, S., and Serrano, M. (2006b). The combination of modified atmosphere packaging with eugenol or thymol to maintain quality, safety and functional properties of table grapes. *Postharvest Biol Technol* 41, 317–327.

Valverde, J.M., Guillen, F., Martinez-Romero, D., Castillo, S., Serrano, M., and Valero, D. (2005). Improvement of table grapes quality and safety by the combination of modified atmosphere packaging (MAP) and eugenol, menthol, or thymol. *J Agric Food Chem* 53, 7458–7464.

Vermeiren, L., Devlieghere, F., Vandekinderen, I., and Debevere, J. (2006). The interaction of the non-bacteriocinogenic *Lactobacillus sakei* 10A and lactocin S producing *Lactobacillus sakei* 148 towards *Listeria monocytogenes* on a model cooked ham. *Food Microbiol* 23, 511–518.

Wandling, L.R., Sheldon, B.W., and Foegeding, P.M. (1999). Nisin in milk sensitizes *Bacillus* spores to heat and prevents recovery of survivors. *J Food Prot* 62, 492–498.

Wei, Q.K., Fang, T.J., and Chen, W.C. (2001). Development and validation of growth model for *Yersinia enterocolitica* in cooked chicken meats packaged under various atmosphere packaging and stored at different temperatures. *J Food Prot* 64, 987–993.

Wimpfheimer, L., Altman, N.S., and Hotchkiss, J.H. (1990). Growth of *Listeria monocytogenes* Scott A, serotype 4 and competitive spoilage organisms in raw chicken packaged under modified atmospheres and in air. *Int J Food Microbiol* 11, 205–214.

Zeitoun, A.A.M. and Debevere, J.M. (1991). Inhibition, survival and growth of *Listeria monocytogenes* on poultry as influenced by buffered lactic-acid treatment and modified atmosphere packaging. *Int J Food Microbiol* 14, 161–169.

Zeitoun, A.A.M. and Debevere, J.M. (1992). Decontamination with lactic-acid sodium lactate buffer in combination with modified atmosphere packaging effects on the shelf-life of fresh poultry. *Int J Food Microbiol* 16, 89–98.

chapter seven

Lessons from other commodities: fish and meat

Morten Sivertsvik

Contents

7.1 Introduction

In the last two decades active packaging technologies have been presented and reviewed by several authors (Labuza and Breene 1989; Labuza 1993, 1996; Ahvenainen and Hurme 1996; Floros et al. 1997; Day 1998; Vermeiren et al. 1999, 2002; Quintavalla and Vicini 2002; Sivertsvik 2003; Suppakul et al. 2003; Cha and Chinnan 2004; Devlieghere et al. 2004; Ozdemir and Floros 2004; Brody 2005; Kerry et al. 2006); however, few of these technologies have been used commercially by the food industry and even fewer for fish and meat products. The reasons are several: loss and thermolability of active components, cost vs. benefit, and incompatible legislation. This chapter will focus on the most promising technologies for meat and fish products and their limitations. This includes the use of atmosphere modifiers such as oxygen scavengers and carbon dioxide emitters, packaging that controls water or with antimicrobial and antioxidative properties, and indicator mechanisms. Modified atmosphere packaging (MAP) is sometimes regarded as an active packaging technology; however, because MAP is a well-established method to extend the shelf life of foods (Sivertsvik et al. 2002b), only its novel approaches, such as *soluble gas stabilization*, will be covered in this chapter.

Muscle foods span a broad spectrum of products, and it is unlikely that one specific novel or active packaging technology will be a success for all, just like not all products observe the same shelf life extension using MA (Sivertsvik et al. 2002a). The potential for an active packaging technology to be a success for a product would depend on the technology's ability to control and inhibit the shelf-life-deteriorating spoilage reactions (e.g., bacterial growth of specific bacteria, oxidative rancidity, or color changes) in the specific product.

The proposed EU legislation COM (2003) 0689 (Arvanitoyannis et al. 2005) defines *active food contact materials and articles* as materials and articles that are intended to extend the shelf life or to maintain or improve the condition of packaged food. They are designed to deliberately incorporate components that would release or absorb substances into or from the packaged food or the environment surrounding the food. *Intelligent food contact materials and articles* are defined as materials and articles that monitor the condition of packaged food or the environment surrounding the food.

These materials and articles shall be manufactured in compliance with good manufacturing practice so that, under normal or foreseeable conditions of use, they do not transfer their constituents to food in quantities that could (1) endanger human health or (2) bring about an unacceptable change in the composition of the food or deterioration in the organoleptic characteristics thereof. For *active* or *intelligent* food contact materials and solutions specific measures may be adopted in accordance with the directive.

Active packaging applications can be grouped according to their main purpose: those that influence or change the in-pack atmosphere; those that remove unwanted components such as water, taint, and off-odors; those that contain or release an active component such as antimicrobial agents or antioxidants; and those that monitor or control in-pack conditions, i.e., smart or intelligent packaging.

7.2 *Atmosphere modifiers*

Many of the most used active packaging technologies are closely related to MA packaging. Together with anaerobic conditions, carbon dioxide, CO_2, is the active gas of MA packaging because it inhibits growth of many of the normal spoilage bacteria (Sivertsvik et al. 2002b). The effect of CO_2 on bacterial growth is complex, and four activity mechanisms of CO_2 on microorganisms have been identified (Farber 1991; Daniels et al. 1985; Dixon and Kell 1989; Parkin and Brown 1982): alteration of cell membrane function including effects on nutrient uptake and absorption; direct inhibition of enzymes or decreases in the rate of enzyme reactions; penetration of bacterial membranes, leading to intracellular pH changes; and direct changes in the physicochemical properties of proteins. Probably a combination of these activities accounts for the bacteriostatic effect (Sivertsvik et al. 2002a).

The CO_2 is usually introduced into the MA package by evacuating the air and flushing the appropriate gas mixture into the package prior to sealing, typically using automatic form–fill–seal or flow-packaging machines. Two other approaches to create a modified atmosphere for a product are to generate the CO_2 and remove the O_2 inside the package after packaging or to dissolve the CO_2 into the product prior to packaging. Both methods can give appropriate packages with a smaller gas/product ratio, and thus decrease the package size, which has been a disadvantage of MAP from the start.

The first approach involves the most commercialized active packaging technology, oxygen scavengers, which are available from several manufacturers (Mitsubishi Gas Chemical Co., ATCO, Sealed Air/Cryovac, Multisorb, and others) in different forms (sachets, packaging film, closures) and with different active ingredients (iron, enzymes, dye). Some of these companies have also developed CO_2 emitters, using the O_2 in the package headspace to produce CO_2 and develop a CO_2/N_2 atmosphere inside a package without the use of gas flushing. Other resources to generate the CO_2 gas inside the packages after closure is the use of dry ice (solid CO_2) (Sivertsvik et al. 1999) or carbonate possibly mixed with weak acids (Bjerkeng et al. 1995).

The commercial use of atmosphere modifiers and O_2 scavengers, in particular for fish and seafood, has been mostly limited to the Japanese market and to dried (seaweed, beef or salmon jerky, sardines, shark's fin, rose mackerel, cod, and squid) or smoked products (Ashie et al. 1996). These ambient stored products have low a_W (<0.85) so the microbial deterioration is not shelf life limiting, and the effect of the O_2 scavengers is to prevent oxidation reactions, discoloration, and mold growth. Other commercial products are fresh yellowtail, salmon roe, and sea urchin, all stored at superchilled conditions packaged with O_2 scavenger to prevent first oxidation and discoloration, but also to some degree bacterial growth (Ashie et al. 1996). Different O_2 scavengers are chosen dependent on the amount of O_2 to scavenge (pack size and material) and product a_W. O_2 scavengers for high a_W foods react faster than scavengers for dry foods, but in general the absorption is slow and exothermic.

For meat, O_2 scavengers have been more successfully commercialized and used in several countries for sliced cooked ham to prevent discoloration (Kerry et al. 2006). Hormel Foods' pepperoni and bacon with Multisorb sachets is one success story. O_2 absorbers also have an effect on the color stability of retail packaged beef (Isdell et al. 1999) however did not affect weight loss.

There is growing interest in centralized preparation of retail-ready meat cuts for distribution to widely dispersed retail stores due to the convenience of having high-quality ready-to-go products that are consistently provided to consumers at lower cost (Jeyamkondan et al. 2000). In order to obtain a long shelf life during chilled transportation in master packages, O_2 scavengers could be incorporated to absorb the residual O_2 and prevent color degradation (Tewari et al. 2001). Similar positive color results have been found in storage trials of pork with O_2 scavengers (Tewari et al. 2002; Sørheim et al. 1995), and one study (Buys 2004) suggests that pork retail packs, bulk packaged in oxygen-depleted atmospheres using O_2 scavengers, were found by consumers to be as acceptable as pork chops stored in oxygen-enriched atmospheres.

Removal of oxygen from package interiors improves shelf life by sub-optimizing the environment for aerobic microbiological growth and for adverse oxidative reactions such as rancidity. Ferrous iron-based oxygen scavengers rely on the presence of moisture for activation, with a water activity of 0.7 required, and 0.85 to 0.9 preferred (Brody 2001).

Oxygen absorbers are designed to reduce oxygen levels to less than 100 ppm in package headspace. In iron-based oxygen scavengers the oxygen is removed by oxidation (rusting) of powdered iron forming nontoxic iron oxide. Oxygen absorbers could be used to create oxygen-free conditions in the headspace of packages with medium barrier properties. The sachet will absorb residual oxygen and oxygen permeated through the packaging material during storage. More inexpensive or environmentally friendly packaging materials with lower oxygen barriers could be used in combination with an oxygen absorber, instead of high-cost barrier materials (Sivertsvik 2003). However, not all oxygen absorbers can be combined with MA packaging. Some of them, like the iron-based Ageless SS-type from Mitsubishi, meant for use with high a_W foods, will unintentionally absorb some of the carbon dioxide present and decrease some of the inhibitory effect of CO_2 bacterial growth. This is caused by a reaction of iron with CO_2 to form ferrous carbonate, and secondarily this ferrous carbonate reacts with O_2. This will also slow down the O_2 absorption (Sivertsvik 2003).

Use of an O_2 scavenger (Ageless SS-100) had only a marginal effect on the microbial growth in packages of fish cakes, fish pudding, and mackerel fillets, and far less than the highly significant effect obtained by MA packaging. However, a significant effect on the microbial activities by using an O_2 scavenger was observed in packages of salmon fillets. The use of O_2 absorbers also inhibited development of rancidity (TBARS) in both mackerel and salmon fillets (Sivertsvik 1997), but to no higher degree than O_2-free MA packaging.

A second approach to create a modified atmosphere is to dissolve CO_2 into the product prior to packaging. Since the solubility of CO_2 increases at lower temperatures and at higher partial and total pressures (Sivertsvik et al. 2004a), a sufficient amount of CO_2 can be dissolved into the product in 1 to 2 h in pure CO_2 prior to retail packaging. This method is called *soluble gas stabilization* (SGS) (Sivertsvik 2000). This is not an active packaging technology by definition, but it is a novel alternative to MA and it has been used successfully alone and in combination with O_2 scavengers. SGS has the potential to prevent package collapse even at low gas-to-product ($^g/_p$) volume ratios (Sivertsvik and Jensen 2005; Sivertsvik et al. 2004a, 2004b) without compromising the quality of the packaged food. The application of SGS has shown promising results on raw fillets of Atlantic salmon (*Salmo salar*) (Sivertsvik 2000, 2003), on sliced meat products (cold cuts) (Jensen 2004), for dairy products (Loss and Hotchkiss 2002; Chen and Hotchkiss 1991), on chicken breast fillets (Rotabakk et al. 2006), and for ready-to-eat shrimp (Sivertsvik and Birkeland 2006).

The effect of SGS treatment, different O_2 absorbers/CO_2 emitters, and combinations of these on the storage quality of salmon fillets revealed the best microbial quality in packages combining SGS with a combined O_2 absorber and CO_2 emitter (Ageless G-100), i.e., the packages with the most CO_2 inside (Sivertsvik 2003). The fastest microbial growth was observed in salmon stored in air without absorbers and in air with Ageless SS-200 and Ageless ZPT-100 O_2 absorbers. Samples packaged with G-100 absorber/emitter gave the best cooked sensory scores, while samples without absorber got the lowest cooked sensory scores. On raw odor the samples packaged with G-100 and without absorbers were evaluated as better than samples with SS-200 and ZPT-100 absorbers. No differences in the color of the samples were observed, in contrast to the reddish color change observed when packaging perch and pike perch fillets with the Ageless G-100 CO_2 emitters (Lyijynen et al. 1997). They observed the same shelf life for fresh perch and pike perch fillets packaged with G-100 as for traditional MAP using anoxic high CO_2 atmosphere, and 2 to 4 days longer shelf life than with overwrap or vacuum packaging. However, the color change and smell of raw liver in the raw fillets in the active packages were observed. This was not observed in the traditional MA packages. No differences between the two packaging technologies were observed after cooking of the fish.

The commercial CO_2 emitters usually contain ferrous carbonate and a metal halide catalyst, although nonferrous variants are available, absorbing the O_2 and producing an equal volume of CO_2. Carbon dioxide could also be produced inside the packages after packaging by allowing the exudates from the product to react with a mixture of sodium carbonate and citric acid inside the drip pad, an approach used successfully for cod fillets (Bjerkeng et al. 1995), increasing shelf life, compared to traditional MAP, even when using a low gas headspace in the package. The Verifrais package manufactured by Codimer, which has been used for extending the shelf life of fresh meats and fish, is a similar concept (Day 1998). This package consists of a standard MAP tray but has a perforated false bottom under which a porous sachet containing sodium bicarbonate/ascorbate is positioned. When exudate from packed meat or fish drips onto the sachet, CO_2 is emitted and counteracts the package collapse due to the CO_2 solubility in the food. Recent findings (Hansen et al. 2004, 2005) suggest that the same spoilage organisms will dominate when using CO_2 emitters as for MA packaging of cod and salmon, and the main benefit of a CO_2 emitter is package size reduction. CO_2 emitters must be used together with a CO_2-enriched MA to avoid package inflation due to the release of CO_2 gas.

7.3 Removal of water or taint

Excess moisture is a major cause of food spoilage, and different humidity absorbers are used to protect dried products' humidity damage; however, these absorbers have a limited effect on fish products. Several companies manufacture moisture drip-absorbent pads, sheets, and blankets for liquid water control in watery foods such as meats, fish, poultry, fruit, and vegetables. Moisture drip absorber pads or false-bottomed trays are commonly placed under packaged fresh meats, fish, poultry, and prepared fruit to absorb unsightly tissue drip discharge. Larger sheets and blankets are used for absorption of melted ice from chilled seafood during air freight transportation (Day 1998). Commercial moisture absorber sheets, blankets, and trays include Toppan Sheet (Toppan Printing, Japan), Thermarite (Thermarite, Australia), and Fresh-R-Pax (Maxwell Chase, U.S.).

An approach to extend shelf life of chilled fresh fish is to decrease the water activity at the surface. The Showa Denko Co. (Tokyo, Japan) has developed a film (Pichit film) in the form of a pillow with entrapped propylene glycol between layers of polyvinyl alcohol (PVA). The PVA film is very permeable to water but is a barrier to glycol. When placed in contact for several hours with the surface of meat or fish by wrapping the film around, it absorbs water and causes injury to spoilage bacteria. This technique can increase the shelf life of ocean fish by 2 to 4 days (Labuza 1993). The action is due to the difference in a_W between the fish (0.99) and the glycol (0.0); thus, the water is rapidly drawn out of the fish surface. This surface dehydration not only inhibits some microbes, but also may injure others without causing change in fish quality (Labuza and Breene 1989). Most likely some glycol also transfers to the food surface, also slowing growth. The Pichit film has been shown to maintain the color of tuna, veal, pork, and beef (Arakawa et al. 1990), since the color in these products is related to the myoglobin content in the meat and dehydration of the surface will lead to increased myoglobin concentration. The effect of the Pichit film on the shelf life of fish has been little exploited, but for salmon fillets the effect of 2 and 4 hours of Pichit pretreatment on microbial growth and sensory spoilage was nonexisting (Sivertsvik 2000).

During storage of packaged muscle foods, microbial metabolites and protein breakdown products, such as amines and aldehydes, accumulate in the headspace of the package, leading to, for example, putrid (H_2S) and fishy (trimethylamine, or TMA) odors. Removal of these components would therefore often enhance the initial perception of the products upon package opening, and also to some degree increase sensory shelf life.

The effectiveness of an innovative foam plastic tray, provided with absorbers for volatile amines and liquids, on the shelf life of different fish ducts packed under a modified atmosphere (40%CO_2:60%N_2) was evaluated in comparison with a standard tray (Franzetti et al. 2001). Fillets of sole (*Solea solea*), steaks of hake (*Merluccius merluccius*), and whole cuttlefish (*Sepia fillouxi*), placed in the two different kinds of trays, were kept at 3°C. The new packaging, associated with a rigorous control of storage temperature, increased the shelf life up to 10 days. In fact, the innovative tray sequestrated the greater part of TMA from the headspace and led to delayed microbial growth, especially of Gram-negative and H_2S-producing bacteria, and in addition it favored the growth of bacterial strains such as *Moraxella phenylpiruvica*, which are not involved in off-flavoring production (especially because of their lipolitic activity [Franzetti et al. 2001]).

A Japanese patent based on the interactions between off-odors and acidic compounds (e.g., citric acid) incorporated in polymers, claims amine-removing capabilities (Vermeiren et al. 1999). Another approach to remove amine odors has been provided by the Anico Co. (Japan). The Anico bags made from a film containing ferrous salt and an organic acid such as citric or ascorbic acid are claimed to oxidize the amine or other oxidizable odor-causing compounds as they are absorbed by the polymer (Vermeiren et al. 1999).

Some commercialized odor-absorbing sachets, e.g., MINIPAX1 and STRIPPAX1 (Multisorb Technologies, U.S.), absorb the odors developing in certain packaged foods during distribution due to the formation of mercaptans and H_2S (Vermeiren et al. 1999). 2-in-1 from United Desiccants (U.S.) is a combination of silica gel and activated carbon packaged for use in controlling moisture, gas, and odor within packaged products. Profresh1 is claimed to be a freshness-keeping and malodor control master batch to polyethylene (PE) and polystyrene (PS). The active component ADI50 (composition not revealed) is claimed to absorb ethylene, ethyl alcohol, ethyl acetate, and H_2S. Whether these commercial odor absorbers are used and feasible for fish products is not known.

Due to the fact that taint removers are designed to mask spoilage reactions by removal of sensorial signals of spoilage, they are unlikely to be allowed by the European legislation (Arvanitoyannis et al. 2005).

7.4 Antimicrobial and antioxidant applications

Some commercial antimicrobial films and materials have been introduced, primarily in Japan. For example, one widely reported product is a synthetic silver zeolite, which has been directly incorporated into food contact packaging film. The major potential food applications for antimicrobial films include meats, fish, bread, cheese, fruit, and vegetables. Several antimicrobial compounds might have the potential to be incorporated into package structures to convert them into active packaging — chlorine dioxide, silver salts, bacteriocins, ozone, and natural spices such as rosemary and its derivatives (Brody 2001) — but few of these have been investigated for use in or on packaging material for fish products. One exception is benzoic acid anhydride on the PE film used on fish fillets (Han 2000).

One antimicrobial packaging application used commercially for semimoist and dried fish products in Japan is ethanol emitters, for example, Ethicap, Antimold 102, Negamold (Freund Industrial), Oitech (Nippon Kayaku), ET Pack (Ueno Seiyaku), and Ageless-type SE (Mitsubishi Gas

Chemical) (Day 1998). These films and sachets contain absorbed ethanol in a carrier material that allows the controlled release of ethanol vapor.

Essential oils have antimicrobial effects, and oils of oregano and cinnamon had the strongest antimicrobial activity, followed by lemongrass, thyme, clove, bay, marjoram, sage, and basil oils. Oregano oil (0.05% v/w) reduced growth of the specific spoilage organism (SSO) *Photobacterium phosphoreum* in naturally contaminated MAP cod fillets and extended shelf life from 11 to 12 days to 21 to 26 days at 2°C (Mejlholm and Dalgaard 2002). Obviously, essential oils can extend the shelf life of MAP muscle food, but because of the volatile and often odorous nature of these components, incorporating them into an active packaging could be a challenge.

Another component with potential as an active packaging ingredient for fresh fish is acetate buffer, which can extend the shelf life of MA-packaged cod fillets by having the fillets sprayed prior to packaging (Boskou and Debevere 2000). Production of total volatile bases and TMA was inhibited in treated fillets for 10 days' storage under modified atmospheres. Inhibition of TMA production can be attributed to growth inhibition of H_2S-producing bacteria, inhibition of the trimethylamine oxide (TMAO)-dependent metabolism of TMAO-reducing bacteria, and the stable pH during storage. The shelf life at 7°C of treated cod fillets, based on cooked flavor score, was almost 12 days, about 8 days more than the shelf life of the control fillets. Potassium sorbate is another preservative shown to increase shelf life of fish products (Fey and Regenstein 1982; Drosinos et al. 1997), and that could be an active ingredient in packaging materials for fishery products.

Incorporating antioxidants, such as vitamins C and E, in the packaging film may have the potential to reduce oxidative reactions such as development of rancid flavor and odor in fatty fish products. The degradation of texture, flavor, and odor of stored seafood is attributed to the oxidation of unsaturated lipids. Processing operations such as salting, cooking, and mincing promote oxidation, while smoking, dehydration, and freezing retard oxidation. The rate and degree of lipid degradation in frozen fish are dependent upon the fish species and muscle type, dark or white. Lipid oxidation proceeds in the following decreasing order: skin, dark muscle, and white muscle. Lipid oxidation within a given species will vary with season and location within the tissue. Metal ions affect oxidation in the following decreasing order: Fe^{2+}, hemin, Cu^{2+}, and Fe^{3+}. Oxidation can be reduced through the use of single or combined antioxidants. Lipid oxidation in fish fillets wrapped by butylated hydroxytoluene (BHT) antioxidant-incorporated PE film was inhibited compared to nonwrapped fish fillets (Huang and Weng 1998). The BHT-incorporated PE film was able to inhibit lipid oxidation in both fish muscle and oil.

Antioxidants have a potential as an active component in packaging materials; however, vacuum packaging has a greater reduction in oxidation than the presence of additives (Flick et al. 1992), and this should favor O_2-scavenging properties before antioxidative properties.

7.5 Indicator applications

Smart packaging, such as time–temperature indicators (TTIs), is a technology that appears to have potential, especially with chill-stored products under anaerobic conditions where microbial safety is not otherwise ensured. Strict temperature control is necessary to ensure microbial safety, and temperature abuse should be avoided because of both safety issues and shortening of shelf life. TTIs could be applied to monitor the temperature and to detect temperature-abused packages (Labuza et al., 1992; Labuza 1993, 1996). Otwell (1997) evaluated the enzyme-based Vitsab TTIs for use in modified atmosphere packaging of seafood. Results demonstrated that the color change from enzyme-based Vitsab labels correlates to spoilage of packaged salmon and other fish. The labels change color before botulism toxin formation. Furthermore, the Vitsab indicator reflects the experience of the contents within the package (Otwell 1997).

A systematic approach for fish shelf life modeling and TTI selection in order to plan and apply an effective quality monitoring scheme for the fish chill chain was developed (Taoukis et al. 1999), modeling the growth of the SSOs of the Mediterranean fish boque (*Boops boops*).

Cox Technologies (Belmont, NC) developed a spoilage indicator, FreshTag, meant to be affixed to the package surfaces. This indicator detected, measured, and signaled the presence of decomposition volatile bases such as ammonia, trimethylamine, and dimethylamine in the headspace of the package (de Kruijf et al. 2002). Results were promising for lean fish but not for fatty fish (Brody 2001), since the shelf life of the fatty fish is limited to not only microbial growth but also oxidative reactions (Sivertsvik et al. 2002a). To our knowledge, FreshTag was never used commercially.

Related to the spoilage indicator is the Toxin Guard™ indicator (Toxin Alert, Mississauga, Canada), a patented system of placing antibody-based tests on polymer packaging films to detect pathogens or other selected microorganisms (e.g., specific spoilage organisms). This indicator can possibly detect growth of pathogenic microorganisms in real time without waiting (Brody 2001). Commercialization of this product is still ongoing.

7.6 The future of active packaging technologies for meat and fish

The future trend in active packaging technologies is likely to develop further toward film-based materials and away from the traditional sachet-based system. Active packaging solutions incorporated in the packaging material itself would increase consumer safety; however, these solutions must be activated upon contact with the food and not lose their active compounds during film production, distribution, and storage prior to use. An example of such a product is Cryovac's oxygen-scavenging film OS2000, used commercially on Nestlé's Buitoni range of fresh pasta. Films with antimicrobial activity are also under final development, and new methods to immobilize the active components on the polymer surface could lead to commercial products. However, the effect of such film on actual foodstuffs is still far less than the effect observed on inhibiting single strands of bacteria in the scientific environment.

Developments in nanotechnology will also impact the development of active packaging technologies, but today this is limited to the use of nanoclays in food packaging. In a recent report, Cientifica forecasted active packaging as the most promising food-related application of nanotechnology and estimated that active packaging will be a $2.7 billion (2.1 billion) market by 2012 (Food Production Daily, 2006). However, these nanotech solutions will only account for a meager percentage of the $402 billion (312 billion) world packaging market.

While active packaging is promising to improve the safety and quality of products, the price tag will significantly hinder its integration into the world market, and these advanced packaging solutions will only be viable for high-value products such as caviar in the near future. Producers of packaging for common products such as milk and meat will not find the benefits of nanotech packaging to be cost effective.

Many of the above-mentioned active packaging technologies do not improve microbial safety of the products above that obtained by traditional MAP. O_2 scavengers and CO_2 emitters give little or no additional shelf life to fresh fish or meat products compared to MAP and vacuum packaging, but the technologies could in some cases replace the traditional packaging technologies. Different antimicrobial components can extend shelf life through inhibiting spoilage organisms. However, the inhibition of the spoilage bacteria reduces bacterial competition that may permit growth and toxin production by nonproteolytic *Clostridium botulinum* or growth of other pathogenic bacteria. The antimicrobial packaging should therefore also be able to inhibit growth of pathogens as well as the specific spoilage organisms to ensure microbially safe products. The risks from botulism in MAP fish have been widely reviewed (Sivertsvik et al. 2002a), and even if the results are not conclusive, there is a potential threat for a packaged fish product to become toxic prior to spoilage, especially at storage temperatures of 8°C or above. TTIs and toxin indicators are active technologies that can be used to ensure the safety of such products. For processed or prepared product changes in the a_W, pH, salt, or heat treatment is used to control the threat, and in the future active packaging technologies could do the same, without adversely changing the flavor, odor, color, and texture of a fresh product.

Active packaging technologies increase the shelf life of fresh meat and fish only marginally and usually not the prime quality. Increased hygiene and reducing the initial counts of specific spoilage organisms, together with reduced storage temperature by 1 or 2°, would be a wiser method to increase shelf life and quality. A technology able to control and prolong the high-quality phase of a fresh product should be preferred, and none of the existing active packaging technologies is able to do this today.

Therefore, active packaging has great potential to be a success for processed and prepared food products with added value, for example, ready-meals, ready-to-eat or ready-to-heat. This is also possibly the segment with the highest growth potential and the segment that could sustain the cost of incorporating such technologies.

References

Ahvenainen R, Hurme E. 1996. Active and Smart Packaging for Meeting Consumer Demands for Qualtiy and Safety. Paper presented at ILSI Conference on Food Packaging, Brussels, Belgium.

Arakawa N, Chu Y-J, Otsuka M, Kotoku I, Takuno M. 1990. Comparative study on the color of animal and tuna muscle after using a contact-dehydrating sheet. *Home Econ Jpn* 41:947–950.

Arvanitoyannis IS, Choreftaki S, Tserkezou P. 2005. An update of EU legislation (directives and regulations) on food-related issues (safety, hygiene, packaging, technology, GMOs, additives, radiation, labelling): presentation and comments. *Int J Food Sci Technol* 40:1021–1112.

Ashie IN, Smith JP, Simpson BK. 1996. Spoilage and shelf-life extension of fresh fish and shellfish. *Crit Rev Food Sci Nutr* 36:87–121.

Bjerkeng B, Sivertsvik M, Rosnes JT, Bergslien H. 1995. Reducing package deformation and increasing filling degree in packages of cod fillets in CO2-enriched atmospheres by adding sodium carbonate and citric acid to an exudate absorber. In *Foods and Packaging Materials: Chemical Interactions*, Ackermann P, Jägerstad M, Ohlsson T, Eds. The Royal Society of Chemistry, Cambridge, U.K., pp. 222–227.

Boskou G, Debevere J. 2000. Shelf-life extension of cod fillets with an acetate buffer spray prior to packaging under modified atmospheres. *Food Addit Contam* 17:17–25.

Brody AL. 2001. Is something fishy about packaging? *Food Technol* 55:97–98.

Brody AL. 2005. Active packaging becomes more active. *Food Technol* 59:82–84.

Buys EM. 2004. Colour changes and consumer acceptability of bulk packaged pork retail cuts stored under O-2, CO2 and N-2. *Meat Sci* 68:641–647.

Cha DS, Chinnan MS. 2004. Biopolymer-based antimicrobial packaging: a review. *Crit Rev Food Sci Nutr* 44:223–237.

Chen JH, Hotchkiss JH. 1991. Effect of dissolved carbon dioxide on the growth of psychrotrophic organisms in cottage cheese. *J Dairy Sci* 74:2941–2945.

Daniels JA, Krishnamurthi R, Rizvi SSH. 1985. A review of effects of carbon dioxide on microbial growth and food quality. *J Food Prot* 48:532–537.

Day BPF. 1998. *Active Packaging of Foods*. Campden & Chorleywood Food Research Association, Chipping Campden, Glos., U.K.

de Kruijf N, van Beest M, Rijk R, Sipilainen-Malm T, Losada PP, De Meulenaer B. 2002. Active and intelligent packaging: applications and regulatory aspects. *Food Addit Contam* 19:144–162.

Devlieghere F, Vermeiren L, Debevere J. 2004. New preservation technologies: possibilities and limitations. *Intl Dairy J* 14:273–285.

Dixon NM, Kell DB. 1989. The inhibition of CO_2 on the growth and metabolism of micro-organisms. *J Appl Bacteriol* 67:109–136.

Drosinos EH, Lambropoulou K, Mitre E, Nychas GJE. 1997. Attributes of fresh gilt-head seabream (*Sparus aurata*) fillets treated with potassium sorbate, sodium gluconate and stored under a modified atmosphere at 0 ±1 degrees C. *J Appl Microbiol* 83:569–575.

Farber JM. 1991. Microbiological aspects of modified-atmosphere packaging technology: a review. *J Food Prot* 54:58–70.

Fey MS, Regenstein JM. 1982. Extending shelf-life of fresh wet red hake and salmon using CO2-O2 modified atmosphere and potassium sorbate ice at 1°C. *J Food Sci* 47:1048–1054.

Flick GJ, Hong GP, Knobl GM. 1992. Lipid Oxidation of Seafood During Storage, ACS Symposium Series 500, Washington. pp. 183–207.

Floros JD, Dock LL, Han JH. 1997. Active packaging technologies and applications. *Food Cosmet Drug Packaging* 20:10–17.

Food Production Daily. 2006. Most Companies Will Have to Wait Years for Nanotech's Benefits. Accessed August 24. Available online at www.foodproductiondaily.com/news/ng.asp?n=69974-nanotechnology-packaging-nano-applications.

Franzetti L, Martinoli S, Piergiovanni L, Galli A. 2001. Influence of active packaging on the shelf-life of minimally processed fish products in a modified atmosphere. *Packaging Technol Sci* 14:267–274.

Han JH. 2000. Antimicrobial food packaging. *Food Technol* 54:56–65.

Hansen AÅ, Eie T, Tamarit MPC, Rudi K. 2004. Explorative Analyses of 16S RDNA Microbial Community in Farmed Salmon Fillets Packed in Modified Atmosphere (MAP) with a CO_2-Emitter. Paper presented at Proceedings of the WEFTA Conference 2004, Lübeck, Germany, September 12–15.

Hansen AÅ, Rudi K, Eie T. 2005. Active Packaging of Pre- And Post-Rigor Farmed Cod (*Gadus morhua)* with Reduced Headspace and a CO_2-Emitter. Paper presented at the 22nd IAPRI Symposium, Campinas, Brazil, May 22–24.

Huang CH, Weng YM. 1998. Inhibition of lipid oxidation in fish muscle by antioxidant incorporated polyethylene film. *J Food Process Preserv* 22:199–209.

Isdell E, Allen P, Doherty AM, Butler F. 1999. Colour stability of six beef muscles stored in a modified atmosphere mother pack system with oxygen scavengers. *Int J Food Sci Technol* 34:71–80.

Jensen, J.S. 2004. Sliced Cold Cuts. *DMRI Newsletter,* June 14.

Jeyamkondan S, Jayas DS, Holley RA. 2000. Review of centralized packaging systems for distribution of retail-ready meat. *J Food Prot* 63:796–804.

Kerry JP, O'Grady MN, Hogan SA. 2006. Past, current and potential utilisation of active and intelligent packaging systems for meat and muscle-based products: a review. *Meat Sci* 74:113–130.

Labuza TP. 1993. Active packaging technologies for improved shelf life and quality. In *Science for the Food Industry of the 21st Century, Biotechnology, Supercritical Fluids, Membranes and Other Advanced Technologies for Low Calorie, Healthy Food Alternatives*, Yalpani M, Ed. ATL Press, Mount Prospect, IL, pp. 265–284.

Labuza TP. 1996. An introduction to active packaging for foods. *Food Technol* 50:68–71.

Labuza TP, Breene WM. 1989. Applications of "active packaging" for improvement of shelf-life and nutritional quality of fresh and extended shelf-life foods. *J Food Proc Pres* 13:1–69.

Loss CR, Hotchkiss JH. 2002. Effect of dissolved carbon dioxide on thermal inactivation of microorganisms in milk. *J Food Prot* 65:1924–1929.

Lyijynen T, Randell K, Skyttä E, Hurme E, Hattula T, Ahvenainen R. 1997. The Shelf-Life of Perch and Pike-Perch Fillets in Different Retail Packages. Paper presented at the 3rd Nordfood Conference, Sustainable Food Production and Competitive Industries, Copenhagen, Denmark.

Mejlholm O, Dalgaard P. 2002. Antimicrobial effect of essential oils on the seafood spoilage micro-organism *Photobacterium phosphoreum* in liquid media and fish products. *Lett Appl Microbiol* 34:27–31.

Otwell S. 1997. Time, temperature, travel: a quality balancing act. *Seafood Int* 57–61.

Ozdemir M, Floros JD. 2004. Active food packaging technologies. *Crit Rev Food Sci Nutr* 44:185–193.

Parkin KL, Brown WD. 1982. Preservation of seafood with modified atmospheres. In *Chemistry and Biochemistry of Marine Food Products*, Martin RE, Flick GJ, Hebard CE, Ward DR, Eds. AVI Publishing Company, Westport, CT, pp. 453–465.

Quintavalla S, Vicini L. 2002. Antimicrobial food packaging in meat industry. *Meat Sci* 62:373–380.

Rotabakk BI, Birkeland S, Jeksrud WK, Sivertsvik M. 2006. Effect of modified atmosphere packaging and soluble gas stabilization on the shelf life of skinless chicken breast fillets. *J Food Sci* 71:S124–S131.

Sivertsvik M. Active Packaging of Seafood: An Evaluation of the Use of Oxygen Absorbers in Packages of Different Seafood Products. Paper presented at the *27th WEFTA Meeting*, Madrid, October 19–22, 1997.

Sivertsvik M. 2000. Use of soluble gas stabilisation to extend shelf-life of fish. In *Proceedings of 29th WEFTA Meeting*, Leptocarya, Pieria, Greece, October 10–14, 1999, pp. 79–91.

Sivertsvik M. 2003. Active packaging in practice: fish. In *Novel Food Packaging Techniques*, Ahvenainen R, Ed. Woodhead Publishing Ltd., Cambridge, U.K., pp. 384–400.

Sivertsvik M, Birkeland S. 2006. Effects of soluble gas stabilisation, modified atmosphere, gas to product volume ratio and storage on the microbiological and sensory characteristics of ready-to-eat shrimp (*Panadalus borealis*). *Food Sci Technol Int* 12:445–454.

Sivertsvik M, Jeksrud WK, Rosnes JT. 2002a. A review of modified atmosphere packaging of fish and fishery products: significance of microbial growth, activities and safety. *Int J Food Sci Technol* 37:107–127.

Sivertsvik M, Jeksrud WK, Vågane Å, Rosnes JT. 2004a. Solubility and absorption rate of carbon dioxide into non-respiring foods. 1. Development and validation of experimental apparatus using a manometric method. *J Food Eng* 61:449–458.

Sivertsvik M, Jensen JS. 2005. Solubility and absorption rate of carbon dioxide into non-respiring foods. 3. Cooked meat products. *J Food Eng* 70:499–505.

Sivertsvik M, Rosnes JT, Bergslien H. 2002b. Modified atmospere packaging. In *Minimal Processing Technologies in the Food Industry*, Ohlsson T, Bengtsson N, Eds. Woodhead Publishing Ltd., Cambridge, U.K., pp. 61–86.

Sivertsvik M, Rosnes JT, Jeksrud WK. 2004b. Solubility and absorption rate of carbon dioxide into non-respiring foods. 2. Raw fish fillets. *J Food Eng* 63:451–458.

Sivertsvik M, Rosnes JT, Vorre Aa, Randell K, Ahvenainen R, Bergslien H. 1999. Quality of whole gutted salmon in various bulk packages. *J Food Qual* 22:387–401.

Sørheim O, Grini JA, Nissen H, Andersen HJ, Lea P. 1995. Pork loins stored in carbon dioxide: colour and microbiological shelf life. *Fleischwirtsch* 75:679–681.

Suppakul P, Miltz J, Sonneveld K, Bigger SW. 2003. Active packaging technologies with an emphasis on antimicrobial packaging and its applications. *J Food Sci* 68:408–420.

Taoukis PS, Koutsoumanis K, Nychas GJE. 1999. Use of time-temperature integrators and predictive modelling for shelf life control of chilled fish under dynamic storage conditions. *Int J Food Microbiol* 53:21–31.

Tewari G, Jayas DS, Jeremiah LE, Holley RA. 2001. Prevention of transient discoloration of beef. *J Food Sci* 66:506–510.

Tewari G, Jeremiah LE, Jayas DS, Holley RA. 2002. Improved use of oxygen scavengers to stabilize the colour of retail-ready meat cuts stored in modified atmospheres. *Int J Food Sci Technol* 37:199–207.

Vermeiren L, Devlieghere F, Beest Mv, Kruijf ND, Debevere J. 1999. Developments in the active packaging of foods. *Trends Food Sci Technol* 10:77–86.

Vermeiren L, Devlieghere F, Debevere J. 2002. Effectiveness of some recent antimicrobial packaging concepts. *Food Addit Contam* 19:163–171.

chapter eight

Modeling modified atmosphere packaging for fruits and vegetables

Yachuan Zhang, Z. Liu, and J.H. Han

Contents

8.1 Introduction

Fruits and vegetables are living, respiring, and edible tissues even after removal from the parent plant.[1] During the respiration, O_2 is consumed and CO_2 and H_2O produced. If a hexose sugar is used as the substrate, the overall equation can be written as follows:

$$C_6H_{12}O_6 + 6O_2 \rightarrow 6CO_2 + 6H_2O + energy \tag{8.1}$$

A low rate of respiration is desirable, for a lower rate lengthens the storage life. The ratio of the volume of CO_2 released to the volume of O_2 absorbed is termed the respiratory quotient (RQ) and is used to evaluate the nature of the respiratory process.[1] RQs are 1, <1, and >1 for carbohydrates, lipids, and organic acids, respectively, in the aerobic respiration process. Very high RQ values (>1.3) usually indicate anaerobic respiration,[1] which is an undesirable process during the storage.

From Equation 8.1, it is easy to see that the augmentation of CO_2 on the right side or the decrease of O_2 on left side would be helpful to retard the respiratory rate, and therefore lead to extension of the storage life of the fruit or vegetable. The principle of modified atmosphere packaging (MAP) is stemmed from this theory. Gas transfers (O_2 and CO_2) between the fruit or vegetable, package, and outside atmosphere follow a diffusion law at steady state. As the O_2 concentration within the plant cells begins to decrease as a result of respiration, the partial pressure of O_2 between the cells and the package interior and exterior atmosphere becomes unbalanced, causing some O_2 to diffuse from the outside atmosphere into the package interior atmosphere. When the rate of O_2 consumption of fresh produce exceeds the rate of permeation of atmosphere O_2 into the package, the O_2 concentration in the package decreases. This, in turn, further slows the respiration rate and equilibrates to the point where the rate of O_2 consumed equals the rate of O_2 permeated through the package film.[1] CO_2 enriches inside the package due to the respiration, leading to the diffusion of CO_2, crossing the package from inside to outside. A true equilibrium is expected to be reached when both O_2 and CO_2 flux due to respiration are equal to the flux due to permeation. However, this true equilibrium is so hard to achieve for polymer film that steady-state conditions inside a package are generally considered to be reached when the rate of O_2 consumption equals the rate of O_2 permeability through the film.[1] Therefore, MAP is a technique for controlling internal atmosphere, taking into account the natural process of respiration of the commodity and the gas permeability of the packaging film.[2] The key function of modeling MAP is to achieve the optimal storage conditions by designing various parameters regarding the storage, such as film permeability, thickness, package weight, etc., and avoiding anoxic conditions or too high O_2 inside the package. In anoxic conditions, a fruit or vegetable will carry out anaerobic respiration. Anaerobic respiration involves the incomplete oxidation of compounds in the absence of O_2 and the accumulation of ethanol, acetaldehyde, and CO_2,[1] consequently leading to fermentative volatile accumulation, off-odor development, and tissue injury.[3] High O_2 may increase oxidative browning and senescence.[4] Each fruit or vegetable needs a different optimal gas formulation,[2] which can be found in Table 8.1. A very effective way of plotting these data has been presented with CO_2 concentration as the ordinate and O_2 concentration as the abscissa (Figure 8.1 and Figure 8.2).

The packaging films used in MAP can be divided into two groups according to the differences in their structure: nonporous films (or continuous films) and porous films (or perforated films). Nonporous films are

Table 8.1 Recommended MAP Conditions of Various Products

Commodity	Temperature range (°C)	O_2 (%)[a]	CO_2 (%)[a]
Fruits			
Apple (whole)	0–5	2–3	1–5
Apple (sliced)	0–5	10–12	8–11
Avocado	5–13	2–5	3–10
Banana	12–15	2–5	2–5
Kiwifruit	0–5	2	5
Mango	10–15	5	5
Pineapple	10–15	5	10
Strawberry	0–5	10	15–20
Vegetables			
Asparagus	0–5	20	5–10
Broccoli	0–5	1–2	5–10
Cabbage	0–5	3–5	5–7
Lettuce (head)	0–5	2–5	0
Lettuce (shredded)	0–5	1–2	10–12
Mushrooms	0–5	21	10–15
Spinach	0–5	21	10–20
Tomatoes (mature)	12–20	3–5	0

[a] Volume or mole percentage; the remainder is nitrogen.

Source: Adapted from Floros, J.D. and Matsos, K.I., in *Innovations in Food Packaging*, Han, J.H., Ed., Elsevier Academic Press, New York, 2005, chap. 10. With permission from Elsevier.

traditionally used in MAP. However, the O_2 and CO_2 permeation rates through nonporous films are often below the produce respiration rate and the package headspace evolves toward a zero-O_2 atmosphere.[5] Therefore, the use of porous films is increasing with the application of MAP. Porous films provide high to very high gas exchange rates. They were introduced for the preservation of fruits and vegetables with very high respiration rates.[5] The gas transportation mechanisms are different between nonporous and porous films. For nonporous films, gas transportation is achieved by the combination of absorption and diffusion processes. For porous films, gas exchange is achieved by gas molecular collisions through the pores. Therefore, the mathematical models used to describe the permeation processes for nonporous and porous films, respectively, are different.[6] Numerous mathematical models have been applied for describing MAP systems, such as the Michaelis' Menten equation, which deals with the respiratory rate, and Fick's first law, which describes the gas diffusion process, etc.[5] These valuable models will be discussed in this chapter. The models generally include kinetics of respiration and permeation, and dependency of the respiration rate and permeability on the temperature, etc. Other factors, such as packed produce weight, respiring surface area, gas permeability of packaging film, film thickness, etc., which affect the equilibrium gas concentrations of

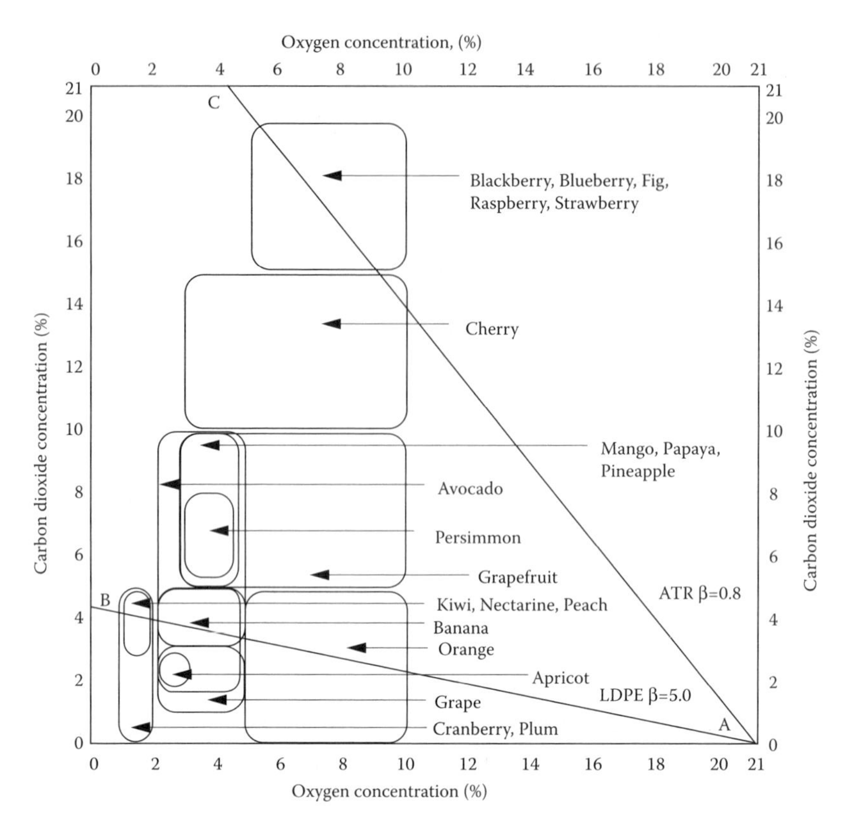

Figure 8.1 Recommended modified atmosphere for storage of fruits. (From Mannapperuma, J.D. et al., in *Proceedings of 5th Controlled Atmosphere Research Conference*, Wenatchee, WA, 1989, p. 225. With permission from Dr. A. Kader.)

packaged produce, are also considered in the establishment of the models.[6,7] Two of the most important parameters are the respiratory rate of the produce and the gas permeability of the packaging film.[8] The optimal MAP conditions (partial pressures of O_2 and CO_2 for the packaged produce at a specific temperature) have to be defined before establishing the MAP models.[3] The objective of this article is to describe these models, including respiration models and gas transportation models, for nonporous and porous films and figure out how to use them to create a MAP system.

8.2 Respiration rate

The first job needed to do for designing and developing a MAP is to determine the respiration rate of the fruit or vegetable under the MAP conditions existing inside the package. Respiration rate data can be obtained from literature for several produce. Respiration rate is a value that can be influenced by various factors. The most important factor is the temperature. The high temperature increases the respiration rate. The influence of temperature

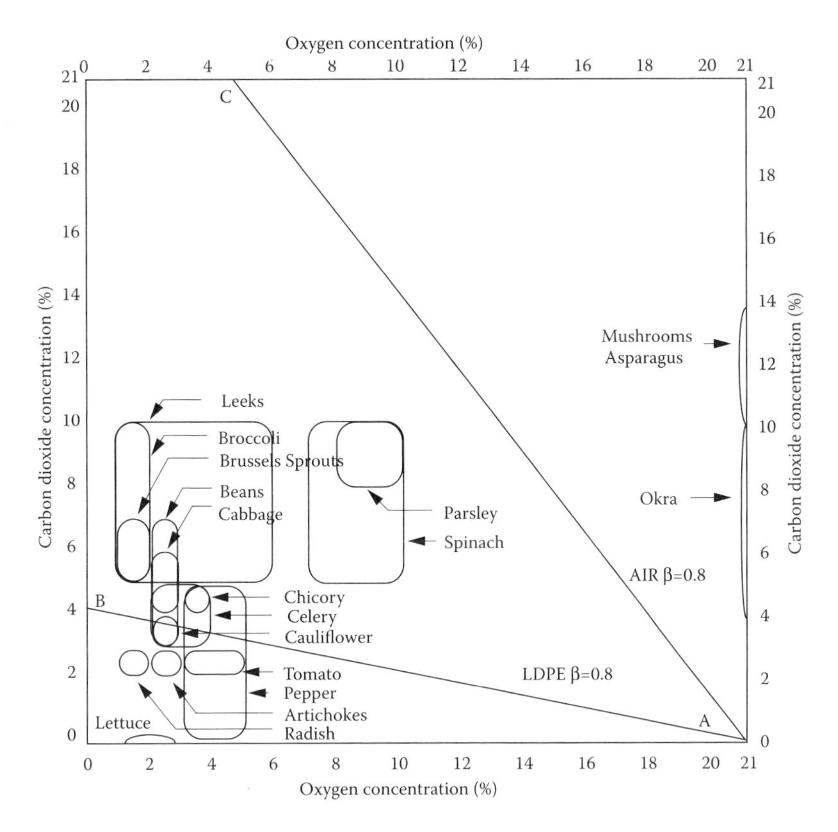

Figure 8.2 Recommended modified atmosphere for storage of vegetables. (From Mannapperuma, J.D. et al., in *Proceedings of 5th Controlled Atmosphere Research Conference*, Wenatchee, WA, 1989, p. 225. With permission from Dr. A. Kader.)

on respiration rate was first quantified with the Q_{10} value, which is the respiration rate increase for a 10°C rise in temperature:

$$Q_{10} = (R_{r2} / R_{r1})^{10/(T_2 - T_1)} \tag{8.2}$$

where R_{r1} and R_{r2} are the respiration rates at temperatures T_1 and T_2, respectively. For various products, Q_{10} values may range from 1 to 4 depending on the temperature range.[1] For example, for vegetables, the Q_{10} values are 2.5 to 4.0 at 0 to 10°C; 2.0 to 2.5 at 10 to 20°C; 1.5 to 2.0 at 20 to 30°C; and 1.0 to 1.5 at 30 to 40°C.[9] The increase in respiration rates declines with an increase in temperature up to 40°C, with Q_{10} becoming less than 1 as the tissue is near its thermal death point (about 50 to 55°C) when enzymes are denatured and metabolism becomes disorderly.[9] The second factor influencing the respiration rate is the O_2 and CO_2 concentration around the produce. Reduction of the O_2 concentration or elevation of CO_2 concentration can slow down the respiration rate. For vegetables, the minimum O_2 content of 1 to

3% is required to maintain aerobic respiration. Low O_2 or high CO_2 also applies some other positive physiological influences on fruits and vegetables. For example, elevated CO_2 levels (from 10 to 15%) significantly inhibit development of Botrytis rot on strawberries, cherries, and other fruits. Also, MAP can be a useful tool for insect control in some commodities.[10]

Respiration rates are commonly obtained by measuring the rate of O_2 consumption and CO_2 production when the produce is sealed inside a closed container. A gas chromatograph is usually used to measure the headspace composition of the gases periodically by withdrawing a certain amount of gas sample with an airtight syringe.[10] Various empirical expressions, such as first order and second order, are fitted to the respiration data using non-least-squares analysis, and the equations that give the best fit to respiration data are then chosen.[2] Besides the empirical expressions, the Michaelis' Menten equation is commonly used to describe the dependency of the respiration rate on O_2 concentration:

$$R_{O_2} = \frac{R_{O_2}^{\max} \cdot p_{O_2}^{pkg}}{k_{1/2} + p_{O_2}^{pkg}} \tag{8.3}$$

where R_{O_2} is the O_2 respiration rate (ml kg^{-1} h^{-1}), $R_{O_2}^{\max}$ is the maximum rate of O_2 consumption (ml kg^{-1} h^{-1}), $k_{1/2}$ is the Michaelis' Menten constant (atm), and $p_{O_2}^{pkg}$ is the internal partial pressure (atm) of O_2. An uncompetitive inhibition equation is commonly used to describe the dependency of the respiration rate on CO_2 concentration:[11]

$$R_{CO_2} = \frac{R_{O_2}^{\max} \cdot p_{O_2}^{pkg}}{k_{1/2} + \left[1 + \frac{p_{CO_2}^{pkg}}{k_i}\right] \cdot p_{O_2}^{pkg}} \tag{8.4}$$

where R_{CO_2} is the CO_2 respiratory rate in units of ml kg^{-1} h^{-1}, $p_{CO_2}^{pkg}$ is the internal partial pressure (atm) of CO_2, and k_i is the inhibition reaction constant.[11]

8.3 Gas transmission in nonporous film

Permeation, absorption, and diffusion are typical mass transfer phenomena occurring in nonporous film. Figure 8.3 shows these three profiles.[12] Therefore, the gas transport properties of packaging films are often described by three common coefficients: diffusivity, solubility, and permeability. The diffusion coefficient (i.e., diffusivity) describes the movement of permeant molecules through a polymer, and thus represents a kinetic property of a polymer–permeant system.

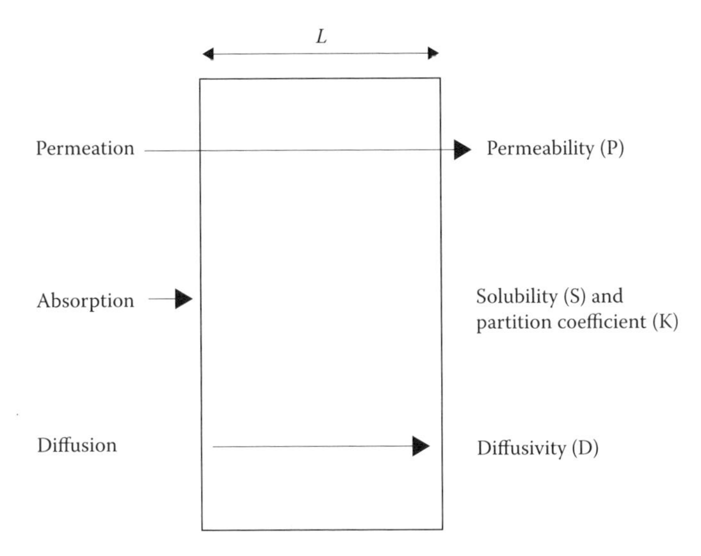

Figure 8.3 Mass transfer phenomena and their characteristic coefficients. (Adapted from Han, J.H. and Scanlon, M.G., in *Innovations in Food Packaging*, Elsevier, New York, 2005, chap. 2, p. 12. With permission from Elsevier.)

Figure 8.4 shows the activated diffusion process[14] used to describe the mass transfer phenomenon of permeant movement in a polymer. Activated diffusion is described as the opening of a void space among a series of segments of a polymer chain due to the oscillations of the segments (an active state), followed by translational motion of the permeant within the void space before the segments return to their normal state. Both the active and normal states are long-lived, as compared with the translational rate of the gas.[8] The diffusion can be described by Fick's law. Fick's first law in one dimension defines the diffusion coefficient (D):

$$J = -D\frac{\partial c}{\partial x} \tag{8.5}$$

where J is the diffusive mass transfer rate of a permeant per unit area, c is the concentration of a permeant, and x is the length or thickness of the film.

Before the permeant diffuses through the film, it must be adsorbed and dissolved into the film surface. The solubility coefficient describes the dissolution rate of the permeant in a polymer.[12] Equation 8.6 shows the linear relationship (Henry's law) between the concentration at the surface of the packaging material and the partial pressure of the gas in atmosphere:

$$c = H^{-1} \cdot p \tag{8.6}$$

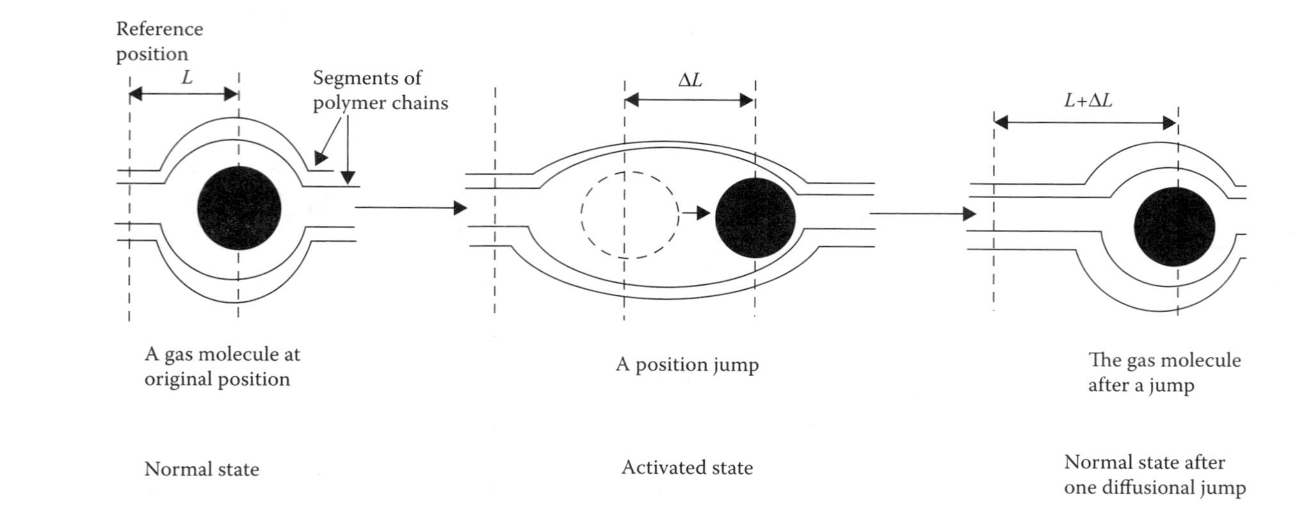

Figure 8.4 The activation process for diffusion. *L* is the reference distance and Δ*L* is the distance the gas molecule moved. (Adapted from Dibenedetto, A.T., *J. Polym. Sci. A*, 1, 3477, 1963. With permission from John Wiley & Sons.)

where c is the concentration of gas at the surface of the film, p is the partial pressure of the gas, and H^{-1} is a constant. The constant H^{-1} can be expressed in the form of Equation 8.7:

$$H^{-1} = \frac{c}{p} = \frac{n}{V_{material} \cdot p} = \frac{n}{V_{STP}} \cdot \frac{V_{STP}}{V_{material} \cdot p} = \frac{S}{V_{STP}} \tag{8.7}$$

where n is the number of moles of the gas that dissolve in the packaging material and $V_{material}$ is the volume of the packaging material into which the gas dissolves. V_{STP} is the volume occupied by 1 mole of the gas under standard temperature and pressure (STP) conditions (0°C and 1 atm). This has a value of 22414×10^{-6} m^3 mol^{-1} for an ideal gas, and thus can be taken as a constant. S is the solubility.[12] Therefore, S can be expressed as

$$S = \frac{n \cdot V_{STP}}{V_{material} \cdot p} \tag{8.8}$$

S has the unit mol m_{gas}^{3} mol^{-1} $m_{material}^{-3}$ Pa^{-1}, which is equivalent to Pa^{-1}, showing the volume ratio of the absorbed gas to the material under standard gas conditions.[12]

If the permeable gas molecules have an affinity to the packaging material matrix, or are immobilized in the microvoids of the matrix polymer at a relatively low pressure, the sorption behavior follows a logarithmic relationship, which is expressed as a Langmuir-type sorption.[13] Therefore, Henry's law (Equation 8.6) can be utilized limitedly at the low gas concentration condition.

The permeability coefficient incorporates both diffusion and solubility properties of the film/permeant system; thus, it provides a gross mass transport property. By integrating Equation 8.5 at the steady-state conditions, Equation 8.9 is expressed as follows:

$$J = D \cdot \frac{\Delta c}{x} \tag{8.9}$$

From Equation 8.7,

$$c = \frac{S \cdot p}{V_{STP}} \tag{8.10}$$

By substituting Equation 8.10 into Equation 8.9,

$$J = D \cdot \frac{S \cdot \Delta p}{V_{STP} \cdot x} = \frac{P \cdot \Delta p}{V_{STP} \cdot x} \tag{8.11}$$

where P is the permeability and Δp is the difference in the partial pressure of the gas on the outside and inside of the packaging film. The partial pressure outside the packaging film would generally be a constant at ambient conditions. From Equation 8.11, the permeability, diffusivity, and solubility of a gas have the relationship shown in Equation 8.12:

$$P = D \cdot S \tag{8.12}$$

Therefore, the SI units of P, D, and S are $m^2\ s^{-1}\ Pa^{-1}$, $m^2\ s^{-1}$, and Pa^{-1}, respectively.[11]

The permeabilities of the commonly used synthetic films are listed in Table 8.2. Usually, the O_2 and CO_2 permeabilities of plastic films increase exponentially with increasing temperature. Previous studies have shown that the relationship agrees with the Arrhenius expression.[15] Therefore, the natural logarithm of the permeability coefficient for both gases depends linearly on the reciprocal of temperature (K):

$$\ln P = \ln P_0 - \frac{E_a}{R} \cdot \frac{1}{T} \tag{8.13}$$

where P is O_2 or CO_2 permeability ($m^2\ sec^{-1}\ Pa^{-1}$), P_0 is a theoretical permeability at infinitively high temperature ($m^2\ sec^{-1}\ Pa^{-1}$), E_a is an activated

Table 8.2 Oxygen and Carbon Dioxide Permeability Coefficients[a] for Various Polymers

Polymer	P_{O_2} (30°C)	P_{CO_2} (30°C)	$\frac{P_{CO_2}}{P_{O_2}}$
Low-density polyethylene	55	352	6.4
High-density polyethylene	10.6	35	3.3
Polypropylene	23	92	4.0
Unplasticized poly(vinyl chloride)	1.2	10	8.3
Cellulose acetate	7.8	68	8.7
Polystyrene	11	88	8.0
Nylon 6	0.38	1.6	4.2
Poly(ethylene terephahlate)	0.22	1.53	7.0
Poly(vinylidene chloride)	0.053	0.29	5.5

[a] Permeability units: $\times 10^{11}$ ml (STP) cm $cm^{-2}\ s^{-1}$ (cm Hg^{-1}).

Source: Adapted from Singh, R.K. and Singh, N., in *Innovations in Food Packaging*, Han, J.H., Ed., Elsevier Academic Press, New York, 2005, chap. 3. With permission from Elsevier.

energy (J mol^{-1}), R is the universal gas constant (8.314 J mol^{-1} K^{-1}), and T is temperature (K).

Since the biodegradable packaging film and coating, made of proteins, starch, lipids, or their combinations, were recently developed, the relationship between permeability and relative humidity (RH) was studied. Water molecule works as a plasticizer in the biodegradable films and coatings. Therefore, moisture content in the films or coatings can significantly influence the permeability of O_2 and CO_2. For the whey protein isolate (WPI) coating system, the relationship between the O_2 permeability and RH is expressed as follows:

$$P_{O_2} = \mathrm{j} \times 10^{kRH} \tag{8.14}$$

where RH is relative humidity (%), P_{O_2} is film O_2 permeability, and j and k are specific constants for different kinds of films. For WPI/sorbitol (ratio 60/40) and WPI/glycerol (ratio 70/30) films, j values are 0.026 and 0.037, respectively, and k values are 0.049 and 0.037, respectively.[15] Equation 8.14 can be used in other hydrophilic packaging films, such as cellulose and various polyvinyl.

In classical packaging design, it is assumed that at steady state, the change in the values of pO_2 and pCO_2 (partial pressure) inside the MAP package at any time could be zero. The following equations are available in describing this steady state:

$$\frac{dpO_2^{in}}{dt} = \frac{P_{O_2}}{x} \cdot A \cdot (pO_2^{out} - pO_2^{in}) - W \cdot R_{O_2} = 0 \tag{8.15}$$

$$\frac{dpCO_2^{in}}{dt} - \frac{P_{CO_2}}{x} \cdot A \cdot (pCO_2^{in} - pCO_2^{out}) - W \cdot R_{CO_2} = 0 \tag{8.16}$$

where A is the total area of the film surface, and W is mass of packaged foods. Outside the package, the total pressure is 1 atm at sea level. Therefore, pO_2 is approximately equal to 0.209 atm and pCO_2 is 0.0003 atm. By inputting the appropriate respiratory models (Equations 8.3 and 8.4) into Equations 8.15 and 8.16, solutions could be obtained for pO_2 and pCO_2 when a particular film and system geometry was used, or for the values of P_{O_2} and P_{CO_2} required to obtain specific pO_2 and pCO_2 conditions at steady state.[11]

The application of Equations 8.15 and 8.16 in practice can be summarized in two aspects. First, Equations 8.15 and 8.16 can be rearranged to make $\frac{AW}{x}$ the subject of each equation and then combined. If gas concentration (c) rather than partial pressure (p) is applied, the relationship between $c_{CO_2}^{in}$ and $c_{O_2}^{in}$ can be derived as follows:

$$c_{CO_2}^{in} = c_{CO_2}^{out} + \frac{P_{O_2}}{P_{CO_2}}(c_{O_2}^{out} - c_{O_2}^{in})\frac{R_{CO_2}}{R_{O_2}} \tag{8.17}$$

where $\frac{P_{O_2}}{P_{CO_2}}$ is a constant for a film and also called $\frac{1}{\beta}$. In the case of low-density polyethylene (LDPE), where β is roughly 5, and assuming $\frac{R_{CO_2}}{R_{O_2}}$ is unity, Equation 8.17 yields a straight line with a slope of $\frac{1}{5}$, shown as line *AB* in Figure 8.1 and Figure 8.2. This means that only the modified atmosphere falling along the line *AB* can be attained by LDPE film. However, $\frac{R_{CO_2}}{R_{O_2}}$ will change the slope ($\frac{1}{\beta}$) if it is not equal to unity. Therefore, Equation 8.17 approximates the relationship between $c_{O_2}^{in}$ and $c_{CO_2}^{in}$.[16]

Second, when a film is selected, P_{CO_2}, P_{O_2}, and x become given parameters. The A of the film package and W of foods in a package are needed to determine $c_{O_2}^{in}$ and $c_{CO_2}^{in}$. Using τ, denoted the ratio of $\frac{W}{A}$, Equations 8.15 and 8.16 can be rearranged to yield the relationship between $c_{O_2}^{in}$ and $c_{CO_2}^{in}$.

$$c_{O_2}^{in} = c_{O_2}^{out} - \frac{xR_{O_2}}{P_{O_2}}\tau \tag{8.18}$$

$$c_{CO_2}^{in} = c_{CO_2}^{out} + \frac{xR_{CO_2}}{P_{CO_2}}\tau \tag{8.19}$$

Assuming the respiration rate did not change with gas concentrations, any changes in τ would simply move the gas compositions inside the package along the *AB* line in Figure 8.1 and Figure 8.2. Increasing τ would decrease the O_2 concentration and increase the CO_2 concentration. Thus, gas composition inside the package changes toward *B* from *A* along the line *AB* in direct proportion to the increase in τ.[16]

The ratio of CO_2 to O_2 permeability (β) of a polymeric film can be a useful criterion for selecting package film.[1] Since values of RQ range from 0.7 to 1.3 for aerobic respiration for most fruits and vegetables, R_{CO_2} can be considered to be roughly equal to R_{O_2}. From Equations 8.15 and 8.16, the ratio can be expressed as follows:

$$\frac{P_{CO_2}}{P_{O_2}} = \frac{p_{O_2}^{out} - p_{O_2}^{in}}{p_{CO_2}^{in} - p_{CO_2}^{out}} \tag{8.20}$$

In normal atmosphere condition, $p_{O_2}^{out}$ is equal to 0.21 atm and $p_{CO_2}^{out}$ is essentially 0 atm. Therefore, Equation 8.20 can be simplified to

$$\frac{P_{CO_2}}{P_{O_2}} = \frac{p_{O_2}^{out} - p_{O_2}^{in}}{p_{CO_2}^{in} - p_{CO_2}^{out}} = \frac{0.21 - p_{O_2}^{in}}{p_{CO_2}^{in}} \tag{8.21}$$

Assuming the total pressure inside the package is equal to the outside atmosphere (for flexible packages, it is true), Equation 8.21 can be used to calculate the ratio of the CO_2 to O_2 permeability to give a desired gas atmosphere inside the package. For example, if it is desired to create an atmosphere containing 2% O_2 and 6% CO_2, then

$$\frac{P_{CO_2}}{P_{O_2}} = \frac{0.21 - p_{O_2}^{in}}{p_{CO_2}^{in}} = \frac{0.21 - 0.02}{0.06} = 3.2 \tag{8.22}$$

Therefore, a film with a CO_2 permeability 3.2 times its O_2 permeability at the temperature of storage should be selected as a packaging material. However, if the permeabilities of the gases are very small, the gas fluxes will be very low according to Equation 8.11. Anoxic conditions can be developed inside the package. So, application of Equation 8.22 in practice requires not only the ratio of $\frac{P_{CO_2}}{P_{O_2}}$, but also the values of the permeability. Useful data on the ratio of $\frac{P_{CO_2}}{P_{O_2}}$ for various films are listed in Table 8.2.[17] Because the permeability can be affected greatly by temperature, the selection of the ratio of $\frac{P_{CO_2}}{P_{O_2}}$ should be considered based on the listed storage temperature accordingly.

8.4 Gas transmission in porous film

In practice, fruits and vegetables after harvest have higher respiration rates than the permeability of the films, resulting in the produce headspace evolving toward a very low O_2 atmosphere.[5] Since CO_2 permeates out four times faster from the package than O_2 invasion for most plastics, the headspace pressure lowers below 1 atm or the package collapses. Therefore, more porous films have recently been introduced for the preservation of fruits and

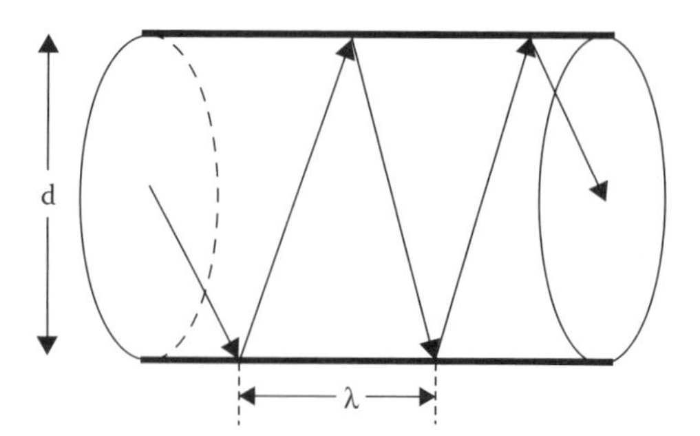

Figure 8.5 Knudsen diffusion. d is the diameter of the pore and λ is the mean free path of the permanent molecules.

vegetables with a very high respiration rate and also for slowly respiring products in a large-sized package.

In a porous film packaging, the gas permeability of film does not affect the gas exchange process due to the gas exchanges through pores. Consequently, the combination of equations such as Fick's law for diffusion or Henry's law for sorption does not describe the transport process. Instead, the mass transfer models of Knudsen, Maxwell (for nonpressurized packaging systems), or Poiseuille (for pressurized systems) are more suitable.[5]

8.4.1 Knudsen diffusion

Knudsen diffusion occurs when the mean free path (λ) is relatively long compared to the pore size, so the molecules collide frequently with the pore wall rather than intramolecular collisions (shown in Figure 8.5). Knudsen diffusion is dominant for pores that range in diameter between 2 and 50 nm,[18] while Maxwell and Poiseulle equations are suitable for larger diameters. The equation for the Knudsen diffusion coefficient is

$$D_K = \frac{d}{3}\sqrt{\frac{8RT}{\pi M}} \tag{8.23}$$

where R and T are the gas constant (8.314 J mol^{-1} K) and temperature (K), respectively, d is the pore diameter, and M is the molar mass of the gas (kg mol^{-1}).

The flux of the gas is proportional to the molar concentration gradient along the pore. The molar flow (flux) of the gaseous substance is described as follows:

$$N_{O_2} = -D_{K,O_2} \cdot A \cdot \frac{dc_{O_2}}{dL} = -\frac{\pi d^2 D_{K,O_2}(p_{O_2}^{out} - p_{O_2}^{in})}{4p^{total} \cdot L} = -\frac{\pi d^2 D_{K,O_2}(c_{O_2}^{out} - c_{O_2}^{in})}{4L} \tag{8.24}$$

$$N_{CO_2} = D_{K,CO_2} \cdot A \cdot \frac{dc_{CO_2}}{dL} = \frac{\pi d^2 D_{K,CO_2}(p_{CO_2}^{in} - p_{CO_2}^{out})}{4p^{total} \cdot L}$$
$$= \frac{\pi d^2 D_{K,CO_2}(c_{CO_2}^{in} - c_{CO_2}^{out})}{4L} \tag{8.25}$$

where D_K is the Knudsen diffusion coefficient, L is the film thickness or the pore length, $c_{O_2}^{out}$ and $c_{O_2}^{in}$ are the O_2 molar fractions outside and inside the packaging film, respectively, $c_{CO_2}^{out}$ and $c_{CO_2}^{in}$ are CO_2 molar fractions, respectively, and A is the cross-sectional area of the pore.

8.4.2 *Maxwellian diffusion*

When the mean free path (λ) of the permanent molecules is relatively short compared to the pore size, the diffusion type belongs to Maxwellian diffusion (shown in Figure 8.6). Maxwellian diffusion is also known as bulk diffusion, which is caused by collisions of gas molecules.[19] This diffusion mode is applicable to Brownian motion, where the movement of each particle is random and not dependent on its previous motion, and the molecules are regarded as a rigid sphere experiencing elastic collision.[20]

$$J_{O_2} = \left(\sum_i^n N_i\right) \cdot c_{O_2} - D_{O_2,air} \cdot \frac{dc_{O_2}}{dL} \tag{8.26}$$

$$J_{CO_2} = \left(\sum_i^n N_i\right) \cdot c_{CO_2} - D_{CO_2,air} \cdot \frac{dc_{CO_2}}{dL} \tag{8.27}$$

where J is the molar flow (flux) of O_2 or CO_2 per unit area in mol s^{-1} m^{-2}, n is the number of components in the mass, D is the diffusion coefficient of O_2 or CO_2 in stagnant air, and c is the molar fraction.

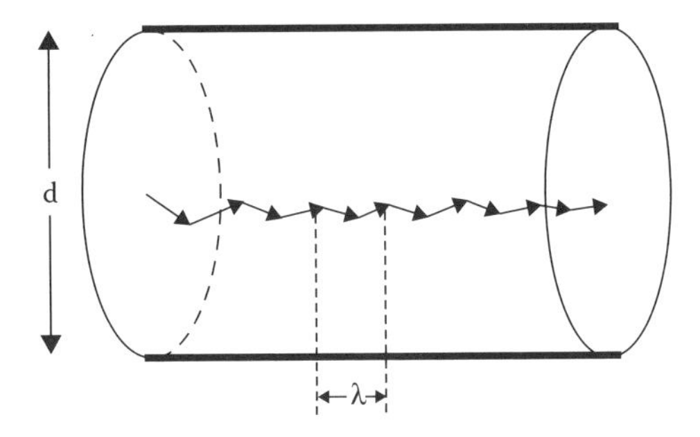

Figure 8.6 Maxwellian diffusion. *d* is the diameter of the pore and λ is the mean free path of the permanent molecules.

The diffusion coefficient of a gas in air ($D_{O_2,air}$ or $D_{CO_2,air}$) can be estimated by the method of Fuller, Schettler, and Giddings (FSG method).[20] It is most accurate for nonpolar gases at low to moderate temperatures. The FSG method is based on the following equation:

$$D_{AB} = \frac{10^{-3} T^{1.75} \sqrt{\frac{M_A + M_B}{M_A \cdot M_B}}}{P\left[\left(\sum v\right)_A^{1/3} + \left(\sum v\right)_B^{1/3}\right]^2} \tag{8.28}$$

where D_{AB} is the diffusion coefficient of compound A in compound B (m^2 sec^{-1}), T is temperature (K), P is pressure (atm), and M_A and M_B are the molecular weight (g mol^{-1}) of compound A and average molecular weight of B, respectively. $\sum v$ is a molar volume based on the method developed by Fuller (m ^{3} mol^{-1}). It can be estimated from volume increments associated with each element in the compound. These increments give the volume (m^3) per mole of atom present. The values are given in Table 8.3.

At steady state, Equations 8.26 and 8.27 can be integrated to give Equations 8.29 and 8.30:

$$J_{O_2} = \phi_{O_2} \frac{D_{O_2,air} \cdot p_{total}}{RTL} \cdot \ln \frac{\phi_{O_2} - c_{O_2}^{out}}{\phi_{O_2} - c_{O_2}^{in}} \tag{8.29}$$

$$J_{CO_2} = \phi_{CO_2} \frac{D_{CO_2,air} \cdot p_{total}}{RTL} \cdot \ln \frac{\phi_{CO_2} - c_{CO_2}^{out}}{\phi_{CO_2} - c_{CO_2}^{in}} \tag{8.30}$$

where $\phi_{O_2} = J_{O_2} \Big/ \sum_i^n N_i$ and $\phi_{CO_2} = J_{CO_2} \Big/ \sum_i^n N_i$, which can be rewritten to give volume flow per pore as

$$N_{O_2} = \phi_{O_2} \frac{\pi d^2 D_{O_2,air}}{4L} \ln \frac{\phi_{O_2} - c_{O_2}^{out}}{\phi_{O_2} - c_{O_2}^{in}} \tag{8.31}$$

$$N_{CO_2} = \phi_{CO_2} \frac{\pi d^2 D_{CO_2,air}}{4L} \ln \frac{\phi_{CO_2} - c_{CO_2}^{out}}{\phi_{CO_2} - c_{CO_2}^{in}} \tag{8.32}$$

Table 8.3 Special Atomic Diffusion Volumes for Use in Equation 8.28

Atomic and Structural Diffusion Volume Increments			
C	16.5	(Cl)[a]	19.5
H	1.98	(S)	17.0
O	5.48	Aromatic or heterocyclic rings	–20.2
(N)	5.69		
Diffusion Volumes of Simple Molecules, $\sum v$			
H_2	7.07	CO_2	26.9
D_2	6.70	N_2O	35.9
He	2.88	NH_3	14.9
N_2	17.9	H_2O	12.7
O_2	16.6	(CCl_2F_2)	114.8
Air	20.1	(SF_6)	69.7
Ne	5.59		
Ar	16.1	(Cl_2)	37.7
Kr	22.8	(Br_2)	67.2
(Xe)	37.8	(SO_2)	41.1
CO	18.9		

[a] Parentheses indicate a value based on only a few data.

Source: Skelland, A.H.P., in *Diffusional Mass Transfer*, Skelland, A.H.P., Ed., John Wiley & Sons, New York, 1974, chap. 3. With permission from John Wiley & Sons, Inc.

where $\phi_{O_2} = N_{O_2} \Big/ \sum_i^n J_i$ and $\phi_{CO_2} = N_{CO_2} \Big/ \sum_i^n J_i$. Therefore, total gas flux can be obtained from multiplying Equations 8.31 and 8.32 by the number of pores in unit area (m^2).

8.4.3 *Transition diffusion*

When pore size is about the mean free path, transition diffusion happens. Transition diffusion has properties of both Knudsen and Maxwellian diffusion. A combination of Knudsen and Maxwellian diffusivity laws is needed since both molecule-to-molecule collision and collision with pore walls contribute to the gas exchange control:[5]

$$N_{O_2} = \phi_{O_2} \frac{\pi d^2 D_{O_2,air}}{4L} \ln \frac{\left(1 + D_{O_2,air} \Big/ D_{K,O_2}\right) \cdot \phi_{O_2} - c_{O_2}^{in}}{\left(1 + D_{O_2,air} \Big/ D_{K,O_2}\right) \cdot \phi_{O_2} - c_{O_2}^{out}} \tag{8.33}$$

$$N_{CO_2} = \phi_{CO_2} \frac{\pi d^2 D_{CO_2,air}}{4L} \ln \frac{\left(1 + D_{CO_2,air} / D_{K,CO_2}\right) \cdot \phi_{CO_2} - c_{CO_2}^{in}}{\left(1 + D_{CO_2,air} / D_{K,CO_2}\right) \cdot \phi_{CO_2} - c_{CO_2}^{out}} \tag{8.34}$$

where D_{K,O_2} and D_{K,CO_2} are the Knudsen diffusivities for O_2 and CO_2, respectively.

It was found that the flow values obtained by Equations 8.25, 8.32, and 8.34 are a function of pore diameter. Equations 8.25 and 8.34 provide the same result of mass flow rage for pore diameters of <0.1 μm. Equation 8.32 is applied for pores of >10 μm. Within the 0.1- to 10-μm range, both Equations 8.25 and 8.32 overestimate the flow and Equation 8.33 is needed.[5] Therefore, special attention should be paid to the pore size of the porous films when selecting an equation to calculate the gas flux.

8.4.4 Hydrodynamic diffusion

Equations 8.24, 8.25, and 8.31 to 8.34 are applicable only when the pressures at both sides of the film are identical. When there is a total pressure gradient along the film, a hydrodynamic flow is established. In this case, total flow is described using Poiseuille's law:[5]

$$J = \frac{\pi d^4 (p_1 - p_2)}{128 \mu L} \tag{8.35}$$

where p_1 and p_2 are the total pressure at each side of the film and μ is the gas viscosity (Pa s).

The composition of the stream is identical to that of the atmosphere with the highest total pressure. Therefore, the flow of compound A through the pore will be equal to

$$N_A = \frac{\pi d^4 (p_1 - p_2)}{128 \mu L} \cdot c_A \tag{8.36}$$

where c_A is the molar fraction of compound A in the gas mixture with the highest pressure. When there is a difference in composition as well as in total pressure, hydrodynamic flow and diffusion flow will coexist and the combination of the equation is

$$N_A = N_{Transition,A} + N_{hydrodynamic,A} = \phi_A \frac{\pi d^2 D_{A,air}}{4L} \ln \frac{(1+ D_{A,air}/D_{K,A})\phi_A - c_{A1}}{(1+ D_{A,air}/D_{K,A})\phi_A - c_{A2}}$$

$$\frac{\pi d^4 (p_1 - p_2)}{128 \mu L} \cdot c_A \tag{8.37}$$

Since the gases permeate through the porous films by pores, the sorption of the gas components into the package materials is negligible in gas transport mechanisms.

Example

Calculate the diffusivity of O_2 in air at 35°C and 1 atm.

Solution:

$$D_{CO_2,air} = \frac{10^{-3} T^{1.75} (\frac{1}{M_{O_2}} + \frac{1}{M_{air}})^{1/2}}{p\left[\left(\sum v\right)_{O_2}^{1/3} + \left(\sum v\right)_{air}^{1/3}\right]^2} \tag{8.38}$$

For air: 1. $v_{air} = 20.1$

2. Average molecular weight of air = 28.97 g mol^{-1}

For O_2, from Table 8.3:

$O = 5.48$, then $O_2 = 10.96$

$v_{O_2} = 10.96$

Molecular weight of O_2 = 32 g mol^{-1}

$$D_{CO_2,air} = \frac{10^{-3}(308)^{1.75}\left(\frac{1}{32} + \frac{1}{28.97}\right)^{1/2}}{1 \cdot \left[(10.96)^{1/3}{}_{O_2} + (20.1)^{1/3}{}_{air}\right]^2} = 16.6 \ \text{cm}^2 \ \text{s}^{-1} \tag{8.39}$$

Generally, $D_{O_2,air}$ and $D_{CO_2,air}$ can be obtained from Table 8.3. They are 16.6 and 26.9 $cm^2\ s^{-1}$, respectively.

8.5 Conclusions

Traditionally, nonporous films were chosen for fruit and vegetable MAP. However, porous films became preferred for produce with high respiration rates because it can increase the permeation of gases greatly. In addition, porous films with bigger pores can effectively reduce the humidity inside the package, which is a concern for the nonporous package. No matter what film is selected, the optimal MAP conditions and the respiration rate at the storage temperature for the given produce should be identified before designing a MAP system. The specific permeability and thickness of films required to establish the MAP can be calculated for a given fruit or vegetable according to the models included in this chapter. Conversely, the weight of the fruit or vegetable can be figured out by the models when the package film with known permeability and surface area are available. Figure 8.1 and Figure 8.2 are valuable guidelines for selecting nonporous packaging films. values can roughly help when choosing a package film. The chemical composition plays less of a role for gas diffusion in porous films, because gas diffusion is mainly decided by the size of the pores and film thickness. With the Knudsen, Maxwell, and transition models, the pore diameter and film thickness required to establish the MAP can be calculated for a given fruit or vegetable, then the corresponding films can be chosen. The high quality of the produce and appropriate temperature are the most important factors for prolonging the storage life of a fruit or vegetable. Only when these two premises are achieved is the modeling MAP meaningful.

References

1. Robertson, G.L., *Food Packaging Principles and Practice*, Marcel Dekker, New York, 1993, pp. 470–506.
2. Floros, J.D. and Matsos, K.I., Introduction to modified atmosphere packaging, in *Innovations in Food Packaging*, Han, J.H., Ed., Elsevier Academic Press, New York, 2005, chap. 10, pp. 159–172.
3. Fonseca, S.C. et al., Influence of low oxygen and high carbon dioxide on shredded Galega kale quality for development of modified atmosphere packages, *Postharvest Biol. Technol.*, 35, 279–292, 2005.
4. Kim, J.G. et al., Effect of initial oxygen concentration and film oxygen transmission rate on the quality of fresh-cut romaine lettuce, *J. Sci. Food Agric.*, 85, 1622–1630, 2005.
5. Del-Valle, V. et al., Modeling permeation through porous polymeric films for modified atmosphere packaging, *Food Addit. Contam.*, 20, 170–179, 2003.
6. Zhu, M., Chu, C.L., and Wang, S.L., Predicting oxygen and carbon dioxide partial pressures within modified atmosphere packages of cut rutabaga, *J. Food. Sci.*, 67, 714–720, 2002.

7. Kim, J.G., Luo, Y., and Gross, K.C., Effect of package film on the quality of fresh-cut salad savoy, *Postharvest Biol. Technol.*, 32, 99–107, 2004.
8. Miller, K.S. and Krochta, J.M., Oxygen and aroma barrier properties of edible films: a review, *Trends Food Sci. Technol.*, 81, 228–237, 1997.
9. Kader, A.A., *Postharvest Technology of Horticultural Crops*, 2nd ed., Division of Agriculture and Natural Resources, University of California, Davis, 1992, pp. 85–92.
10. Escalona, V.H. et al., Changes in respiration of fresh-cut butterhead lettuce under controlled atmospheres using low and superatmospheric oxygen conditions with different carbon dioxide levels, *Postharvest Biol. Technol.*, 39, 48–55, 2006.
11. Van de Velde, M.D., Van Loey, A.M., and Hendrickx, M.E., Modified atmosphere packaging of cut Belgian endives, *J. Food Sci.*, 67, 2202–2206, 2002.
12. Han, J.H. and Scanlon, M.G., Mass transfer of gas and solute through packaging materials, in *Innovations in Food Packaging*, Han, J.H., Ed., Elsevier Academic Press, New York, 2005, chap. 2, pp. 12–23.
13. Cussler, E.L., *Diffusion, Mass Transfer in Fluid Systems*, 2nd ed., Cambridge University Press, New York, 1997, pp. 308–329.
14. Dibenedetto, A.T., Molecular properties of amorphous high polymers. An interpretation of gaseous diffusion through polymers, *J. Polym. Sci. A*, 1, 3477–3487, 1963.
15. Cisneros-Zevallos, L. and Krochta, J.M., Internal modified atmospheres of coated fresh fruits and vegetables: understanding relative humidity effects, in *Innovations in Food Packaging*, Han, J.H., Ed., Elsevier Academic Press, New York, 2005, chap. 10, pp. 159–172.
16. Mannapperuma, J.D. et al., Design of polymeric packages for modified atmosphere storage of fresh produce, in *Proceedings of 5th Controlled Atmosphere Research Conference*, Wenatchee, WA, 1989, p. 225.
17. Singh, R.K. and Singh, N., Quality of packaged foods, in *Innovations in Food Packaging*, Han, J.H., Ed., Elsevier Academic Press, New York, 2005, chap. 3, pp. 24–44.
18. Malek, K. and Coppens, M.D., Knudsen self and Fickian diffusion in rough nanoporous media, *J. Chem. Phys.*, 119, 2801–2811, 2003.
19. Hosticka, B. et al., Gas flow through aerogels, *J. Noncryst. Solids*, 225, 293–297, 1998.
20. Skelland, A.H.P., Ed., *Diffusional Mass Transfer*, John Wiley & Sons, New York, 1974, pp. 49–80.

chapter nine

Interaction of food and packaging contents

Kay Cooksey

Contents

9.1 Introduction

Food products are mainly judged by quality and safety. Packaging can play a role in preserving quality factors such as flavor, aroma, texture, color, overall appearance, and nutritional properties. Furthermore, interactive packaging may be designed to enhance these properties. The type of food, chemical composition, size, storage conditions, expected shelf life, moisture content, aroma/flavor, and appearance are just a few of the characteristics that must be taken into consideration when selecting the right material for a food product. A continuing trend in food packaging is the design of packages to extend the shelf life of foods while maintaining fresh-like quality. This places a high demand on selecting materials that not only provide the needed properties to maintain the quality of the food, but also do so at a cost-effective price. The permeability of the packaging material is one of the most critical features of the package for affecting the quality of the food product. Materials can be selected to provide a very long shelf life, but one must ask whether there is a need for the best barrier. Furthermore, does the extension of the shelf life justify the cost of the material and the quality of the food? Therefore, knowing the important factors for material selection based on permeability is an essential part of the package design process.

The safety of the product is also affected by the packaging. This can include protection from food spoilage or pathogenic microorganisms. Migration of compounds from the packaging into the product, causing off-odors, off-flavors, or other compounds, that may pose a risk if migration occurs under conditions that cause regulatory concerns. Another interaction that could affect the quality and/or safety of the product is sorption of product components into the packaging. Some of these interactions can eventually affect the overall integrity of the package and product.

9.2 Definitions

A discussion of permeability involves many terms that are commonly misunderstood or confused. Therefore, a definition of terms is necessary to better understand the factors that affect permeability.[1]

- Permeability is the transfer of molecules from the product to the external environment, through the package, or from the external environment through the package to the product (Figure 9.1).
- Sorption is the movement of molecules contained by the product into but not through the package (Figure 9.2).
- Migration is the movement of molecules originally contained by the package into the product (Figure 9.3).

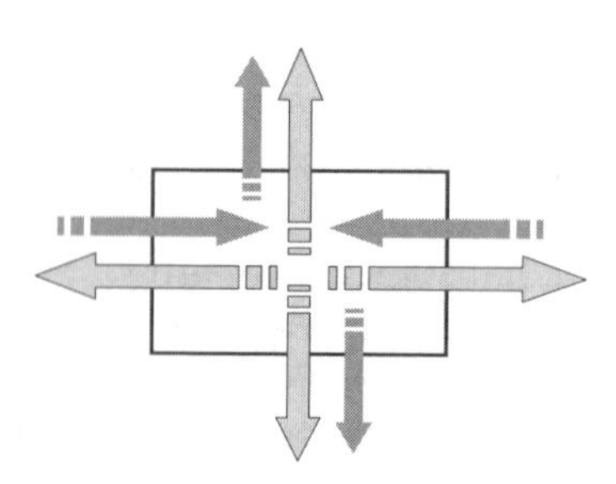

Figure 9.1 Permeability involves transfer of molecules into or out of the package.

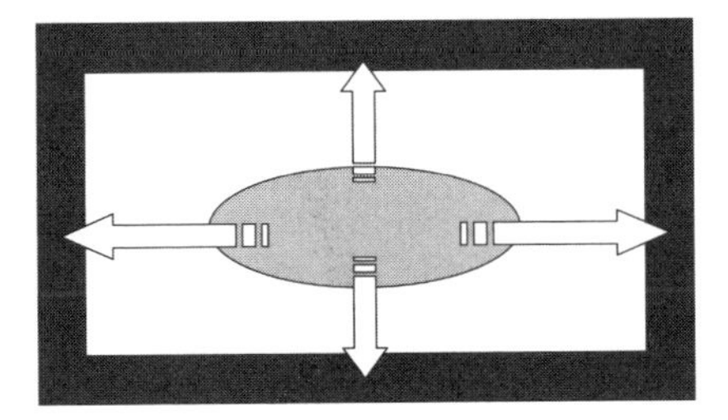

Figure 9.2 Sorption is the movement of molecules from the product to the package.

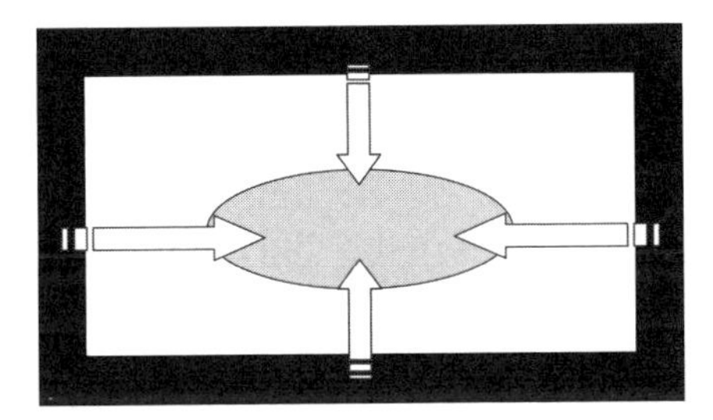

Figure 9.3 Migration is the movement of molecules from the package to the product.

These three components make up the majority of interactions that take place. Although permeability does not directly measure each of these interactions, it does have an influence on each. For example, a material such as low-density polyethylene is a good water barrier but will scalp certain flavor and aroma compounds from foods. Plasticizers may be added to polymers such as polyvinyl chloride to affect their flexibility and permeability, but migration of the additives is also affected.

The equation used to express permeability is

$$P = (D)(S)$$

where P is the permeability coefficient, D is the diffusion coefficient, and S is the solubility coefficient. The diffusion coefficient is a measure of how rapidly penetrant molecules are moving through the barrier, in the direction of lower concentration or partial pressure. D is a kinetic term that describes how fast molecules move in a polymer matrix. The solubility coefficient is the amount of transferring molecules retained or dissolved in the film at equilibrium conditions. S is a thermodynamic term that relates to how many molecules dissolve in a polymer matrix.

When an equal concentration of molecules are moving through a polymer at a constant rate, a steady state is reached, known as Fickian behavior. When measuring water and oxygen permeability, this is common behavior for most packaging polymers such as low-density polyethylene and polyester. However, some do not have steady-state behavior. Polymers such as

Table 9.1 Units Used to Express Permeability and Transmission Rates

Term	Units
Permeability	cc*mil/100 in.^{2}*24 h*ATM @ %RH and temp.
Transmission rate	cc/100 in.^{2}*24 h*ATM @ %RH and temp.

Source: Cooksey, K. et al., *Food Technol.*, 53, 60, 1999.

nylon and ethyl vinyl alcohol diffuse at different rates depending upon penetrant concentration and time. Polymers with this behavior are referred to as non-steady-state or non-Fickian polymers. The chemical nature of the polymers and that of the permeants determine whether the polymer will behave in a Fickian or non-Fickian manner.

It is also important to understand the difference between the terms *permeability rate* and *transmission rate*. According to ASTM,[2] transmission rate (TR) is defined as the movement of a permeant in unit time through a unit area under specified conditions of temperature and relative humidity (Table 9.1). The thickness of the material is not incorporated into the definition but is implied to be the thickness of the test film sample. Permeability (P) is defined as the movement of a permeant through a unit area of unit thickness induced by a unit vapor pressure difference between two specific surfaces under specific conditions of temperature and humidity at each surface (Table 9.1). In other words, permeability is the arithmetic product of transmission rate and thickness.

9.3 Factors affecting product/package interactions

9.3.1 Food Factors

Food composition factors that are affected by product/package interactions are texture, flavor, color, nutritional value, and overall acceptability. Composition of the foods, such as flavor/aroma compounds, enzymes, pH, pigments that affect color, water activity, and microbiological factors, determines how the food should be packaged to best maintain safety and quality. Processing is typically part of the preservation of foods that also affects package design. Properties of foods, shelf life, and desired quality are used to select the processing method. Common processes used with foods include refrigeration, freezing, blanching, canning, pasteurization, irradiation, dehydration, concentration, and fermentation. When designing a package, it is important to know which method of preservation is used and what role packaging will play in the process. It may be integral to the method or may simply maintain the characteristics created during processing. Other factors to consider are residues from washing, surface area created by slicing or grating, chemical composition of preservatives, and possible pesticide residues, all of which can further complicate interactions. One of the most important factors to consider during the design phase is the distribution environment and types of physical or temperature abuse that the product/package might experience. This can drastically accelerate or alter interactions if not properly anticipated and should be included in the testing and design of the total package.

9.3.2 Polymer morphology

Polymers can be classified as crystalline, amorphous, or semicrystalline. The polymer chains in the crystalline regions are linear and more tightly packed, therefore creating a more tortuous path for various penetrants to flow through. Amorphous polymer chains are branched and have more void space, which allows for freer movement throughout the polymer. Specifically, crystalline (glassy) films such as polystyrene have lower permeability rates than films that are semicrystalline, such as polyethylene. Crystalline polymers such as polyester are generally considered to be better barriers than semicrystalline and glassy (amorphous) polymers. However, crystallinity can vary even for a particular polymer such as polyester, depending upon processing conditions. For example, polyethylene terephalate (PET) film oriented at 90°C has 22% crystallinity compared to PET film oriented at 115°C, which has 31% crystallinity. Permeation of ethyl acetate was decreased by a factor of 4× in PET film with 31% crystallinity.[3]

9.3.3 Polarity

Polarity affects solubility and therefore diffusivity. The best way to describe this phenomenon is the phrase "like dissolves like." In other words, polar substances prefer to dissolve and stay associated with polar substances. Limonene (nonpolar) sorption is higher for polyethylene (nonpolar) than hexanal, which is more polar than limonene.[4] Another example was presented by Van Willige and others,[5] who reported exposure of oriented polypropylene (OPP) (nonpolar) and PET (moderately polar) to two moderately polar compounds (aldehyde and carvone) at 40°C for 14 days. The compounds were more readily absorbed by PET than OPP. The presence of a functional group and polarity of a permeant can also affect sorption. For example, carvone ($C_{10}H_{14}O$) and limonene ($C_{10}H_{16}$) both have 10 carbons in their chain, but limonene is less polar than carvone; therefore, limonene permeates faster through nonpolar polymers than carvone.[3]

The length or size of the molecules involved have a strong influence on permeation of flavor and aroma compounds. Compounds of eight or more carbons are sorbed more easily than shorter-chain molecules. More highly branched molecules are sorbed more than linear molecules, and larger molecules are more likely to condense on the surface of a film than smaller molecules.[6]

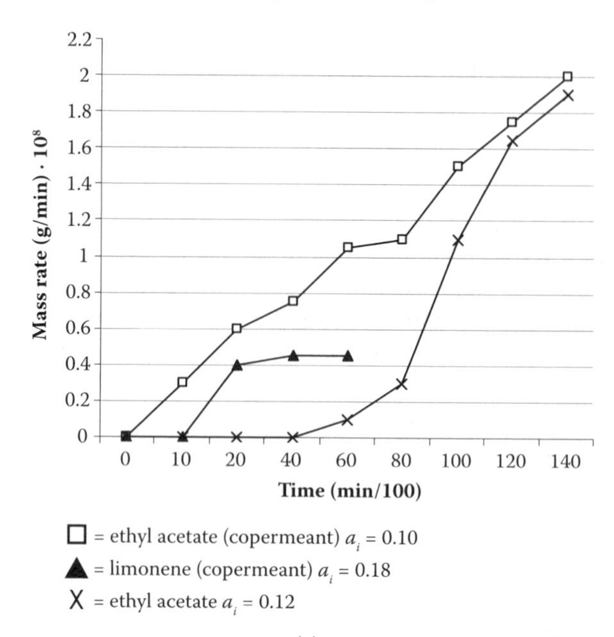

Figure 9.4 Effects of copermeant on permeation. (a) □ = ethyl acetate (copermeant) $a_i = 0.10$; ▲ = limonene (copermeant) $a_i = 0.18$; X = ethyl acetate $a_i = 0.12$. (b) □ = ethyl acetate (copermeant) $a_i = 0.10$; ▲ = limonene (copermeant) $a_i = 0.29$; X = ethyl acetate $a_i = 0.12$.

9.3.4 *Presence of copermeant*

The presence of a copermeant and its concentration can also affect the permeance of flavor compounds. In the example provided by Giacin and Hernandez,[3] the permeations of ethyl acetate and limonene are measured using biaxially oriented polypropylene. In Figure 9.4a, the presence of limonene as a copermeant increased the permeation of ethyl acetate alone, thus providing a synergistic effect. When the concentration of limonene copermeant was increased (Figure 9.4b), the synergistic effect was not observed. When higher levels of limonene were present, it became a more aggressive permeant and suppressed the permeation of ethyl acetate.

9.3.5 *Relative humidity*

Relative humidity has a very significant effect on the permeability of flexible films. When specifying conditions for testing, it is important to consider the conditions to which the material will be exposed. In some cases, as relative humidity increases, permeability increases (Table 9.2). Hygroscopic materials such as nylon, polyvinyl alcohol, polyvinyl acetate, and uncoated cellophane are most significantly affected by the presence of moisture. The presence of

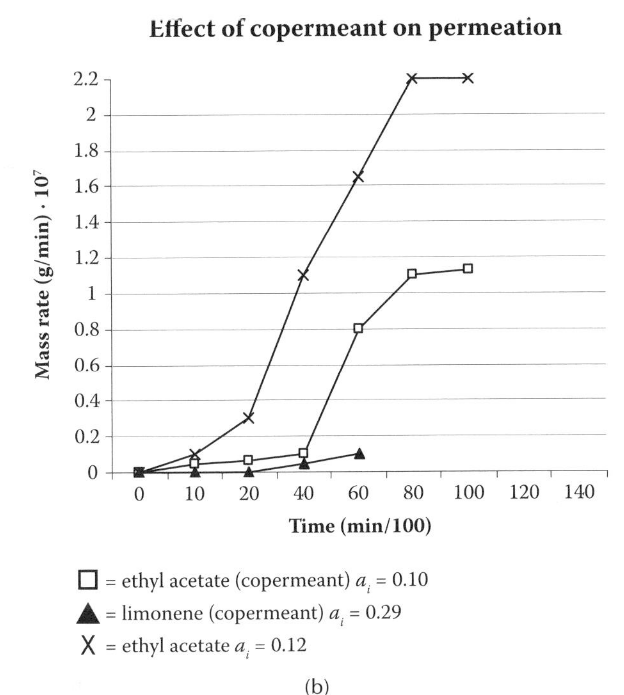

(b)

Figure 9.4 (Continued.)

Table 9.2 Effect of Relative Humidity on Oxygen Permeability (10^{11} ml*cm/cm^2*sec*cmHg) of Selected Films Measured at 25°C

Polymer	0% RH	100% RH
Polyvinyl alcohol	0.0006	1.5
Uncoated cellophane	0.0078	12.0
Nylon 6	0.06	0.3
Polyvinyl acetate	3.3	9
Acrylonitrile–styrene copolymer	0.06	0.06
High-density polyethylene	6.6	6.6
Low-density polyethylene	28.8	28.8

Source: Robertson, G.L., *Food Packaging Principles and Practice*, Marcel Dekker, New York, 1993.

hydroxyl groups (–OH) is responsible for the phenomena observed for most of these polymers. Polyethylene (high and low density), known to be an excellent water barrier, is unaffected by humidity levels. Acrylonitrile copolymer is also unaffected by relative humidity.[1] A relative humidity of 50% is a common condition used for testing to simulate ambient conditions; however, the relative humidity inside most food packages can easily be 100%. Therefore, it is important to specify the percent relative humidity when

Table 9.3 Transmission Rate (g*mm/m^2/day) of Selected Organic Compounds through Low-Density Polyethylene

Permeant	0°C	21.1°C	54.4°C
Acetic acid	0.14	1.22	25.9
Benzaldehyde	0.15	2.67	81.0
Ethyl alcohol	0.75	6.5	149
Phenol	0.04	0.2	9.4
Propyl alcohol	0.03	0.2	8.8
Toluene	22.7	199	2270
Xylene	14.2	101	1420

Source: Robertson, G.L., *Food Packaging Principles and Practice*, Marcel Dekker, New York, 1993.

requesting testing of materials or when evaluating permeability data from suppliers to accurately predict performance under actual conditions of use.

9.3.6 *Temperature*

An increase in temperature also causes an increase in permeability. This is not surprising since the laws of kinetics apply. As temperature increases, molecules have more energy and will move more easily through a polymer matrix. According to Delassus,[7] permeability increases proportionally to an increase in temperature, but the linear relationship between temperature and permeability rate is affected by the Tg (glass transition temperature) of the polymer. The general rule is oxygen permeability increases about 9% per °C when above Tg and about 5% per °C when below Tg. To observe the effect of temperature on flavor and aroma compounds, see Table 9.3. All permeants and polymers exhibit the same behavior and the transmission rate increases more significantly between 21.1 and 54.4°C than between 0 and 21.1°C.

9.3.7 *Plasticizers*

Additives such as plasticizers can increase the permeability rate for gas, water, and flavor/aroma compounds. Small amounts (less than 1%) do not have a significant effect on permeability,[7] but some polymers, such as polyvinyl chloride (PVC), rely heavily on plasticizers for flexibility. PVC can contain as much as 50% additives (including plasticizers) to achieve the flexibility desired for most food packaging applications, which can increase oxygen permeability by as much as 10 times.[7]

9.3.8 *Orientation*

Orientation can increase the degree of packing, which generally decreases transmission rate (Table 9.4). The magnitude of reduction in transmission rate differs depending on the polymer. Acrylonitrile styrene copolymer is an

Table 9.4 Oxygen Transmission of Polymers in Relationship to Effect of Orientation

Polymer	% orientation	Oxygen transmission (cc/100 in.^{2}*day*atm)
Polypropylene	0	150
	300	80
Polystyrene	0	420
	300	300
Polyester	0	10
	500	5
Acrylonitrile 70%/styrene 30% copolymer	0	1.0
	300	0.9

Source: Delassus, P., in *The Wiley Encyclopedia of Packaging Technology,* 2nd ed., Brody, A. and Marsh, K.S., Eds., John Wiley & Sons, New York, 1997.

excellent barrier, and orientation only reduces its barrier properties by 10%, but polyolefins such as polypropylene and polystyrene have 45 and 29% reduction in transmission rate after orientation, respectively. The oxygen transmission rate of polyester was reduced by 50% with 500% orientation.[7]

9.4 Interactions with gas atmospheres

Modified atmosphere packaging has been commonly used for packaging a variety of products. Each gas used in the package is selected and carefully balanced for a specific purpose. Oxygen may be used at low levels to control microbial growth, color, rancidity, and enzymatic activity that could lead to deterioration. Carbon dioxide is known to have a direct and effective antimicrobial effect but may negatively affect other food characteristics depending upon the level. Nitrogen may be used to displace oxygen and act as a package filler. The solubility of these gases in foods as well as their permeation through the package must be considered to properly design a controlled or modified food package.

9.5 Examples of product/package interactions

Numerous studies show the variety of interactions that can take place between food products and the package under a variety of conditions. The following are just a small representation of some of the more commonly studied areas. Limonene is a good example of an aggressive flavoring agent important for imparting the citrus characteristic for fruit flavor/aroma products. The study by Mannheim et al.[8] involved measuring sorption of limonene in glass vs. polyethylene packaging. After 24 days of storage, the orange juice contained in glass packages lost very little d-limonene compared to the juice contained in the aseptic carton. As shown in Figure 9.5, the loss of the d-limonene in the carton was nearly equal to the amount gained in the carton material during storage. This indicates that the carton material scalped the d-limonene away from the orange juice due to the polar nature

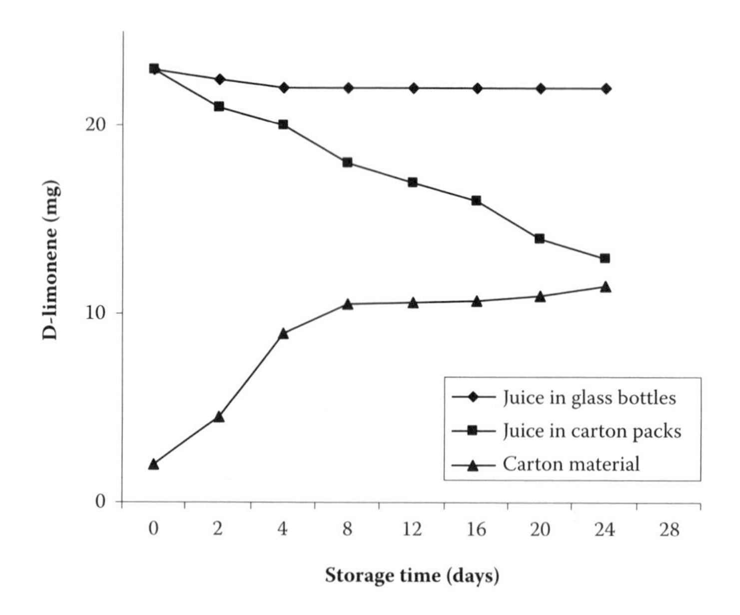

Figure 9.5 d-Limonene in juice contained within glass bottles and carton packs and in carton material stored at 24°C, 40% RH. (From Mannheim, C.H. et al., *J. Food Sci.*, 52, 737, 1987.)

of the food contact material (low-density polyethylene). While juice contained in the carton may have a lower d-limonene concentration, theoretically, a sensory panel might indicate that the amount lost did not affect their preference for the juice. If it did affect preference, a producer might elect to add flavoring to make up for the expected loss of flavor over time, which might be a less expensive option than using glass packaging.

Another study involving limonene was performed by Hirose et al.,[9] who studied the effects of limonene on the mechanical and permeation properties of polyethylene copolymer (Surlyn) and low-density polyethylene. They found that tensile strength was decreased, percent elongation was increased, and oxygen permeability was increased for LDPE and Surlyn. Seal strength was decreased for LDPE but not for Surlyn. These effects on the properties of the film were attributed to the strong affinity limonene had toward the materials, particularly LDPE, and its plasticizing effect on the materials within 3 days of exposure to the limonene.

According to a study by Miranda,[10] some materials are good barriers toward a variety of solvents while others are consistently poor for all solvents. Table 9.5 shows that the multilayer material containing ethylene vinyl alcohol (EVOH) was rated as having low permeation of all solvents and the glassine had high permeation of all solvents. Some materials were good barriers depending upon the solvent. Nylon served as a good barrier to almost all of the solvents, but when glassine was included in the multilayer structure, its barrier performance toward almost all of the solvents was poor.

Table 9.5 Permeation by Single-Component Solvents in Various Materials

	Allyl sulfide	Acetic acid	Methyl ethyl ketone	Ethyl acetate
1.25 mil PE/EVOH/EVA	Low	Low	Low	Low
1.25 mil PE/nylon/EVA	Low	Medium	Low	Low
1.35 mil PVDC/PP/PVDC	Low	High	High	Medium
1.8 mil PVDC/glassine/EVA	Medium	High	High	High
1.8 mil glassine	High	High	High	High

Note: Low = <0.1 g/m^2*day*100 ppm; medium = 0.1–1.0 g/m^2*day*100 ppm; high = >1.0 g/m^2*day*100 ppm.

Source: Miranda, N., *J. Packaging Prof.*, 1, 12, 2000.

Table 9.6 Permeation of Various Aromas in Various Materials as Determined by a Sensory Panel

	LDPE (50 μm)	PP/PE (20 μm/ 50 μm)	PET/PE (20 μm/ 50 μm)	EVOH/PE (15 μm/ 50 μm)	PVDC/ Nylon (15 μm)
Orange essence	<1 day	<1 day	>1 day, <7 days	>7 days	<1 day
Strawberry essence	<1 day	<1 day	>1 day, <7 days	>1 day, <7 days	<1 day
Curry powder	<1 day	>1 day, <7 days	<1 day	>1 day, <7 days	<1 day
Garlic powder	<1 day	<1 day	<1 day	>1 day, <7 days	<1 day
Coffee powder	<1 day	>1 day, <7 days	>1 day, <7 days	>7 days	>1 day, <7 days
Geraniol	<1 day	>1 day, <7 days	>1 day, <7 days	>7 days	>7 days
Methyl ionone	<1 day	>1 day, <7 days	>1 day, <7 days	>7 days	>7 days

Source: Miranda, N., *J. Packaging Prof.*, 1, 12, 2000.

Table 9.6 shows the results of a study involving a variety of aromatic compounds that were packaged and sealed in pouches made of a variety of materials. Sensory panelists were asked if they could smell the aroma in the package. Panelists could detect all of the odors within less than 1 day in the LDPE monolayer pouches, which supports the fact that it is known to be a poor aroma barrier. The material that had the best barrier (undetectable for greater than 7 days) for four of the seven aroma compounds had EVOH as one of the layers. For the remaining three compounds it was moderately effective. When designing a material as an aroma barrier, it is important to balance cost with performance. While EVOH is more expensive than LDPE, it will perform better, and it may be more cost effective to use a thin layer of EVOH with LDPE to enhance its aroma barrier performance.[10]

A variety of polymers were exposed to a tropical mix flavor and stored for up to 35 days. Films were washed with water for 3 minutes and then offered to the sensory panelist. Panelists were asked to respond based on whether they could clearly detect the odor, could faintly smell or smell part of the aroma, or could not detect it at all (Table 9.7). LDPE was the thickest material used in the study, yet it clearly sorbed the tropical mix aroma while EVOH did not pick up the odor throughout the study. Again, this demonstrates that the barrier properties of the materials are more strongly influenced by their chemical composition and interaction than thickness.[10]

9.6 Conclusions

When determining what material works best for specific food applications, it is first important to understand the terminology used to describe materials. If prediction using equations is necessary, be sure to choose the equation that best fits the data provided and the information desired. Consider factors such as polymer characteristics and chemical interaction with characteristics of the permeant. Finally, determine the conditions (relative humidity and temperature) under which the material will be used. Taking all these known factors into consideration will not guarantee the best possible material for the application, but it will definitely provide for a more informed decision that could save time and money in the future.

Table 9.7 Absorption of Tropical Mix Flavor of Various Materials as Determined by a Sensory Panel

	1 h	2 h	15 h	1 day	4 days	35 days
EVOH-32 (15 μm)	Not detected	Not detected	Not detected	Not detected	Not detected	Not detected
EVOH-44 (15 μm)	Not detected	Not detected	Not detected	Not detected	Not detected	Not detected
BO nylon (15 μm)	Not detected	Not detected	Faint/ partial	Faint/ partial	Faint/ partial	Faint/ partial
BOPP (20 μm)	Faint/ partial	Faint/ partial	Faint/ partial	Faint/ partial	Faint/ partial	Faint/ partial
PVDC (25 μm)	Not detected	Not detected	Faint/ partial	Faint/ partial	Faint/ partial	Faint/ partial
LDPE (50 μm)	Clearly detected	Clearly detected	Clearly detected	Clearly detected	Clearly detected	Clearly detected

Source: Miranda, N., *J. Packaging Prof.*, 1, 12, 2000.

References

1. Robertson, G.L. *Food Packaging Principles and Practice*. Marcel Dekker, New York, 1993.
2. ASTM. *Annual Book of Standards*, Vol. 15.09. ASTM, Philadelphia, 1994.
3. Giacin, J. and Hernandez, R. Permeability. In *The Wiley Encyclopedia of Packaging Technology*, 2nd ed., Brody, A. and Marsh, K.S., Eds. John Wiley & Sons, New York, 1997.
4. Hernandez-Munoz, P., Catala, R., Hernandez, R.J., and Gavara, R. Food aroma mass transport in metallocene ethylene-based copolymers for packaging applications. *Journal of Agricultural and Food Chemistry*, 46, 5238, 1998.
5. Van Willige, R.W.G., Schoolmeester, D.N., Van Ooij, A.N., Linssen, J.P.H., and Boragen, A.G.J. Influence of storage time and temperature on absorption of flavor compounds from solutions by plastic packaging materials, *Journal of Food Science*, 67, 2023, 2002.
6. Ducruet, V.J., Rasse, A., and Geigenbaum, A.E. Food and packaging interactions: use of methyl red as a probe for PVC swelling by fatty acid esters, *Journal of Applied Polymer Science*, 62, 1475, 1996.
7. Delassus, P. Barrier polymers. In *The Wiley Encyclopedia of Packaging Technology*, 2nd ed., Brody, A. and Marsh, K.S., Eds. John Wiley & Sons, New York, 1997.
8. Mannheim, C.H., Miltz, J., and Letzter, A. Interaction between polyethylene laminated cartons and aseptically packed citrus juices. *Journal of Food Science*, 52, 737, 1987.
9. Hirose, K., Harte, B.R., Giacin, J.R., Miltz, J., and Stine, C. Sorption of d-limonene by sealant films and effect on mechanical properties. In *Food and Packaging Interactions*, Hotchkiss, J., Ed. American Chemical Society, Washington, DC, chap. 3, 1988.
10. Miranda, N. Evaluating barrier to complex flavors, aromas. *Journal for Packaging Professionals*, 1, 12, 2000.
11. Cooksey, K., Marsh, K.S., and Doar, L.H. 1999. Predicting permeability and transmission rate for multilayer materials. *Food Technology*, 53, 60, 1999.

chapter ten

Natural nontoxic insect repellent packaging materials

Shlomo Navarro, Dov Zehavi, Sam Angel, and Simcha Finkelman

Contents

10.1 Introduction

Materials such as polymers, paper or paperboard, textiles, and metal foils are all used in producing packaging material for packaged foods. A major drawback of such packaging materials is that pests leading to infestation of the packaged foods can penetrate them. The degree of pest infestation of packaged foods depends upon the pest species involved, the time of exposure to invading pests, and the prevailing environmental conditions.

In many instances, synthetic pesticides have been the only effective measure available for controlling pest infestation of stored foods. However, most synthetic pesticides have significant adverse effects on humans and the environment, and accordingly, their use has been substantially excluded from packaged foodstuffs.

The use of nontoxic crude extracts of neem or turmeric, capable of repelling insects, to protect packaged foodstuffs from insect infestation was proposed and a patent was granted (Navarro et al., 1998a). This approach was developed based on the assumption that many plants inherently produce various chemicals that protect them against insects, and extracts from these plants may affect the metabolism of insect species other than those attacking the plant from which the chemical was derived. The search for naturally occurring substances is an important approach for the development of an ecologically sound plant protection strategy suitable for adoption by the food industry. The authors acknowledge the visioned approach of Shmuel Shatski, general manager of BioPack, for providing the funds for the development of insect-repelling packaging materials quoted in this publication.

In this chapter the authors delineated effective substances of plant extracts that could be used to develop improved food packaging materials that are insect repellent and can prevent insect penetration into the packages. The insect-repelling substances that were developed are nontoxic to humans, environmentally compatible, and suitable for protecting food and the like from insect infestation.

10.2 Insect damage and types of penetration into packaging materials

10.2.1 Insects that cause damage to food packages

There are two types of insects that attack packaged products: penetrators, insects that can bore holes through packaging materials, and invaders, insects that enter packages through existing holes, such as folds and seams and air vents (Highland, 1984; Newton, 1988). *Sitophilus* spp., *Rhyzopertha dominica* (F.), *Plodia interpunctella* (Hübner), *Lasioderma serricorne* (F.), and *Stegobium paniceum* (L.) are some of the stored product insects that are capable of penetrating food packaging. However, *Tribolium* spp., *Cryptolestes ferrugineus* (Stephens), and *Oryzaephilus* spp. cannot penetrate intact packages but as invaders enter through existing holes in the package (Highland, 1991).

Beetles and moths comprise the majority of stored grain insect pests. Ambient temperature and moisture content of the commodity have a major influence on the rate of insect development. The rate of beetle development is generally more affected by temperature than by commodity moisture content (Hagstrum and Milliken, 1988). Moth development is more dependent on ambient humidity above the grain and moisture in the grain.

Stored product insects are mainly of tropical and subtropical origin and have spread to temperate areas via international trade. Because insects cannot control their body temperature, their rates of development and reproduction increase with rising temperature. Consequently, most of them become inactive at low temperatures (10 to 15°C) and will die after prolonged periods at very low temperatures (0 to 5°C). Most species are unable to hibernate or enter an inactive phase termed diapause, though some such as *P. interpunctella* and *Trogoderma granarium* Everts do diapause.

For each insect species there is a minimum and maximum temperature at which they are able to develop (at certain low temperatures, oviposition and larval growth ceases; at specific high temperatures, egg sterility occurs and mortality increases). Conversely, there is a temperature range at which oviposition and insect development are optimal. The lower and upper limits and optimal temperatures of most of the important stored product species have been studied and are well known.

Survival of *Tribolium castaneum* (Herbst) from egg to adult is highest between 25 and 27.5°C and decreases rapidly below and above this temperature (Howe, 1960). Temperatures below 15°C generally arrest all insect development sufficiently to prevent damage, though not to cause mortality. For most insects, sustained temperatures above 40°C and below 5°C are lethal.

Each stored product pest species has different food requirements. Studies have been made to identify the nutritional requirements of different species in order to breed them on artificial diets. Clearly these requirements affect the ability of insects to develop on different stored products and their ability to compete with other species. Consequently, for each stored product, there is a range of insect pests.

All stored product insects are negatively phototrophic, which means that they stay away from sunlight. Because of their phototrophic behavior, they are generally not visible to the casual observer.

10.2.1.1 Order Coleoptera (beetles)

Adults are typified by forewings modified to form rigid wing covers (elytra) that meet along the back and generally cover the abdomen. The hind wings are membranous, folded beneath the elytra, and used for flying. The beetles have biting mouth parts, and the upper plate of the first segment of the thorax covers the other segments to form a shield (pronotum). These are the tanks of the insect world. Metamorphosis is complete, and larvae may be active with well-developed or grub-like and sessile legs. Of about 250,000 species known to man, more than 200 are associated with stored products, but only a few constitute the major stored product pest species.

Stored product beetles are typified by being small enough to be able to move between units of food products and live in the airspaces. Their armored bodies make them very adaptable to this environment. Some are primary pests capable of entering undamaged products. Generally their larvae are soft and sessile. Others are secondary pests that feed on broken or damaged grains, chaff, and dust, but cannot penetrate sound grain. Their larvae are active and well protected (wiry). A third group is scavengers of dead insects and mold feeders, but they do not attack grain kernels.

The most common beetles that penetrate flexible packages are *R. dominica*, *Sitophilus oryzae* (L.), *L. serricorne*, and *S. paniceum*. The most common invaders are *T. castaneum*, *Tribolium confusum* J. du Val, *Cryptolestes* spp., and *Oryzaephilus surinamensis* (L.).

10.2.1.2 Order Lepidoptera (moths)

The adults of Lepidoptera are typified by fragile wings covered with scales, which often have delicate colorings and markings. The head has sucking mouth parts (proboscis) or mouth parts are absent and long filiform antennae. The adults are generally characterized by their short lives. The larvae are caterpillars with biting mouth parts that inflict damage. Young larvae are excellent invaders, and old larvae have the capability to penetrate packaging laminates.

Most common moths that invade or penetrate flexible packages are larvae of *Ephestia cautella* (Wlk.) and *P. interpunctella*.

10.2.2 Insect damage to plastic packages and types of damage

Insects infesting stored foods are one of the most common household problems. The many different kinds of insects that invade stored dried foods are often referred to as pantry pests. Insects contaminate more food than they consume, and most people find the contaminated products unfit for consumption. Stored food insect pests are often discovered when they leave an infested food to crawl or fly about the premises. They often accumulate in pots, pans, or dishes or on window sills.

Nearly all dried food products are susceptible to insect infestation; examples include cereals and their products (flour, cake mix, cornmeal, rice, spaghetti, crackers, and cookies), pulses, nuts, cocoa and coffee beans, confectionary-like chocolate, dried fruits, spices, powdered milk, seeds, and cured meats. Nonfood items that may be infested include all types of feed, dry pet food, ornamental seed and dried plant displays, ornamental corn, dried flowers and potpourri, garden seeds, and rodent baits.

The quality hazards to stored food due to insects are (1) package perforation (Figure 10.1), (2) devouring of the agricultural product (Figure 10.2), (3) substantial damage to the product, (4) contamination of the product (Figure 10.3), and (5) esthetical objections.

10.2.3 Propensity of damage by insects

A stored food product may become infested at the processing plant or warehouse, in transit, at the store, or in the consumer's home. Most of the insects attacking stored foods are also pests of stored grain or other commodities and may be relatively abundant outdoors. Food products that are left undisturbed on the shelves for long periods are particularly susceptible to infestation. However, foods of any age can become infested.

Stored food insects are capable of invading the food package or penetrating unopened paper, cardboard, and plastic-, foil-, or cellophane-wrapped packages. They may chew their way into packages or crawl in through folds and seams. Insects within an infested package begin multiplying and can spread to other stored foods or food debris that has accumulated in corners, cracks, and crevices, and eventually the entire cupboard.

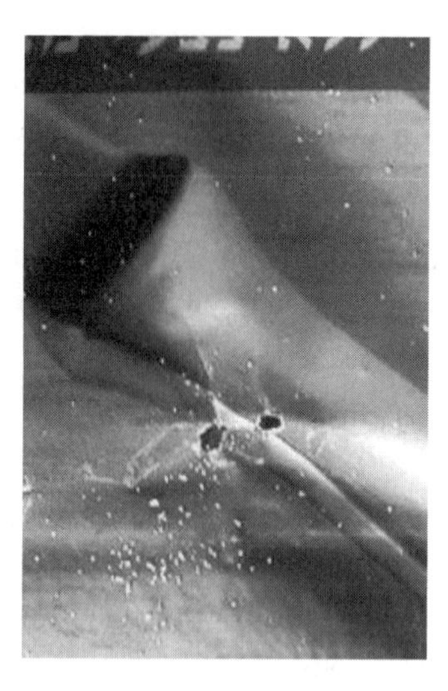

Figure 10.1 A package damaged by the adult of the cigarette beetle, *L. serricorne*.

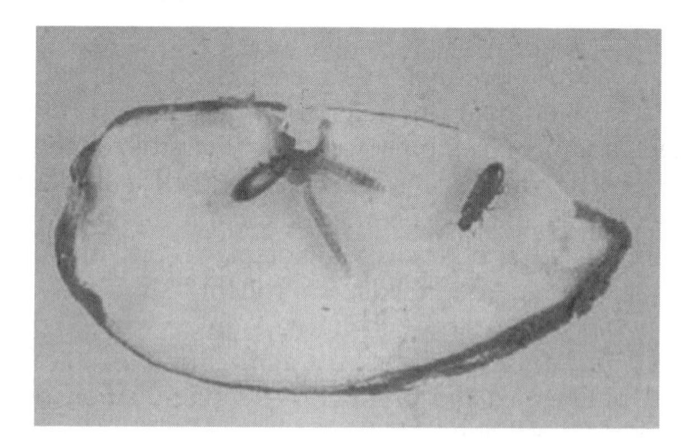

Figure 10.2 Almond damaged by the adult of the red flour beetle, *T. castaneum*.

All stages (egg, larva, pupa, and adult) may be present simultaneously in infested products.

Most food infestations of storage pests maintain themselves on spills in the crevices of cupboards and drawers or in opened packages of food stored for long periods.

Stored food insects have been a major factor in food losses and the most difficult to combat. Insects are almost always present in food stores. The insects originate either from residual infestations hidden within the storage structural materials or from stored product insects that lay their eggs on the product.

The insect pests of stored grain have environmental requirements that greatly affect their abundance and consequently their potential danger for causing damage. The most important environmental factors are temperature

Figure 10.3 Chocolate damaged by the tropical warehouse moth larvae *E. cautella.*

and moisture (climate), food requirements, and competition with other living organisms.

For stored product pests the influence of external climate is reduced by the fact that the "climates" within a warehouse or grain storage silo may be very different from that outside. Thus, insects that are unable to withstand outdoor winter conditions in temperate climates may be able to survive and develop in relatively warm grain masses in warehouses, storages, or the heated buildings of food processing factories, even in relatively cool or cold climates.

10.2.4 Resistance of plastic films to insect penetration

Several factors need to be considered in developing a package for a product. The food package needs to be designed not only to maintain the quality of the product but also to attract the consumer. Cheap packaging can lead to infestation by insects and microorganisms. Food manufacturers are aware that if a consumer finds an insect in a package, it can cause a lasting and often irreversible impression, ultimately resulting in the loss of that customer. Although most insect pests under the category of invaders enter packages through existing openings that are a result of poor seals, the penetrators have the capacity to enter almost all existing packaging films. According to Mullen (1997), most infestations are the result of invasion through seams and improper closures. Minute openings in packaging materials due to sealing failures during manufacture or handling attract pests and are often large enough to permit entry of first instars of most stored product insects.

Most of the packaging materials for fresh fruits and vegetables as well as stored dry and semidry food commodities are cellulose or plastic or a combination of plastic and paper. Some plastic materials are rigid while others are flexible films.

Flexible packaging films vary in resistance to penetration. Bags made of laminated foil and paper were found to be more resistant than bags made of cellophane, polyethylene, multilayer paper, and fiberboard box (Kvenberg, 1975). The study was based on eight stored product insects, where the adult and larvae of the cadelle, *Tenebrioides mauritanicus*, and the cigarette beetle, *L. serricorne*, penetrated the test material.

In a method developed by Highland and Wilson (1981) to test insect penetration of flexible packaging films, 18 types of polymer films and fibers, used by 25 different companies, were exposed to adults of the lesser grain borer, *R. dominica*. Polyurethane and polyester films were most resistant. Polypropylene and fluorinated carbon polymer films varied in resistance to penetration. All Kraft papers, plasticized polyvinyl chloride (PVC), cellulose, polyvinyl alcohol, Saran, polyethylene, and ethyl vinyl acetate copolymer films were the most susceptible to penetration.

A recent study Riudavets et al. (2007) assessed the ability of *R. dominica*, *S. oryzae*, and *O. surinamensis* to penetrate different plastic films. They studied polypropylene, polyethylene, and polyester, and a multilayer film (paper, polyethylene, aluminum, and polyethylene). Damages observed in each material were evaluated under binocular microscope. All three species were able to penetrate the films tested. *R. dominica* was the species with the highest penetration ability. The intensity of damage produced by all three species was higher in polyethylene than in polypropylene and polyester. In the multilayer film, *R. dominica* showed a similar penetration ability independently of the film side exposed to the insect, since the aluminum foil was the layer acting as a barrier to avoid the penetration of this species (Figure 10.4 to Figure 10.6).

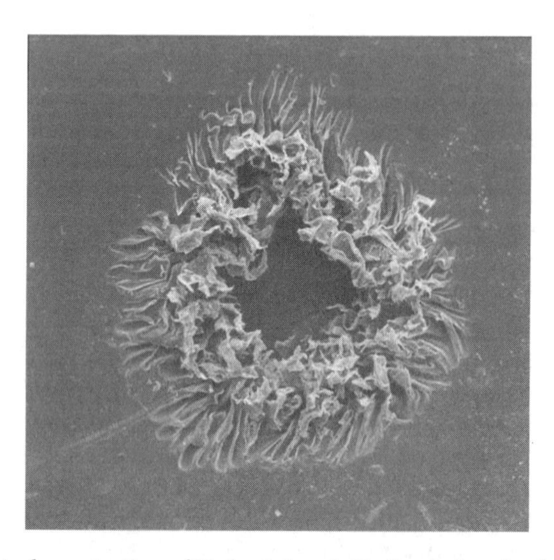

Figure 10.4 Typical penetration of *R. dominica* adults through polyethylene laminates under microscope. (Riudavets 2005, personal communication.)

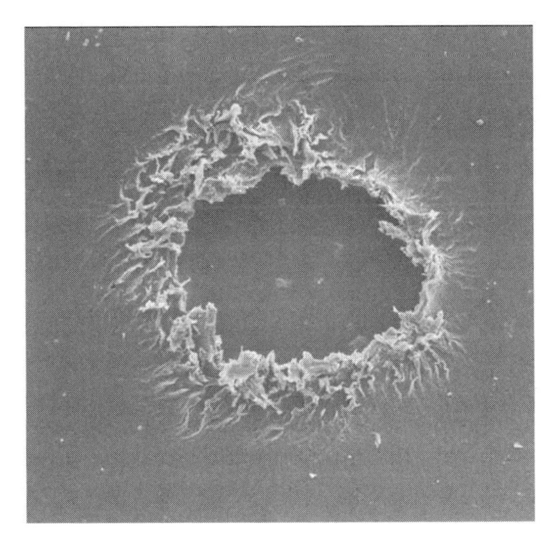

Figure 10.5 Typical penetration of *R. dominica* adults through polypropylene laminates under microscope. (Riudavets 2005, personal communication.)

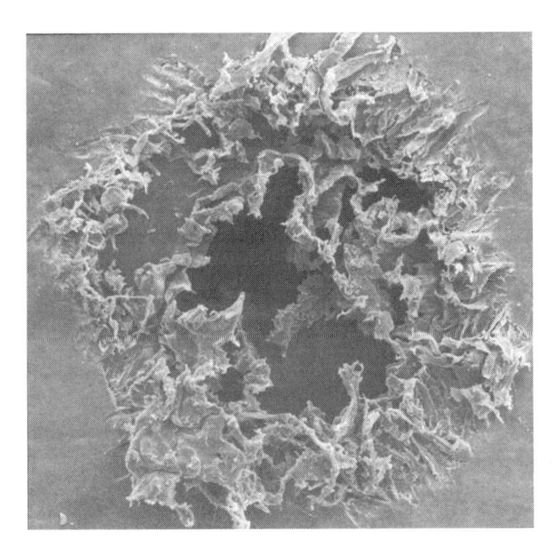

Figure 10.6 Typical penetration of *R. dominica* adults through multilayer paper–aluminum laminates under microscope. (Riudavets 2005, personal communication.)

10.2.5 Economic impact of insect damage to packaging material

Packaging of food products has an important role as a barrier for insect pests, and plastic films are among the most effective packaging materials. World expenses in packaging materials and equipment reach 240,000 million euros per year (Hanlon et al., 2000). Cardboard, paper, and plastic films are the most important materials used. However, during recent years there has been an increase in the use of plastic and a decrease in the use of paper and cardboard. Today there is also an increase in the use of new materials, such as bioplastics. World consumption of plastic films for packaging is approximately 100 million tons, with more than 30 different types of materials. The most common plastic materials used are polyethylene, polypropylene, and polyester. To combine the characteristics of these plastic films with other materials such as aluminum or cardboard, complex packaging materials and multilayer plastic films have been developed, and their use is increasing.

Food products are packaged to protect them against external attacks due to handling, hits, or other mechanical actions or against the effects of macro- and microorganisms. Among macroorganisms, insects are one of the most important affecting the packaged final products. In the marketing of durable raw and processed agricultural products, their quality and wholesomeness must be maintained from the time they are packaged until they reach the consumer. This is achieved by use of different kinds of packages whose purposes are to offer convenience, to render them attractive for promoting their sale, and above all to provide a physical barrier against external adverse influences, including the ingress of insect pests.

Control of insect pests of durable food commodities relies heavily upon hygiene and, to a limited extent, on the use of fumigants on raw materials, but almost only on hygiene and physical means after the food commodities are processed. Application of a selective and limited number of contact insecticides is permitted in the food processing plants, but their use directly on processed food is not permitted due to their toxicity. Therefore, the possibility of infestation and contamination of the commodity by insect pests during the post-manufacturing stages remains a major problem.

The introduction of insect pest management techniques into the food industry enables the integration of several chemicals and nonchemical preventative and control measures in the storage and processing plant to prevent contamination by insects of the commodity before it is packaged.

The extensive use of packaging in modern food distribution systems provides a potentially effective tool in the management of insect pests. New and modified food handling procedures, increasingly stringent sanitation standards, and increasing international trade impose a need for systems that will protect food from infestation, from the time it is packaged until the package is opened by the consumer. Knowledgeable selection of packaging materials can help produce packages that resist infestation (Highland, 1991).

10.3 Natural substances for protection of food packages from insects

10.3.1 Traditional insect control practices

In today's food industry, one of the aims is to eliminate the use of noxious pesticides against food-infesting pests and to replace them with natural, nontoxic, environmentally friendly compounds.

The utilization of plant materials to protect field crops and stored commodities against insect attack has a long history. Many of the plant species concerned have also been used in traditional medicine by local communities and have been collected from the field or specifically cultivated for these purposes. Leaves, roots, twigs, and flowers have been admixed as protectants, with various commodities in different parts of the world, particularly India, China, and Africa.

The neem tree (*Azadirachta indica*) is native to the Indo-Pakistan subcontinent and grows abundantly in this region. Neem trees are plentiful in South Asia and other developing countries where farmers are aware of its properties. In rural India an age-old practice is to mix dried neem leaves and turmeric powder with stored grain to keep away insects. To combat insect pests, food grain stored in gunny bags is mixed with dried neem leaves. Those who store wheat in mud bins rub fresh neem leaves on the inside walls of the bins. In the districts of Nawabshah and Khairpur, in Pakistan, palli is commonly used for storage. Some farmers plaster the bin walls and the top with mud having crushed neem leaves. In Rahim Yar Khan District, neem extract is sprinkled on the wheat straw packed at the bottom of Palli before pouring in the grain (InPho Newsletter, 2006).

Turmeric, *Curcuma longa* L., is a tropical herb of the Zingiberaceae family indigenous to southern Asia. The aromatic yellow powder from its mature rhizomes was used in Asian countries for many centuries as a yellow vegetable table dye for silks and cottons. It is still used in foods as a condiment, particularly as an essential ingredient of curry powder, in medicine as a stomachic, carminative, anthelmintic, laxative, and cure for liver ailment, and also as an ant repellent in India (Su et al., 1982).

10.3.2 Botanical insect repellent extracts

Over the last 30 years, intensive and pioneering research has been conducted on neem and its derivatives, turmeric, and various other plant materials. As a consequence, the potential role of botanicals in the fields of antifeedants, repellents, toxicants, and growth regulators has been established (Islam, 1986). Antifeedants inhibit insect boring, while repellents prevent insects from invading food packages through even the tiniest of openings, or even to approach packages containing repellent substances. Repellents and antifeedants act at some distance from the commodity (Gebbinck, 1999) and are not considered insecticides.

Numerous plant substances have been isolated and tested on stored product insects, and among these, azadirachtin (extracted from neem) appears to be particularly promising as a potential stored product protectant (Subramanyam and Hagstrum, 1996). Neem is listed as an approved pesticide for organic agriculture in the U.S. The active ingredient in neem, azadirachtin, may be responsible for its insecticidal activity. However, neem is not an approved product in the U.S. or Europe for use in or in contact with food. Mixing neem extracts with other materials can boost their power.

Among these so-called promoters are sesame oil, pyrethrums, and piperonyl butoxide (National Research Council, 1992).

Pyrethrum is extracted from *Chrisanthemum cinerariifolium*. Pyrethrum and pyrethrins synergized with piperonyl butoxide were approved in the U.S. for use as an insect repellent on the outer layers of food packages or with adhesives (Highland, 1991). The repellency of pyrethrins was found the primary mode of action against insect penetration and invasion (Laudani and Davis, 1955).

Methyl salicylate, an insect repellent, is registered in the U.S. for use in food packaging to control stored product insects (Radwan and Allin, 1997).

Insect repellents are used to prevent insects from entering packages by modifying their behavior (Highland, 1984; Mullen, 1994; Watson and Barson, 1996; Mullen and Mowery, 2000). DEET, neem, and protein-enriched pea flour are repellent to many stored product insects when tested by exposure on filter paper or in preference chambers (Khan and Wohlgemuth, 1980; Xie et al., 1995; Fields et al., 2001). Rajab et al. (1988) authored a number of publications describing the isolation identification and structure of limonoid insect antifeedants.

Whalon and Malloy (1998) used a mixture of plant extracts from eucalyptus, orange peel, cinnamon, neem, turmeric, and sweet flag to apply to lacquers on food packages and claimed that the coatings repelled Indian meal moths from invading the packages, over an 8-week period.

According to the FAO Compendium on Post Harvest Operations (InPho Newsletter, 2006), neem and pyrethrum are 2 of 130 plants known to control insects that have been used commercially to date. Both neem and pyrethrum are unstable and could not be used for long-term storage of grain (Cox, 2002).

Cox (2002) and Hou et al. (2004) reviewed the use of natural insect repellents and deterrents on stored food products. These reviews as well as various previous papers have pointed out that stored grain insect pests can be controlled using nontoxic plant-based repellent and antifeedant substances to prevent the insects from coming in contact with the food, not to kill them.

According to the above reviews, although there has been considerable interest recently in the use of plant-derived repellents (Mordue and Blackwell, 1993; Ignatowicz and Wesolowska, 1994; Xie et al., 1995; Lui and Ho, 1999; Obeng-Ofori et al., 2000; Weaver and Subramanyam, 2000), no successful use was made of repellents to protect stored grain from insect attack on a commercial scale. Extracts from *Ocimum* spp. (Labiatae) have been shown to be repellent to *R. dominica*, *T. castaneum*, *S. paniceum*, and all three species of *Sitophilus* (Desphande and Tipnis, 1974; Bekele et al., 1996; Obeng-Ofori and Reichmuth, 1997). If suitable plant chemical repellents are identified, they could be used to provide protective bands around grain bulks or they could be incorporated into packaging materials, such as sacking and paper, to inhibit invasion by pests. Methyl salicylate has recently received regulatory approval in the U.S. for use as a repellent in food packaging (Mullen and Pedersen, 2000).

Wild angelic produces bisabolangelone, which is a good feeding deterrent for *Sitophilus granarius* (L.), *T. confusum*, and *T. granarium*, and terpenoid lactones, secondary metabolites of many plants, have shown antifeedant properties. They could therefore find use in preventing insect damage in stored grain (Cox, 2002). Glucosinolates found mostly in the Cruciferae plant family are toxic to livestock but not to humans; they repel cabbage root fly. Ash tree foliage contains substances that can deter gypsy moth larvae. According to Cox (2002), to date, no tests have been carried out using oviposition or antifeedant inhibitors to repel grain storage insects.

Cox (2002) described a number of chemicals coined semiochemicals, produced by insects, which can act as insect repellents. When aphids are attacked by predators, they release a pheromone hormone, sesquiterpane hydrocarbon (E)-beta farnasene, which causes dispersal of the aphids. However he stated that the production of repellents from plants would be less expensive than their synthesis from complex semiochemicals.

10.3.3 Repellency and antifeedant tests on turmeric oil, neem, and pyrethrum

There are several publications in the literature on the biological activity of turmeric rhizome extracts in repelling and penetration prevention of storage insects.

Su et al. (1982) isolated two compounds from *C. longa* and identified their spectral characteristics as *ar*-turmerone and turmerone. They gave class IV and class III repellency, respectively, to *T. castaneum*, after 8 weeks of exposure. Although *ar*-turmerone and turmerone evoked repellency, no mention was made of the ability of these compounds to prevent penetration or perforation by the insects. The importance of the concentration of the two compounds in the crude extracts was not investigated. Turmerone was reported to be unstable upon exposure to air and slowly aromatizes to *ar*-turmerone (Alexander and Rao, 1973). Turmeric oil extract at 680 $\mu g/cm^2$ applied to filter paper produced class IV (67%) repellency against *T. castaneum* 4 weeks after application (Jilani and Su, 1983). Studies by Jilani et al. (1988) indicated that turmeric oil, sweet flag oil, and neem oil not only repel *T. castaneum* (the red flour beetle), but also interfere with its normal reproduction and development. Jilani and Saxena (1990) reported that turmeric oil and sweet flag oil were significantly more repellent during the first 2 weeks than neem oil and neem-based insecticides, but thereafter their repellency decreased more rapidly than that of neem oil and neem-based insecticides. *R. dominica* adults made significantly fewer and smaller punctures in filter paper discs (7 cm diameter) treated with the test materials than in the control discs. The paper described the extraction process to obtain oils from turmeric and sweet flag rhizomes, but a complete analysis of the extracts was not given.

Tripathi et al. (2002) found that *C. longa* oil reduced oviposition of *T. castaneum* by 72% using filter paper. They also found 81% antifeedant activity against *R. dominica*, *S. oryzae*, and *T. castaneum* at 40 μg/cm^2. In all the works cited above, no attempt was made to uncover a specific fraction or compound capable of repelling or inhibiting insect puncturing.

Navarro et al. (1998a) tested the repellency and penetration prevention (antifeedant effect) of neem extracts, natural pyrethrum extracts (50%), and turmeric petrol ether extract on papers against *R. dominica* and *T. castaneum* adults. The neem extracts included NeemAzal T/S (1% azadirachtin AI, other related limonoids, and neem oil), azadirachtin (30% purity), and neem oil.

Repellency tests with both insects showed that the most effective neem extract was NeemAzal T/S. It was tested at a dose of 50 μg/cm^2 and resulted as class IV (of five classes). Azadirachtin, at the same dose, was substantially less effective (class II). Neem oil at a dose of 800 μg/cm^2 (class III) had the same activity as turmeric extract (at the same dose), both less active than NeemAzal T/S. The repellent effect of pyrethrum could not be detected, as it is actually an insecticide. At the low dose of 5 μg/cm^2, the *T. castaneum* insects showed substantial repellency (class III), but the *R. dominica* insects were moribund. For both insect species, there was no significant synergistic effect of piperonyl butoxide (PBO) on repellency in any of the tested extracts. The pyrethrum results were in accordance with other literature reported results, for example, McDonald et al. (1970), who reported that for a mixture of pyrethrum and PBO the integrated repellency effect is the sum of the two individual repellencies.

The penetration prevention (antifeedant effect on penetration) was tested in a way similar to that of the nonchoice test (see Section 10.4.2.3), with application of different dosages and testing after several time intervals, up to 75 days from application on the papers. Exposure times to *R. dominica* adults were 24 and 48 hours.

The penetration prevention results of neem extracts showed a protective effect in a dose-dependent manner. NeemAzal T/S at 31 to 500 μg/cm^2 reduced significantly the penetration at all the dosages for up to 75 days after treatment. It prevented any penetration at the highest dosage after 60 days. Neem oil and azadirachtin showed penetration prevention efficacy for short periods and were found to be less effective than NeemAzal T/S. Neem oil application resulted in at least partial penetration prevention at all the tested dosage levels of 160 to 2560 μg/cm^2 and up to 30 days after application. The complete residual (no penetration by insects) effect of neem oil was only obtained at the highest dosage after a 1-day delay. However, the tests showed that dosages of 1280 and 2560 μg/cm^2 gave significant protection until 30 days after treatment. Azadirachtin, at dosages of 31 to 500μg/cm^2, was tested 1 day after application on paper. Although penetration was significantly reduced, it was still apparent after confined exposure for 48 hours.

Pyrethrum extract at dosages of 2.5 to 160 μg/cm^2 after 1 and 15 days' time delay showed a reduced penetration at all the dosages except 2.5 and 5 μg/cm^2. High dosages that prevented all penetration also resulted in insect mortality. Turmeric extract showed a protective effect of up to 75 days using high dosages (1280 and 2560 μg/cm^2) (see Section 10.4.2).

Results of Navarro et al. (1998a) and those reported in the literature (Jilani and Su, 1983; Jilani and Saxena, 1990; Malik and Naqvi, 1984) are in agreement about the basic findings that neem extracts and turmeric extracts have substantial repellency and antifeedant effects against storage insects. On the other hand, there are almost no records in the literature on the long-term residual penetration prevention effectiveness of those extracts, while the active compounds slowly degrade. In this context, the results of Navarro et al. (1998a) showed that the most promising material tested was NeemAzal T/S, which gave complete protection at the highest concentration for 60 to 75 days. In storage conditions degradation processes caused by sunlight are less of a concern. This is supported by the observations of other investigators (National Research Council, 1992; Daniel and Smith, 1990; Makanjuola, 1989; Mordue and Blackwell, 1993).

The diversity of results regarding neem extracts may be explained by:

1. Differences in the experimental settings and test conditions.
2. Isman et al. (1990) reported that azadirachtin content varied widely between different neem oil samples. They showed that there is a clear trend in which bioactivity of neem oil is related to its azadirachtin content.
3. Recognition that azadirachtin is not the only neem component contributing to the repelling and penetration prevention efficacies of neem extracts. The combined effect of several components imparts their characteristic activity. For the neem extracts Isman et al. (1990) compared the bioactivity test results of pure azadirachtin to oil spiked with azadirachtin and showed that the bioactivity of azadirachtin is enhanced by the presence of the oil. They concluded that other potentially active constituents are present in these oils that can act as synergists to activate azadirachtin. They also pointed to the possibility that a botanical preparation may enhance the stability of azadirachtin and other active ingredients. Mordue and Blackwell (1993) reported that limonoid mixtures may be more effective than azadirachtin alone, that neem oil has insecticidal properties by itself, unrelated to its azadirachtin content, and that crude formulations may contain volatile repellent components. These results are in accordance with those of Navarro et al. (1998a). These facts led Navarro et al. (1998a) to the conclusion that the specific effectiveness of any particular neem oil or turmeric extract should be tested and confirmed before being applied in practice.

10.3.4 *Insecticidal activity and toxicity of natural essential oils and their derivatives*

Besides the above plant extracts, which can be used as insect repellents or antifeedants, a number of plant extracts have been used as insecticides. Champon (2000) used mustard oil, lemon extract, and vegetable oil and others as a soil treatment, insecticide, fumigant, and structural fumigant. Coats et al. (2001) state that monoterpenoids, found in essential oils of mints, pine, cedar, citrus, eucalyptus, and spice compounds in this class, are commercially available to control fleas on pets and carpets, insects on house plants, and fumigation in honeybee colonies. Rajamannan and Okioga (1997) developed a pesticide from the plant *Tagetes minuta* that is claimed to be able to kill nematodes, wire worms, and insects. Hsu et al. (2001) claimed garlic oil extract combined with an essential oil has insecticidal effects on a number of insect pests as well as a fungicidal effect. Reeves and Shanker (1970) reported that a crude extract of garlic caused 100% mortality in five species of *Culex* and *Aedes* mosquito larvae when used in doses of 12 ppm or more. Borzatta et al. (2001) describes a process for the synthesis of alkylbenzodioxole derivatives from essential oils, especially sassafras oil, which contains 75% of such derivatives, which can be used to prepare insecticides such as piperonyl butoxide.

Although plant source repellents and antifeedants are not intended to kill the insects, curcuma oil, for example, if used at concentrations of 2000 $\mu g/cm^2$, can become nematocidal. Lee et al. (2001) found that concentrations of 1000 and 500 ppm *ar*-turmerone caused 100 and 64% mortality, respectively, in *Nilaparvata lugens* female adults. Against *Myzus persicae* female adults and *Spodoptera litura* larvae, *ar*-turmerone was insecticidal at 2000 ppm (Koul, 1987). At a concentration of 2.1 $\mu g/cm^2$ *ar*-turmerone was almost ineffective against *S. oryzae*, *Callosobruchus chinensis* (L.), and *L. serricorne* as well as larva of *P. interpunctella*.

10.4 Development of a natural nontoxic insect repellent for insect packaging materials

10.4.1 Background

Turmeric oleoresins, or turmeric extracts, are obtained by solvent extraction of the turmeric (*C. longa* L.) dried powdered rhizome. Depending on the extraction solvent, the extraction process, and the turmeric type and cultivar, the oleoresin contains various proportions of curcuminoids (coloring matter), volatile essential oil (imparts the flavor to the product), and nonvolatile fatty and resinous materials. Different polar and nonpolar solvents can be used as extractors. For example, 21 Code of Federal Regulations (CFR) 73.615 has allowed the following solvents for the extraction: acetone, ethyl alcohol, ethylene dichloride, hexane, isopropyl alcohol, methyl alcohol, methylene chloride, and trichloroethylene. Since the curcuminoids are the main compounds of commercial interest in turmeric rhizome and the turmeric oleoresin is valued mainly for its curcuminoids content, a polar solvent is usually prefered for the rhizome extraction. The commercial methods of extraction will vary by manufacturer and are proprietary information.

Turmeric essential oil is obtained by distillation or by CO_2 supercritical fluid extraction (Gopalan et al., 2000) of the powdered rhizome. Steam distillation appears to be the main commercial process. Commercial turmeric essential oil is also called crude oil. The essential oil can also be separated from oleoresins by extraction with hexane (Jayaprakasha et al., 2001) or other lipophilic solvents (as is sometimes done in the separation process of curcuminoids from oleoresins) and is called pure oil. Such separation and the liquid extraction of the rhizome may lead to the loss of volatile compounds of the oil during evaporation of the solvent. Furthermore, if methyl or ethyl alcohol is used as the solvent, artifacts may be produced by reactions of esterification or transesterification, etherification, and acetal formation. The solid residue that remains after the total removal of the essential oil from turmeric oleoresin by liquid extraction is defined as a solid residue of turmeric oleoresin (TOSR).

Navarro et al. (2005), in U.S. Patent Application 20050208157-A1 and European Patent Application 04101309.5, described the biological activity and the chemical composition of turmeric oil, turmeric oleoresins, enriched turmeric oil fractions, and solid residue of turmeric oleoresin in repelling stored product insects. The following section describes their main findings regarding the development of a natural, nontoxic insect repellent for insect packaging materials. Their work on turmeric oil fractions was carried out using crude essential oils from known sources that had similar biological activity.

Two main repelling effects — repellency and antifeedancy — were responsible for the repelling activity. Navarro et al. (2005) showed that several compounds are responsible for this biological activity. These effects are composed of partial contributions of several components, each with its own specific activity. Among these are components that have reverse effects — attraction or feedant. It was also established that components present at low concentrations could contribute substantially to the repellency and antifeedant effect. Their findings were the basis of two patents on pest-impervious packaging materials (Navarro et al., 1998a, 2005).

10.4.2 Bioassays

10.4.2.1 Insect repellency and penetration prevention bioassays

Three types of fast semiquantitative bioassays — repellency, nonchoice, and choice tests — were developed to evaluate the repellent and antifeedant effects of plant extracts on stored food insects.

In these tests, 80 g/m^2 white printing papers were impregnated with test samples of plant extracts at varying concentrations and representative insects were exposed to them. The effects on the insects were found to be nonlinear with concentration. It was concluded that the tests should be performed at a concentration range to which the insects would be most sensitive. At relatively high concentrations of the oils the biological activity could reach saturation and adversely affect insect sensitivity.

10.4.2.2 Repellency test

The repellency bioassay was devised to test the efficacy of preventing insects from moving onto paper treated with turmeric oil or other biologically active compounds or mixtures.

The propensity of the tested sample to repel insects was determined against adults of the red flour beetle, *T. castaneum*, using the methods by Laudani et al. (1955) and McDonald et al. (1970). Accordingly, filter paper strips were treated with acetone solutions of samples, usually at 50 $\mu g/cm^2$ (or any other predetermined concentration). The acetone was then completely evaporated. Two untreated strips were attached lengthwise, edge to edge, on either side of the treated paper strip. Four glass rings, 2.5 cm high and 6.4 cm i.d., were placed over the joined edges of the papers, on each side of the treated strips, where one half of each ring was found on the treated paper and the other half was found placed over the nontreated control paper. Ten adults of each species were placed into each glass ring. The numbers of insects found on the treated and untreated paper halves were recorded at 10 fixed times, from 1 to 24 hours after exposure. The average count of all four ring replicates, each containing 10 reading periods, was converted to percent repellency values (Navarro et al., 2005) (Figure 10.7).

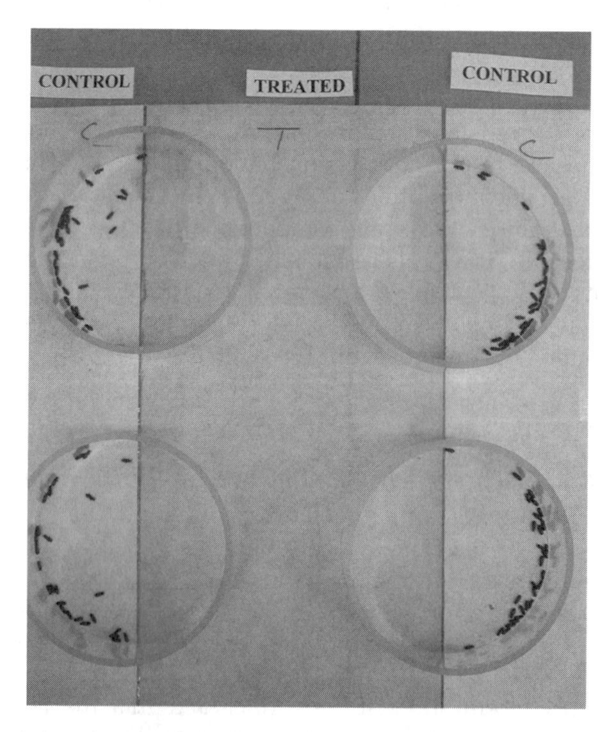

Figure 10.7 Adults of *T. castaneum* concentrated in the nontreated arena in a repellency test.

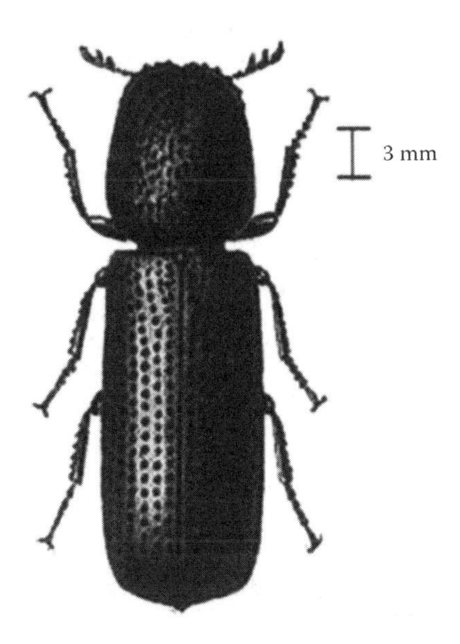

Figure 10.8 Adult *R. dominica* used in penetration tests.

10.4.2.3 Nonchoice bioassay test for penetration

Navarro et al. (1998b, 2005) tested the efficacy of preventing adults of the lesser grain borer, *R. dominica* (Figure 10.8), from boring holes into white office printer paper discs treated with turmeric extracts. The average number of holes in the control compared to the average number of holes in the treated discs was expressed as penetration prevention efficacy (PPE).

In preparing the test, the paper discs were treated with acetone solutions of sample extracts at dosages of 50 and 640 μg/cm^2 in the routine evaluation tests (other dosages of up to 2560 μg/cm^2 were also tested in special experiments) and used in the bioassay the next day. The tested paper discs and the controls were pressed with a supportive piece of wire mesh between two open-ended glass cylinders (Figure 10.9). Ten insects were then placed inside each top cylinder. The tests were carried out in 10 replicates, usually for 24 hours, in the dark at 27°C and 65% RH (Navarro et al., 2005).

10.4.2.4 Choice bioassay test for penetration

Navarro et al. (2005) used the same experimental device (Figure 10.9) and conditions in the choice test as used in the nonchoice test. However, in the choice test, the paper discs were divided in half; one side was impregnated with the sample under test dissolved in acetone, while the other half treated with acetone alone was used as a control. In this test a dosage of only 50 μg/cm^2 of the test sample was applied on the paper. The test insects were entrapped on the paper and could choose to perforate the control, the impregnated part of the paper, or the border between the two halves of a

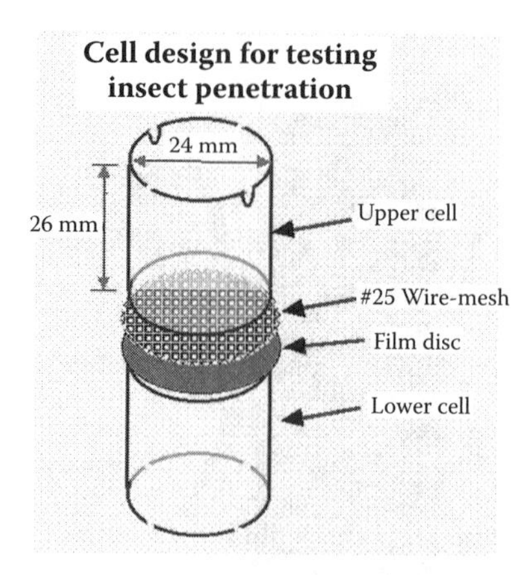

Figure 10.9 Device used in penetration tests.

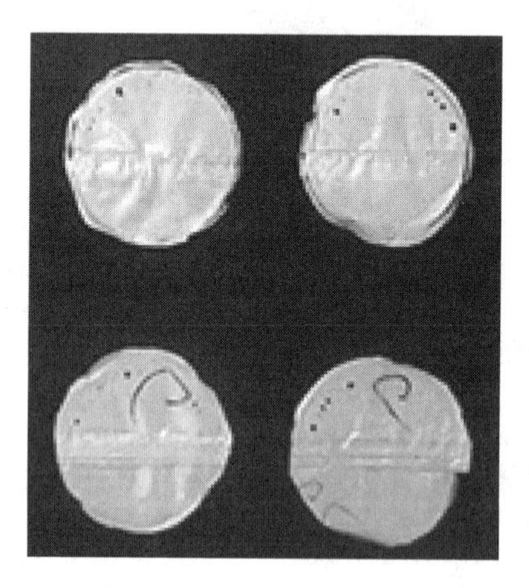

Figure 10.10 Penetration holes made by *R. dominica* adults on the nontreated area in the choice test for penetration.

disc, or not to bore holes anywhere on the two disc halves (Figure 10.10). At the end of each exposure period, the number of perforations appearing in the two parts of the disc were counted. A comparative analysis was performed using the Student's t test for residual effect, and the differences between control and dosages applied were determined using Dunnet's test (Anon., 1989).

10.4.2.5 Interpretation of results

The repellency test directly measures the ability of the tested sample to repel insects, while the other two bioassays revealed a combined effect of the antifeedant components and the repelling components in the tested sample on the insects. The choice test indicates the combined additive effects of both antifeedant and repelling components to prevent perforation, which differs from the nonchoice test where the repelling components enhance perforation and the antifeedant components prevent perforation. From the data of all three bioassays a conclusion could be drawn as to the antifeedant effect of the sample. For example, in the choice test, if the insects did not perforate the impregnated part of the paper but only the untreated part, and the results of repellency showed that the tested sample had low repellent efficiency, it was then concluded that the sample contained antifeedant components. When the sample had a high repellency, the choice test showed a high efficiency even in the presence of low antifeedant effect. In such a case, the nonchoice test would give a negative efficacy result. This negative efficacy could be reduced only when higher antifeedant components were present in the sample. When the effect of antifeedancy was higher than repellency, a positive efficacy value was obtained.

10.4.2.6 Tests for insect penetration of packaging films

The literature describes a number of methods for determining insect penetration. Higland and Wilson (1981) described a test device for penetration of *R. dominica* that consists of five pieces of machine assembly made of aluminum tubing. Wohlgemuth (1979) used for penetration tests a device consisting of several aluminum plates, the test foil, and a cover. Gerhardt and Lindgren (1954) studied two penetration test methods. One consisted of test bags exposed to insects. The second method consisted of two small plastic cups, with a 5-cm-diameter hole in their lids. One cup contained a small amount of food and the other 50 adult insects, and the film to be tested for penetration was inserted between the two cups. Mullen (1994) developed a rapid method to determine the effectiveness of insect-resistant packaging. This technique was based on exposing *P. interpunctella* larvae to plastic pouches. Other authors used pouches made of test films exposed to insects for penetration (Bowditch, 1997; Cline, 1978; Mullen, 1994).

Navarro et al. (1998b) developed a reliable and simple method, less expensive than testing with devices using metal sections. The test device (Figure 10.9) was the same as in the nonchoice test (Section 10.4.2.3) above, with the tested packaging film replacing the paper, and it enabled evaluation of the material following extremely short exposure times. Penetration by *R. dominica* adults increased with increasing exposure periods. Exposure periods of *R. dominica* adults (Figure 10.8) to the test films were 24, 48, or 72 hours or 7 days. The thickness of the films played an important role. The resistance of films used could usually be evaluated after 24 hours of exposure. A possible explanation for the speed at which insects penetrated could be the wire mesh used adjacent to the test film. Apparently, insects could use the wire mesh for support to press their mouth parts against the tested laminate material.

For long-exposure tests on samples of plastic laminates impregnated with repellents and antifeedants, adding of new food every 24 hours and removing dead insects and replacing them with live ones every few days was necessary.

10.4.3 Variability of the plant extracts in inducing insect repellence and penetration prevention

Navarro et al. (2005) tested several turmeric rhizome extracts from various sources for repellency and penetration prevention in the repellency and nonchoice tests. The extracts included laboratory Soxhlet extracts of powdered turmeric rhizomes from various sources, commercial oleoresin, and commercial essential oils prepared by steam distillation. Repellency tests were carried out on two test species, adults of *R. dominica* and *T. castaneum*, exposed for 24 and 48 hours separately. The applied turmeric extract dosage was 800 $\mu g/cm^2$ on the paper. The papers were kept for 4 days before exposure and the tests were run and insects counted for 5 days. In the nonchoice test, unlike the routine test, the turmeric extracts were applied at dosages of 50 to 2560 $\mu g/cm^2$ on the papers and the tests were carried out after 1, 15, 30, 45, 60, and 75 days of turmeric oil application on the papers. The petroleum ether extracts of turmeric rhizomes (oleoresins by definition, assumed to be essential oils by composition) induced repellency of 50 to 60% (averaged over 5 days) with both the test insects, thereby showing that the turmeric extracts contained highly effective insect-repelling substances. Such high repellency was not found in other turmeric essential oils tested from 20 other various origins.

The penetration prevention results showed a protective effect in a dose-dependent manner. The petrol ether extracts, applied at a dosage of 640 $\mu g/cm^2$, showed a substantially reduced penetration for up to 60 days. Higher turmeric extract dosages of 1280 and 2560 $\mu g/cm^2$ resulted in extended periods of protection of over 75 days. The nonchoice test shows that turmeric extractions (petroleum ether or steam distillation) resulted in a high penetration prevention efficacy (PPE) of 80 to 100% of these oils. The other turmeric extracts provided medium (40 to 80%) or low (0 to 40%) PPE in the nonchoice test.

Because the nonchoice test results are also influenced by the repellency effect, which acts in the reverse direction in this test, it is expected that all of the cited extracts have a rather strong antifeedant effect, as all of them showed a strong repellency. Nonetheless, in comparing the effect of different turmeric extracts on preventing insect penetration ability, it was discovered that the source of turmeric rhizomes, as well as the particular extraction method employed, substantially affected the biological activity of turmeric extracts. Rather high turmeric extract doses were initially applied on the papers in the various tests. This doubtless created saturation effects on the tested pests. High doses are impractical in most cases for use in pest-impervious packaging materials. Although these bioassays were run on paper, results point to the potential for long-term residual effectiveness of turmeric extracts in penetration prevention when integrated in packaging materials Navarro et al. (1998a, 2005).

10.4.4 Conclusions

Navarro et al. (2005) demonstrated that turmeric essential oils and several turmeric oil fractions have both repellency and antifeedant effects on test storage insects. The solid residue of turmeric oleoresin has a strong antifeedant effect. It was also demonstrated that the integrated activity contributed by several compounds is responsible for the high activity of the essential oil and its fractions as a repellent and antifeedant. The turmeric oleoresin and its components provide a versatile system with varied components to become integrated in and compatible with food packaging materials and, as a result, lead to very effective and diverse pest-impervious packaging materials.

As a result of diverse rhizome sources and extraction methods, there is a tremendous variation in composition of commercial turmeric essential oils and oleoresins. It should be emphasized that each commercial batch of turmeric oleoresin, essential oil, TOSR, or oil fraction for the production of pest-impervious packaging material should be carefully and comprehensively checked for its biological activity and its chemical composition. Therefore, in U.S. Patent Application 20050208157-A1 and European Patent Application 04101309.5 (Navarro et al., 2005), among the 60 claims, essential claim 1 was on "a composition-of-matter comprising a substance usable in producing packaging material and at least one compound selected from the group consisting of *ar*-turmerone, sesquiterpene alcohols and a turmeric oleoresin solid residue."

10.5 Possibilities of impregnating packaging materials with repellents

10.5.1 Concepts associated with laminate composition

Various types of materials are used for dried food commodities packaging, including plastic polymers, paper, cardboard, textiles, and metal foil (usually aluminum). All of these are subject to boring attacks by insects.

Laminates are composed of several kinds of film layers, sandwiched together by adhesives under pressure, with heat. Some laminates can be produced without heat. Packaging materials composed of laminates are constantly gaining in popularity. These laminates are composed mostly of polymers, polymers and paper, or polymers and aluminum foil. Polymers used in producing laminates may include polyethylene, polypropylene, polyester, and many others. Coating substances to packaging materials include lacquers, varnishes, and paints, with or without an outer polymer film layer.

Usually, packaging materials selected are capable of conferring optimal mechanical and chemical protection to the packaged product, while being cost effective and therefore simple to produce. Laminate physical properties such as strength, elasticity, transparency, and permeability of water and gases are well determined. Additives in the laminate materials that come into contact with food must not migrate into the food or affect the organoleptic properties of the food they protect. The seams of plastic laminate bags must remain tightly closed for the intended shelf life of the laminate, and the laminate sheeting must be free of pinholes.

Considering the complexity of laminate compositions, there are potentially several options and a variety of methods for producing pest-impervious packaging materials by fabricating or modifying the packaging material to include suitable repellents, with or without carriers:

1. Coating ordinary packaging materials with a pest-impervious coating composition. The coating material can be produced by dissolving the active compounds in the solution, suspension, emulsion, or melt of the coating composition, by solvent compounding or by any other suitable method.
2. Adding, dissolving, or dispersing the active compounds in the adhesive, lacquer, or paint or any other additive between the layers of laminates used as ordinary packaging materials.
3. Polymer-based packaging materials can be generated by mixing the active compounds with the polymer as a melt, by solvent compounding, through processes such as extrusion, molding, foaming, casting, or dipping.
4. Paper-based packaging materials can be generated by adding active compounds to a paper pulp emulsion. Paper, paperboard, or textile substances can be generated by impregnation.
5. The margins of the packaging material that serve for welding or gluing of the packaged product are impregnated or coated with the active materials.

All these methods are well established and well described, for example, by Appendini and Hotchkiss (2002).

The level of the active compounds in the packaging material should be sufficiently high so as to effectively and reliably render resistance to insect pests. On the other hand, it should be sufficiently low so as not to substantially weaken or otherwise substantially alter the physical properties of the packaging material, such as its strength and its elasticity. The packaging materials should be safe and nontoxic. Ultimately, cost effectiveness considerations will determine whether these materials will be commercialized.

10.5.2 Coating applications

Whalon and Malloy (1998) in U.S. Patent 5,843.215 describe the use of mixtures of plastics with plant substances with proven capacity to repel insects and protect food packages from insect invasions. Whalon and Malloy (1998) prepared paint varnish coatings for food packaging materials to deter Indian meal moths. These coatings contained various mixtures of limonene from etheral oils, eucalyptal from eucalyptus, perillaldehyde from mandarin orange peel, loncadol from cinnamon oil, neem oil from neem leaves and seeds, turmerone from turmeric, aserone from sweet flag, and cinnamon oil from cinnamon. The coatings were applied to cartons and tested with surrounding moths over an 8-week period. The patent claimed that the treated varnishes repelled the insects while nontreated varnishes did not. The repellency was due to the volatility of the plant extracts in the surface varnishes. Volatile compounds in surface applications tended to lose their potency rather quickly with time. Whalon and Malloy's patent did not indicate the shelf life of these varnish applications as pest-impervious packaging material or the capability to deter other insects.

10.5.3 Pest-resistant laminates containing plant extracts

Feasibility studies and preliminary tests have shown that turmeric extracts can be effectively included in various conventional packaging materials without affecting their physical quality while conferring pest resistance (Navarro et al., 2005). In these experiments, turmeric oil extracts at various concentrations were successfully dissolved in lacquer, glue, and pigments used in industrial manufacturing of packaging films. The turmeric-amended lacquer remained smooth and well spread over the sheets and the turmeric-amended glue dried properly. In other experiments, turmeric essential oil was successfully embedded in commercial PVC sheets. The turmeric extract did not affect the physical appearance of PVC, and the tests resulted in substantial insect penetration prevention efficiency. In addition, turmeric extracts were dissolved in PSA lacquer and then brushed over bags and boxes of breakfast cereal. In this case too, the tests resulted in substantially improved insect penetration prevention efficiency.

The feasibility studies and bioactivity tests for pest resistance in the preliminary experiments and in the development phase of laminates production were carried out at the Food Science Laboratory at the Agricultural Research Organization (ARO), Volcani Center during 2002 to 2004 (Navarro et al., 2005). The bioassays for repellency (exposure time of insects, 24 hours) and penetration (exposure time, 24 hours or a week), described earlier (Sections 10.4.2.2 and 10.4.2.6), were used to test the experimental impregnated laminates. The results were indicative and could guide the manufacturer in a comparable way for the development phase.

In the experimental laminate tests, good repellency results could be obtained, similar to the results obtained with tests on the paper. In the tests for penetration, the insects often made scratches in the plastic without puncturing even the outer layer of the laminate. At times, the outer layer of the laminate was scratched open but the insect failed to penetrate the laminate. This might have been due to the insect sensing the antifeedant compounds after uncovering the surface and then withdrawing.

10.5.4 *Shelf life tests of treated laminates*

Food shelf life is the length of time that corresponds to a guaranteed optimal quality or tolerable loss in quality of the packaged food. The extended-period expiration date of the packaged food relates to food safety. The shelf life of packaging materials, pest-impervious or regular untreated material, should be at least as long as the length of the food expiration date.

The aging of packaging materials refers to the variation and deterioration of their properties of interest, those related to efficacy, over time. Aging and shelf life studies are an indispensable tool to anticipate the behavior of the packaging materials throughout their commercial life under certain conditions of temperature and storing, and allow the manufacturer to detect and correct formulation problems not anticipated in the first stage of development and prior to release for use.

Determining the effects of aging on a packaging material or package/product with a long shelf life (typically 0.5 to 1 year) in real time is a lengthy process that could severely delay market introduction. The possibility of conducting an accelerated aging test that simulates the effects of real time is most welcomed. Data obtained from accelerated aging testing should represent a conservative estimate of shelf life and are tentative until real-time aging studies on the packaging material or product/package combination are completed.

Accelerated aging techniques are based on the assumption that for stored food packages the temperature is the main aging parameter and that the influence of other parameters, like humidity and light, is negligible. It is assumed that the chemical reactions involved in the deterioration of materials follow the well-known Arrhenius equation for reaction rate dependence on temperature. In a simplified protocol for accelerated aging and as a rule of thumb, it is stated that a 10°C increase or decrease in the temperature of a homogenous process results in an approximately two-time or half-time change in the rate of a chemical reaction (equivalent of saying that the aging factor is 2). Using this rule, it can be calculated, for example, that a shelf life of 38 days at 55°C simulates 1 year of shelf life at an ambient temperature of 22°C. However, since this formula is only an approximation and is based on rate kinetics of a single chemical reaction, with numerous assumptions regarding the reaction order kinetics and the activation energy, and not on real multicomponent reactions in packages or packaging materials, the direct extrapolation of this model to the aging of packaging materials must be used with caution (Hemmerich, 1998). Nolan (2006) stated that in research on homogeneous plastic materials, a higher aging factor (2.5 to 3.0) has been found. However, since package systems are usually made up of several different materials, a conservative aging factor of 1.8 to 2.0 is typically used. This aging factor is also used in the medical device package testing industry. This aging factor usually results in a built-in safety factor, ensuring that enough time under test has been achieved to satisfy the *estimate* of shelf life. Since the simplified protocol for accelerated aging enables testing at one elevated temperature, the selected temperature should be a temperature that avoids unrealistic failure conditions, such as deformation due to melting. For many packaging materials 55°C will be the highest suitable temperature for accelerated aging tests. It should be emphasized that real-time aging must be performed in conjunction with any accelerated aging study of a new packaging material to correlate the results found during accelerated aging.

The aging and shelf life testing of pest-impervious materials or food packages faces the problem of lack of information and experience on the behavior of such materials with regard to the deterioration with time of the repelling capability of insects. There are numerous possible combinations of using the turmeric oil, its fractions, or the solid residue of oleoresins in pest-impervious packaging materials, depending on the biologically active composition used and the way of application. Also, there are many potential mechanisms for deterioration of the repelling capability with time, as a result of any incompatibility between the components of the plastic packaging material and the additive composition. These mechanisms include physical mechanisms (diffusion and evaporation of biologically active components) and chemical mechanisms (induced thermal decomposition and reactions of biologically active components with laminate components). It was assumed by the manufacturer (and by now approved in several cases by real-time testing) that the simplified protocol for accelerated aging could be used for testing the shelf life of the various pest-impervious packaging materials with regard to the characteristic qualities and efficacies as packaging materials and packages (including welding quality).

Nonetheless, a simplified protocol for accelerated aging testing was considered unsafe in guaranteeing a conservative result for the shelf life of the packaging material's penetration prevention efficacy, even with an aging factor of 1.8 to 2.

For pest-impervious packaging materials or food packages designated for a relatively short storage time, the ultimate way of testing would be real-time shelf life testing. For long-shelf-life packages, fully accelerated aging testing was considered, which could also result in some information regarding the real mechanism of aging. Nonetheless, the implementation of full testing faced several difficulties and barriers:

1. A fully accelerated aging testing should enable the determination of the kinetics reaction order of the total aging processes and the dependence of the reaction rate constant on temperature. Therefore, the tests should be carried out at least at three elevated temperatures, with several time intervals at each temperature. At the end of each

time interval, the biological activity should be measured. As the penetration prevention test is destructive, the expected number of samples in the tests is rather high.

2. A need exists for a semiquantitative, reliable, and suitable test for the penetration prevention capability (antifeedant and repellency combined effects) of the pest-impervious packaging materials or the food packages. It is quite common for packaging materials for dry food for long storage, like flexible laminates, to be persistent for 2 weeks when food packages massively and constantly are exposed to storage insects, before any apparent penetration is revealed. As described in Section 10.6, a minimum testing time of 4 weeks is needed in order to get indicative and statistically reliable results for effective pest-impervious food packages. Each test or test series should be accompanied with comparable controls of regular food packaging and could take almost 2 months.
3. It was necessary to decide what percentage of penetration prevention would be the limit between success and failure of the pest-impervious packaging material, in the extreme conditions of the tests. No correlation has been established between the results of the tests with massive exposure conditions and the real storing conditions of food packages. Tentatively and as a command decision, it was decided that the shelf life tests would be carried out until the results show 50% deterioration in the high penetration prevention efficacy in freshly prepared pest-impervious packaging material.

Considering these difficulties and the enormous amount of work needed for executing a fully accelerated aging test, accelerated aging tests were not implemented. Instead, when time permitted, real-time testing was performed over long time intervals, covering the practical shelf life necessary for several food commodities. Likewise, dry food manufacturers were encouraged to test pest-impervious packaging materials by themselves, in order to be persuaded if such materials could solve the specific problems they face with their commodities.

10.5.5 *Safety and fragrance aspects of plant extract additives in food packaging materials*

Given the obvious importance of producing safe and wholesome food, it is important that food packaging not affect the food with which it comes into contact. The important issue is the potential migration of unsafe ingredients, monomers, or additives from plastic into food. The untreated regular packaging materials are obviously safe and certified materials.

Turmeric extracted from *C. longa* is a very well known food additive and is used as a spice, seasoning, and flavoring. As such, turmeric essential oil, oleoresins, and natural extractives (including distillates) appear on the FDA's generally recognized as safe (GRAS) list (21 CFR 582.20) and on the FEMA

(Flavor and Extracts Manufacturers Association in the U.S.) GRAS list (182.10, 182.20) without any limitation of use in foods. These lists of food additives are generally recognized as safe by a consensus of scientific opinion. Standard migration tests based on prescribed food stimulants, which included overall migration testing and specific migration tests, ensured the safeness of the pest-impervious laminates proposed for packaging dry food commodities. The migration tests were run by a certified laboratory, subject to national (Israel) provisions and according to EU and U.S. regulations.

Turmeric essential oil is the fragrant essence of turmeric rhizome. It has the same fresh, spicy-woody aroma, also characterized as musky earthy aroma, as the powdered turmeric spice. *ar*-Turmerone, turmerone, and other main components of the essential oil have considerable contribution to the aroma. The inclusion of turmeric oil in packaging materials imparts a distinctive repelling efficacy, and at the same time imparts a characteristic odor to the packaging material. According to the composition of the pest-impervious packaging materials, it can be odorous with varying intensity on the outer and inner sides of the laminate that is in contact with the packaged food. As long as the penetration prevention efficacy is apparent (at least while the repellency effect exists), the packaging material is odorous on the outer side. The odor issue was taken into account as an important parameter during the development stage of the pest-impervious laminates and in the final pest-impervious laminates compositions. Using recommended laminates, the human olfactory system only perceived the odor at a very short distance from the food packages. This odor was considered pleasant or at least was not rejected in consumer tests. There is no such migration when a laminate with an outer active coating or paint varnish is used. When the turmeric oil was added to the adhesive or internal lacquer layer, the odor penetration could be diminished by a relatively thick plastic layer on the inner side of the laminate, or avoided by an aluminum foil layer on the inner side of the laminate. The accompanying odor problem could be solved by replacing turmeric oil with nonodorous turmeric oil fractions or the solid residue as the biologically active constituents in pest-impervious packaging materials.

10.6 Laboratory and field test results with nontoxic insect repellent packaging materials

10.6.1 Organoleptic test results with rice, pasta, nuts, and sunflower seeds exposed to packages constructed with treated laminates

The inclusion of turmeric oil in food packaging may cause customer objection should the integrity of the food product be influenced by the turmeric oil aroma. Since not all food products are consumed as they are stored in their package, initial tests were carried out with macaroni and polished rice inside treated packages and compared with untreated packages. All the treated and untreated packages were stored under controlled conditions of room temperature at 25°C for 3 months. At the end of the storage period, an official laboratory taste analysis test was performed. Test reports indicated that although the atmosphere inside the treated packages at the opening had a strong typical turmeric oil aroma, this was hardly detectable in the uncooked product in the package. Furthermore, the typical aroma of turmeric oil was not detectable by the test panels (with appropriate replicates) in any of the tests carried out following cooking of the macaroni and polished rice.

A series of commercial-scale tests was carried out with roasted groundnuts, almonds, and sunflower seeds. These roasted products were kept for 6 months in turmeric oil-treated packages and in untreated packages as control and then subjected to taste panels. The packages were kept in ordinary storage conditions from June 2004 to January 2005. On a careful examination before the opening of the packages, the typical turmeric oil aroma could be detected on the outside of the packages. This aroma was not detectable from a distance but only when the packages were held close to the nose. The same aroma could not be observed upon opening the package due to the interference of the strong roasted nuts and sunflower seeds aroma. The packages were tested by a team of the packaging company, who indicated that none of the roasted products (including the controls) had an objectionable aroma that would cause the product to be rejected.

10.6.2 Treated laminates as house fly repellents

Field tests were carried out with a pest-impervious laminate sheet to evaluate the efficacy of repelling the house fly, *Musca domestica* L. Treated laminate sheet was placed on a standard dry fruit box (80 cm long, 60 cm wide, 8 cm high). A petri dish (8.5 cm diameter) that contained a thin layer of a fly attractant liquid, Fly Buster, was placed on the center of the box. The fly attractant served to promote adult flies to approach the petri dish and to trap them in the liquid. Another petri dish with Fly Buster was placed on a box covered with an ordinary polyethylene liner (without treatment). Both boxes were laid on the ground under a fichus tree, 2 m apart, in August, when the daily temperatures fluctuated within the range of 22 and 30°C with ambient RH of 60 to 90%.

The Fly Buster remained in the petri dish for three successive weeks. Each test was run for 24 hours. At the end of each test, the insects in the petri dishes were counted and removed from the liquid. Results of four replicates showed that on the average, 19 adult flies were caught in the petri dish of control and only 5.5 flies caught in the petri dish of the treated laminate sheet. Flies were not observed most of the time during the day around the dish on the treated laminate, whereas around the dish on the nontreated (control) sheet flies were apparent most of the time. These preliminary results show clearly that the inclusion of turmeric oil in liners can provide a good solution to prevent flies from landing on surfaces.

10.6.3 Testing the repellency of plant volatile oils on other insect pests

10.6.3.1 Repellency of volatile oils against three mosquito vectors

Tawatsin et al. (2001) evaluated the effect of volatile oils extracted by steam distillation from four plant species (turmeric [*C. longa*], kaffir lime [*Citrus hystrix*], citronella grass [*Cymbopogon winterianus*], and hairy basil [*Ocimum americanum*]) in mosquito cages and in a large room for their repellency effects against three mosquito vectors, *Aedes aegypti, Anopheles dirus,* and *Culex quinquefasciatus*. The turmeric, citronella grass, and hairy basil oils, especially with the addition of 5% vanillin, repelled the three species under cage conditions for up to 8 hours. The oil from kaffir lime alone, as well as with added 5% vanillin, was effective for up to 3 hours. The standard repellent deet provided protection for at least 8 hours against *Ae. aegypti* and *Cx. quinquefasciatus* and only 6 hours against *An. dirus*. However, deet with the addition of 5% vanillin protected against the three mosquito species for at least 8 hours. The results of the large-room evaluation confirmed the results obtained under cage conditions. This study demonstrates the potential of turmeric, citronella grass, and hairy basil oils as topical repellents against both day- and night-biting mosquitoes. The three volatile oils can be formulated with vanillin as mosquito repellents in various forms to replace deet (N,N-diethyl-3-methylbenzamide), the most common chemical repellent currently available.

10.6.3.2 Repellency of volatile oils against ants and clothing moths

In our preliminary studies, Finkelman and Navarro tested several turmeric oils as insect repellents on a wide range of insect groups that are associated with food, stored products, processed and unprocessed packaged food products, and household pests. The groups of insects tested successfully were from among the Coleopteran Bostrichidae (*R. dominica*), Tenebrionidae (*T. castaneum*), and Anobiidae (*L. serricorne*), among the Lepidoptera Phycitidae (*E. cautella* and *P. interpunctella*), and among the Dipterans (*M. domestica*), and from among the Formicids were several house ant species.

The special characteristics of some botanical essential oils to specifically repel clothing moths are mentioned in the literature. This field is relatively new and attractive in view of the carcinogenic nature of naphthalene and para-dichlorobenzene. Their unpleasant smell has created significant interest in the scientific community.

Acknowledgments

The authors thank Eng. R. Dias and Ms. Miriam Rindner of the Department of Food Science, Prof. Fadel Mansur of the Department of Plant Protection, Israel Agricultural Research Organization, and Dr. Ahmet Guray Ferizli of the Department of Plant Protection, University of Ankara, Turkey, for their dedicated work during the research phases of the plant extracts. We also thank Mr. Shmuel Shatsky, General Manager, Mr. Eitan Amichai of BioPack and Eng. Ronny Mozes of Global Roto-Sheika, Caesarea, Israel, for their cooperation during the development of the insect-resistant laminates.

References

Adams, R.P. (Ed.). (2001). *Identification of Essential Oil Components by Gas Chromatography/Quadruple Mass Spectrometry,* Allured Pub. Co., and other GC/MS databases: Flavor2 (HP Flavors library) and NIST02 (NIST Rev. D.04.00, October 2002) Carol Stream, IL.

Alexander, J. and Rao, G.S. (1973). The chemistry of *ar*-turmerone, atlantone and related ocyclic sesquiterpenoids, ketones. *Flavour Ind.* 4:390.

Anon. (1989). *JMP User's Guide.* SAS Institute, Cary, NC, p. 464.

Appendini, P. and Hotchkiss, J.H. (2002). Review of antimicrobial food packaging. *Innovative Food Sci. Emerging Technol.* 3:113–126.

Bekele, A.J., Obeng-Ofori, D., and Hassanali, A. (1996). Evaluation of Ocimum suave as a source of repellents, toxicants and protectants in storage against three major stored product insect pests. *Int. J. Pest Manage.* 42:139–142.

Borzatta, V., Brancaleoni, D., and Battistini, C. (2001). Process for Synthesis of 5-Alkylbenzodioxoles. U.S. Patent 6252092.

Bowditch, T.G. (1997). Penetration of polyvinyl chloride and polypropylene packaging films by *Ephestia cautella* (Lepidoptera:Pyralidae) and *Plodia interpunctella* (Lepidoptera:Pyralidae) larvae, and *Tribolium confusum* (Coleoptera:tenebrionidae) adults. *J. Econ. Entomol.* 90:1028–1031.

Champon, L.S. (2000). All Natural Soil Treatment and Insecticide Composition Containing Plant Extract Heat Components. U.S. Patent 6,051,233.

Cline, L.D. (1978). Penetration of seven common flexible packaging materials by larvae and adults of eleven species of stored-product insects. *J. Econ. Entomol.* 71:726–729.

Coats, J.R., Peterson, C.J., Tsao, R., Eggler, A.L., and Tylka, G.L. (2001). Biopesticides Related to Natural Sources. U.S. Patent 6,207,705.

Cox, P.D. (2002). Potential for using semiochemicals to protect stored products from insect infestation. A review. *J. Stored Products Res.* 40:1–25.

Daniel, S.H. and Smith, R.H. (1990). The repellent effect of neem (*Azadirachta indica* A. Juss) oil and its residual efficacy against *Callosobruchus maculatus* (Coleoptera: Bruchidae) on cowpea. In *Proceedings of the 5th International Working Conference on Stored-Products Protection*, Bordeaux, France, 1992, pp. 1589–1596.

Desphande, R.S. and Tipnis, H.P. (1974). Insecticidal activity of *Ocimum basilicum. Pesticides* 11:11–12.

Fields, P.G., Xie, Y.S., and Hou, X. (2001). Repellent effect of pea (*Pisum sativum*) fractions against stored-product insects. *J. Stored Products Res.* 37:359–370.

Gebbinck, E.K. (1999). Synthesis of Model Compounds Derived from Natural Clerodane Insect Antifeedants, Dissertation, Wageningen, http://www.agralin.nl/wda/abstracts/ab2666.html.

Gerhardt, D. and Lindgren, D.L. (1954). Penetration of Packaging Films. *California Agriculture*, June, pp. 3–4.

Gopalan, B., Goto, M., Kodama, A., and Hirose, T. (2000). Supercritical carbon dioxide extraction of turmeric (*Curcuma longa*). *J. Agric. Food Chem.* 48:2189–2192.

Hagstrum, D.W. and Milliken, G.A. (1988). Quantitative analysis of temperature, moisture, and diet factors affecting insect development. *Ann. Entomol. Soc. Am.* 81:539–546.

Hanlon, J.F., Kelsey, R.J., and Forcinio, H.E. (2000). *Handbook of Package Engineering.* CRC Press, Boca Raton, FL, pp. 1–30.

Hemmerich, K.J. (1998). Accelerated Aging: General Aging Theory and Simplified Protocol for Accelerated Aging of Medical Devices. *Medical Plastics and Biomaterials Magazine*, July 1998, http://www.devicelink.com/mpb/archive/98/07/002.html.

Highland, H.A. (1984). Insect infestation of packages. In *Insect Management for Food Processing*, Baur, F.J. (Ed.). American Association of Cereal Chemists, St. Paul, MN, pp. 311–320.

Highland, H.A. (1991). Protecting packages against insects. In *Ecology and Management of Food-Industry Pests*, Gorham, J.R. (Ed.). Association of Official Analytical Chemists, Arlington, VA, pp. 345–350.

Highland, H.A. and Wilson, R. (1981). Resistance of polymer films to penetration by lesser grain borer and description of a device for measuring resistance. *J. Econ. Entomol.* 74:67–70.

Hou, X., Fields, P., and Taylor, W. (2004). The effect of repellents on penetration into packaging by stored product insects. *J. Stored Products Res.* 40:47–54.

Howe, R.W. (1960). The effects of temperature and humidity on the rate of development and the mortality of *Tribolium confusum* Duval (Coleoptera, Tenebrionidae). *Ann. Appl. Biol.* 48:363–376.

Hsu, H.K., Zhou, J., and Chang, C.H.L. (2001). Natural Pesticide. U.S. Patent 6,231,865.

Ignatowicz, S. and Wesolowska, B. (1994). Potential of common herbs as grain protectants: repellent effect of herb extracts on the granary weevil, *Sitophilus granarius*. In *Stored Product Protection. Proceedings of Sixth International Working Conference on Stored Product Protection*, Canberra, pp. 790–794.

InPho Newsletter. (2006). FAO Compendium on Post Harvest Operations. http://www.fao.org/inpho.

Islam, B.N. (1986). Use of some extracts from Meliaceae and Annonaceae for control of Rice Hispa, *Dicladispa armigera*, and the pulse beetle, *Callosobruchus chinensis*. In *Proceedings of the 3rd International Neem Conference*, Nairobi, Kenya, pp. 217–242.

Isman, M.B., Koul, O., Luczynski, A., and Kaminski, J. (1990). Insecticidal and antifeedant bioactivities of neem oils and their relationship to azadirachtin content, *J. Agric. Food Chem.* 38:1406–1411.

Jayaprakasha, G.K., Negi, P.S., Anandharamakrishnan, C., and Sakariah, K.K. (2001). Chemical composition of turmeric oil: a byproduct from turmeric oleoresin industry and its inhibitory activity against different fungi. *Z. Naturforschung.* 56c:40–44.

Jilani, G. and Saxena, R.C. (1990). Repellent and feeding deterrent effects of turmeric oil, sweetflag oil, neem oil, and a neem-based insecticide against lesser grain borer (Coleoptera: Bostrychidae). *J. Econ. Entomol.* 83:629–634.

Jilani, G., Saxena, R.C., and Reuda, B.P. (1988). Repellent and growth-inhibiting effects of turmeric oil, sweetflag oil, neem oil, and Margosan-O on red flour beetle (Coleoptera: Tenebrionidae). *J. Econ. Entomol.* 81:1226–1230.

Jilani, G. and Su, H.C.F. (1983). Laboratory studies on several plant materials as insect repellents for protection of cereal grains. *J. Econ. Entomol.* 76:154–157.

Khan, M.A. and Wohlgemuth, R. (1980). Diethyltoluamide as a repellent against stored-products pests. *Anzeiger Schadlingskunde Pflanzenschutz Umweltschutz* 53:126–127.

Koul, O. (1987). Antifeedant and growth inhibitory effects of calamus oil and neem oil on *Spodoptera litura* under laboratory conditions. *Phytoparasitica* 15:169–180.

Kvenberg, J.E. (1975). Invasion and penetration of consumer packages in short-term storage by stored-product insects. In 1st International Working Conference on Stored-Product Entomology, Savannah, GA, pp. 627–634. http://bru.gmprc.ksu.edu/proj/iwcspp/.

Laudani, H. and Davis, D.F. (1955). The status of federal research on the development of insect-resistant packaging. *Tech. Assoc. Pulp Paper Ind.* 38:322–326.

Laudani, H., Davis, D.F., and Swank, G.R. (1955). A laboratory method of evaluating the repellency of treated paper to stored-product insects. *Tech. Assoc. Pulp Paper Ind.* 38:336–341.

Lee, H. and Shin, W. et al. (2001). Insecticidal activities of *ar* turmerone identified in *Curcuma longa* rhizome against *Nilaparvata lugens* (Homoptera: Delphacidae) and *Plutella xylostella* (Lepidoptera: Yponomeutidae). *J. Asia Pac. Entomol.* 4:181–185.

Lui, Z.L. and Ho, S.H. (1999). Bioactivity of the essential oil extracted from *Evodia rutaecarpa* against the grain storage insects *Sitophilus zeamais* and *Tribolium castaneum*. *J. Stored Prod. Res.* 35:317–328.

Makanjuola, W.A. (1989). Evaluation of extracts of neem (*Azadirachta indica* A. Juss) for the control of some stored product pests. *J. Stored Prod. Res.* 25:231–237.

Malik, M.M. and Naqvi, S.H.M. (1984). Screening of some indigenous plants as repellents or antifeedants for stored grain insects. *J. Stored Prod. Res.* 20:41–44.

McDonald, L.L., Guy, R.H., and Spiers, R.D. (1970). *Preliminary Evaluation of New Candidate Materials as Toxicants, Repellents, and Attractants against Stored-Product Insects*, Marketing Research Report. USDA, Washington, DC, 8 pp.

Mordue, A.J. and Blackwell, A. (1993). Azadirachtin: an update. *J. Insect Physiol.* 39:903–924.

Mullen, M.A. (1994). Rapid determination of the effectiveness of insect resistant packaging. *J. Stored Prod. Res.* 30:95–97.

Mullen, M.A. (1997). Keeping bags at bay. *Feed Manage.* 48:29–33.

Mullen, M.A. (2006). Insect-resistant packaging. In *Insect Management for Food Storage and Processing*, 2nd ed., Heaps, J.W. (Ed.). AACC International, St. Paul, MN, pp. 35–38.

Mullen, M.A. and Mowery, S.V. (2000). Insect-resistant packaging. *Int. Food Hygiene* 11:13–14.

Mullen, M.A. and Pederson, J.R. (2000). Sanitation and exclusion. In *Alternatives to Pesticides in Stored Product IPM*, Subramanyam, B. and Hagstrum, D.W. (Eds.). Kluwer Academic Publishers, London, pp. 45–92.

National Research Council. (1992). *Neem: A Tree for Solving Global Problems.* National Academy Press, National Research Council, Washington, DC, 141 pp.

Navarro, S., Dias, R., Ferizli, A.G., and Mansur, F. (1998a). Insect Repelling Food Packaging Materials. Israel Patent 125,130/2 and Australia Patent 4,530,499.

Navarro, S., Ferizli, A.G., Dias, R., and Mansur, F. (1998b). A Device for Testing Resistance of Packaging Films to Penetration by Storage Insects, unpublished internal report of the ARO.

Navarro, S., Finkelman, S., Zehavi, D., Dias, R., Angel, S., Mansur, F., and Rindner, M. (2005). Pest-Impervious Packaging Material and Pest-Control Composition. U.S. Patent 20050208157-A1 and European Patent Application 04101309.5.

Newton, J. (1988). Insects and packaging: a review. *Int. Biodeterioration* 24:175–187.

Nolan, P.J. (2006). Ask the Expert Testing and Inspection, Medical Product Manufacturing News. http://www.devicelink.com/mpmn/archive/06/01/testing_expert.html.

Obeng-Ofori, D., Jembere, B., Hassanali, A., and Reichmuth, C. (2000). Effectiveness of plant oils and essential oil of *Ocimum* plant species for protection of stored grains against damage by stored product beetles. In *Stored Product Protection. Proceedings of the Seventh International Working Conference on Stored Products Protection*, Beijing, pp. 799–808.

Obeng-Ofori, D. and Reichmuth, C.H. (1997). Bioactivity of eugenol, a major component of *Ocimum suvae* (Wild.) against four species of stored product Coleoptera. *Int. J. Pest Manage.* 43:89–94.

Radwan, M.N. and Allin, G.P. (1997). Controlled-release Insect Repellent Device. U.S. Patent 5,688,509.

Rajab, M.S. and Bentley, M.D. (1988). Tetranortriterpenes from *Melia volkensii. J. Nat. Prod.* 53:840–844.

Rajab, M.S., Bentley, M.D., Alford, A.R., and Mendel, M.J. (1988). A new limonoid insect antifeedant from the fruit of *Melia volkensii. J. Nat. Prod.* 51:168.

Rajab, M.S., Bentley, M.D., Alford, A.R., Hassanali, A., and Chapya, A. (1988). A new limonoid from *Turraea robusta. Phytochemistry* 27:2353.

Rajab, M.S., Bentley, M.D., and Fort, R.C. (1988). Biomimetic formation of a nimbin class limonoid. *J. Nat. Prod.* 51:1292–1293.

Rajamannan, A.H. and Okioga, D.M. (1997). Pesticide Product Derived from the Plant *Tagetes minuta*. U.S. Patent 5,662,915.

Reeves, E.L. and Shanker, V.A. (1970). An Ingredient of Garlic Oil Is a Potent Mosquito Larvacide. *Chemical and Engineering News*, February 16, 1970.

Riudavets, J. and Salas. (2006). Evaluation and characterization of damage produced by insect pests in packaging films by insect pests. In *Proceedings of the Conference of the International Organization for Biological and Integrated Control of Noxious Animals and Plants (IOBC)*. West Palaearctic Regional Section (WPRS) (OILB SROP) Working Group on Integrated Protection of Stored Products. Gent. Belgium *Bulletin Vol.* 30(2):127–132.

Su, H.C.F., Horvat, R., and Jilani, G. (1982). Isolation, purification, and characterization of insect repellents from *Curcuma longa* L. *J. Agric. Food Chem.* 30:290–292.

Subramanyam, B. and Hagstrum, D.W. (1996). Resistance measurement and management. In *Integrated Management of Insects in Stored Products*, Subramanyam, B. and Hagstrum, D.W. (Eds.). Marcel Dekker, New York, pp. 331–397.

Tawatsin, A., Wratten, S.D., Scott, R.R., Thavara, U., and Techadamrongsin, Y. (2001). Repellency of volatile oils from plants against three mosquito vectors. *J. Vector Ecol.* 26:76–82.

Tripathi, A.K., Prajapati, N.V., Bahi, J.R., Bansal, R.P., Khanuja, S.P.S., and Kumar, S. (2002). Bioactivities of the leaf essential oil of *Curcuma longa* (var. Ch-66) on three species of stored product beetles (Coleoptera). *J. Econ. Entomol.* 95:183–189.

Watson, E. and Barson, G. (1996). A laboratory assessment of the behavioural response of *Oryzaephilus surinamensis* (L.) (Coleoptera: Silvanidae) to three insecticides and the insect repellent N,N-diethyl-m-toluamide. *J. Stored Prod. Res.* 32:59–67.

Watters, F.L. (1966). Protection of packaged food from insect infestation by the use of silica gel. *J. Econ. Entomol.* 59:146–149.

Weaver, D.K. and Subramanyam, B. (2000). Botanicals. In *Alternatives to Pesticides in Stored Product IPM*, Subramanyam, B. and Hagstrum, D.W. (Eds.). Kluwer Academic Publishers, London, pp. 303–320.

Whalon, M.E. and Malloy, G.E. (1998). Insect Repellent Coatings. U.S. Patent 5,843,215.

Wohlgemuth, R. (1979). Protection of stored foodstuffs against insect infestation by packaging. *Chem. Ind.* 5:330–334.

Xie, Y.S., Fields, P.G., and Isman, M.B. (1995). Repellency and toxicity of azadirachtin and neem concentrates to three stored-product beetles. *J. Econ. Entomol.* 88:1024–1031.

chapter eleven

RFID temperature monitoring: trends, opportunities, and challenges

Bill Roberts

Contents

11.1 Background

The arrival of radio frequency identification (RFID) technology as a logistical tool for the perishable food supply chain has presented the industry with a learning curve that is both steep and long; for supply chain visibility, there is still no magic bullet. Nonetheless, RFID tools are increasingly being understood, and leveraged, in the context of specific applications driven by well-defined value propositions. This chapter examines an emerging RFID application that will extend the range of logistical options to include monitoring and interpretation of time–temperature history of perishable goods. Here we summarize some of the perspectives gained from experience in developing this new type of solution.*

11.2 Setting the stage

The following enablers and drivers are believed to be setting the stage for a large-scale move toward RFID in transit temperature-monitoring processes:

Enablers

1. Application-specific integrated circuit (ASIC) technology
2. RFID air interface standards

Drivers

1. Avoidable spoilage losses associated with time–temperature exposure
2. RFID data capture/management infrastructure developments

Regarding the two enablers cited:

- ASIC technology can be seen as the simplification of electronic devices arising from highly economical, custom-designed silicon chips. Without ASIC chip designs, for example, pallet-level monitoring of goods such as produce might not become economically feasible.
- The second enabler, RFID standards,** represents a means to achieve cost/performance enhancements with respect to critical data capture. RFID protocols offer a uniquely cost-effective option for transferring modest amounts (up to about 100 kilobytes) of data between a tag and a reader, appropriately placing most of the performance burden on the reader so as to enable very low cost tag designs. As a wireless communication process, RFID carries the promise of automated reading without need for close reader proximity or direct line of sight. Relative to temperature-monitoring devices, however, RFID standards in place today are not adequate to drive widespread adoption. While most RFID tags do not need batteries, the RFID tags associated

* Sealed Air Corporation has introduced a line of RFID temperature-monitoring tags and accessories under the TurboTag™ trade name. See www.turbo-tag.com.
** Most RFID subject matter is contained within the ISO 18000 family of standards.

with temperature-monitoring applications probably will.* It is currently being undertaken within EPCglobal** and ISO*** to update RFID standards for improved communications with battery-powered tags. Other standards may be needed to support universality of interpretation of complex data associated with temperature-monitoring tags and other high-functionality RFID tags.****

Regarding the two drivers cited above:

- The most important driver is the continued potential for return on investment in new or improved temperature-monitoring solutions. This arises from losses that persist across the supply chain, attributable to insufficient control over time–temperature history. To the extent that better information about this history can lead to greater avoidance of these losses, it becomes the value proposition for better methods of temperature monitoring.
- In view of the enabling aspect of RFID standards, as already discussed, business decisions are made by major retailers and others to invest in networked RFID readers and associated IT systems. This emerging RFID infrastructure is not truly enabling to temperature monitoring per se: essentially all current temperature-monitoring systems operate outside of any industry-wide data capture infrastructure, and presumably could continue to do so effectively. The reason RFID infrastructure becomes a driver, then, is that the existence of this extensive new infrastructure will foster a desire among end users for data capture solutions that make effective use of it, avoiding a need for separate systems that handle temperature data.

The picture that emerges is that because of the enablers and drivers cited above, very large numbers of temperature-monitoring RFID tags will sell at very low cost, being processed through existing RFID reader infrastructure, resulting in significantly reduced spoilage across the perishables supply

* The need for batteries arises from the use of a data-logging process that must continue to operate outside the RF energy field that transiently powers passive RFID tags. The term *active* is typically applied to RFID tags that use battery power to support communications. However, the need for a battery to support data logging goes above and beyond communications. Currently, no clear naming convention differentiates between these two reasons for having a battery.

** See www.epcglobalinc.org for general information. Standards discussions within EPCglobal for battery and sensing tags have reached the Joint Requirements Group (JRG) stage, which is a predecessor to an action group for standards development.

*** See www.iso.org for general information. An ISO committee has been formed as the SC31 WG4 SG3 *ad hoc* on External Inputs to RFID Devices, chaired by Richard Rees. This committee is providing input that will ultimately influence the ISO 18000 family of standards governing all RFID tags.

**** The Smart Active Labels Consortium (SAL-C; see www. sal-c.org) has taken an early role in highlighting the problem of handling complex custom data from logging RFID tags. A submission from SAL-C to ISO formed a basis for the *ad hoc* committee currently under way within ISO (previous footnote).

chain. The remainder of this chapter will attempt to elaborate on aspects of the process of getting from today's reality to tomorrow's vision.

11.3 Existing solutions

To have a perspective on RFID temperature monitoring, one must first understand something about what existing monitoring methods offer.* Four main types of temperature-monitoring systems currently in use (time–temperature indicators [TTIs], strip chart recorders, data loggers, and transponder networks) are discussed below.

11.3.1 Time–temperature indicators (TTIs)

These are the lowest-cost alternative. They generally consist of a label with an active coating or element that gives rise to an easily visible change over time. This change occurs more rapidly at higher temperatures, and the change may be abrupt or gradual. The visual observation of a TTI label thus can serve as a quick means of estimating spoilage of a product, provided that there is a good match between the characteristics of the TTI and the goods that are monitored.

TTIs generally do not give instant warning in the case of brief spikes in temperature, and they cannot be used to "play back" the actual time–temperature history for diagnosis of the conditions that led to a positive (spoilage) reading. Furthermore, most TTIs cannot detect excessive low-temperature excursions such as unwanted freezing events, and cannot be used to precisely quantify a degree of spoilage prior to expiration. Also, it can be difficult to precisely align the TTI and product spoilage characteristics since TTIs are manufactured using a limited range of predefined formulations, so as to achieve reasonable economies of scale in their production.

By virtue of their low cost and ease of reading, TTIs have been the monitoring method of choice for delivery into the hands of the consumer as a smart feature on a package. Because of their data capture limitations,** TTIs have found limited use for cold-chain distribution quality assurance processes.

TTIs are activated either by bringing them out of very cold storage (under which conditions they would remain in their "as manufactured" state) or by some final conversion step that enables the visual change development to begin (e.g., bringing two reactants or elements together to initiate a slow diffusion process or slow chemical reaction).

* Examples of products currently available would be too numerous to cite properly in this format. It is suggested that readers will have no difficulty locating vendors of any of these products by modern Internet searching methods using the descriptions provided.

** By way of exception to this statement, see www.bioett.com for a TTI system that relies on digitization of a nonvisual (RF analog) output and, by virtue of multiple readings collected within a host system, can act as a supply chain-monitoring tool.

11.3.2 Strip chart recorders

These are a kind of visual device that, in contrast to TTIs, is engineered precisely to play back the temperature history of goods monitored. Their visual output is a strip of paper that contains a recorded graph of temperature vs. time. The fact that these devices are interpreted by visual inspection makes them quite versatile (different people may look for different things on the graph, such as upper or lower temperature excursions, peak widths, peak heights, total durations, etc.).

These devices enable users to simultaneously make a pass/fail assessment and see the history that gave rise to this status. However, it is generally not possible to perform digital manipulations on these data, e.g., a determination of a performance metric such as mean temperature or archiving of the time–temperature history on a computer (other than as a scanned image perhaps).

Strip chart recorders are initiated by a mechanical means, such as pushing a switch or pulling out a pin.

11.3.3 Data loggers

These devices are the digital answer to strip chart recorders. They are solid-state computing devices with no moving parts. They excel at precise digital analysis of a temperature history and can be programmed to deliver computed results directly to end users, typically in the form of a visual indication such as a liquid crystal display (LCD) or light-emitting diode (LED) indicator light. As their name implies, they also deliver a stream of measured time–temperature data for further off-line analysis, archival storage, and generation of printed records. Most data loggers combine a simple visual output with a wired computer communication interface. Some newer data loggers may include wireless interfaces rather than wired ones, in some cases even following RFID protocols. Some may be readable by handheld computers (PDAs).

Most data loggers specialize in limit checking: delivering visual alarms whenever predefined upper or lower temperature limits are passed during a logging session. Calculations such as mean kinetic temperature and average temperature can also be performed. However, it is generally not possible to use data loggers to execute shelf-life-based monitoring processes from integration of time–temperature history, reminiscent of TTIs, since they typically do not perform the needed calculations over the course of the logging episode. Recent developments do, however, include some shelf-life-monitoring data loggers.*

Data loggers are typically initiated by a push button or pull tab or by a command received via the computer interface (possibly wireless).

* For examples of emerging products in this area, see www.infratab.com and www.clinisense.com.

Given their rapid rate of advancement and diversification, making a distinction between new RFID tags and old data loggers can become a source of confusion. For present purposes, the following semantic formalism is suggested: If a data logger having an RFID interface utilizes ASIC technology, meaning that it has been reduced to a single silicon chip (or perhaps two chips), it may be considered a *temperature-monitoring RFID tag*. Otherwise, the term *RFID data logger* is suggested. In essence, this is a distinction based on size and price, which, although somewhat arbitrary, may be very significant in terms of commercial viability in newer applications.

11.3.4 Transponder networks

The operators of temperature-controlled warehouses, production facilities, and distribution vehicles assume responsibility for maintaining temperature control of an environment within which perishable goods are held. They have systems in place to achieve this temperature control, but these systems may not be totally reliable. Operators may therefore seek parallel systems that monitor the performance of their temperature controls and specialize in communication, alerting the appropriate persons to take corrective action in the event of undue temperature excursion, as quickly as possible. Wireless transponder networks have become a very popular choice for these kinds of applications. These are wireless device networks that rely on location-specific battery-powered temperature sensors that can actively transmit over a fairly long range (several hundred feet) to a base station that is itself connected by a wired or wireless computer networking infrastructure, bringing status information to operators or to archival data capture systems for assessments of trends in system performance.

Transponder networks, which often include buffering of time–temperature data as a safeguard against network downtime, are easy to confuse with data loggers, but they typically operate in a complementary fashion for monitoring of locations rather than goods, continually rather than for a specific time. They are ideally suited for checking system status and may or may not be suited to logistical decision making about individual shipments.* A compelling argument in favor of continued use of monitoring associated with the goods themselves may ultimately arise from the importance of shelf life estimation as a means to achieve decreased spoilage losses in the supply chain. This topic is elaborated below.

* The inference of a temperature history about goods in transit/storage from a location-monitoring system requires a system that makes the translation reliably between location-sensing data and product temperature, which can become unwieldy or unworkable if all relevant locations are not monitored at all times, or if linking of shipments to locations and time periods is not sufficiently precise.

11.4 *The case for shelf life monitoring*

A standard approach to logistics through a distribution center might be characterized as FIFO with temperature limits. FIFO refers to "first in, first out" logistics, meaning that sequencing of arrivals is preserved as sequencing of departures, maintaining as constant as possible the time spent at the distribution facility. This approach fits the methods currently available for monitoring and managing goods (note that data loggers, chart recorders, transponder networks, and perhaps some TTIs all offer a means for establishing compliance with temperature limit rules). The implicit decision with FIFO is to process all goods as though they are of equal time sensitivity, once they are seen to be in compliance with the temperature limit rules. An extension to this FIFO approach would be to define the "in" time as the time of production of goods rather than time of arrival at a distribution center. Thus, goods that had been produced first would be the first ones "out," compensating somewhat for variations in transit time to the distribution center. The ultimate extension of FIFO would be to further define the "out" time as time of arrival at a store shelf, thus accounting for variability in transit time inherent in different delivery routes from the distribution center, holding as constant as possible the total time between production and sale of perishable goods.

What underlies the FIFO approach is the assumption that all goods that are in compliance are to be treated equally in terms of their time sensitivity. This makes actual time–temperature histories irrelevant in most cases, relegating their role to the exception handling (rejection) processes associated with failed goods, and possibly some off-line trend assessments. The reality for many products is that, within the limit-compliant zone, different temperature histories can have a marked and predictable effect on rate of spoilage.*

Temperature history serves to accelerate or decelerate the loss of shelf life relative to actual time. This gives rise to the concept of FEFO logistics, meaning "first expiring, first out." The term *expiring* in place of *in* (recall FIFO), is telling, because expiration is an anticipated future event, not a past occurrence. After all, it is the anticipated occurrence of spoilage that really drives logistics on perishables, not the occurrence of particular events in the past. If we can better anticipate spoilage, we can better manage these logistics. Conversely, if we use data loggers without attempting to calculate remaining shelf life, we have not captured a large portion of the potential value of the loggers in optimizing our logistics to minimize spoilage.

* Much has been published, and is continually being better understood, regarding the correlation between time–temperature exposure and shelf life. As a lead reference see Wells, J.H. and Singh, R.P., A kinetic approach to food quality prediction using full-history time-temperature indicators, *J. Food Sci.*, 53, 1866, 1988.

11.5 The case for pallet-level monitoring

Having made a qualitative case for shelf life (re)estimation at critical logistic decision points, in support of FEFO logisitics ("first expiring, first out"), another four-letter acronym should be added to the mix: GIGO ("garbage in, garbage out"). In other words, the creation of systems for estimating which products will expire first will rely on the existence of high-quality inputs. Imagine trying to estimate remaining shelf life for an entire truckload of goods that comprise different products that have had a wide range of time–temperature histories. If done with logged data associated with the overall truckload, the result could well be worse than no information at all, on account of a lack of accuracy.

Shelf life is determined by two general information types, or inputs:

1. *Time–temperature history* after production (what is measured by the data logger)
2. *Temperature sensitivity rules* for the product (stored by the data logger or the reader system and used to translate time–temperature history into useful outputs — shelf life in this case).

Both of these inputs must be accurate enough to achieve a logistically useful output. It is not possible to comment on all sources of inaccuracy for all inputs here, but one particular source of error affecting either or both inputs greatly may be related to the level of monitoring (e.g., item level, pallet level, or truckload level). The more granular, the more possible it is to ensure accuracy of shelf life inputs and outputs.

Obviously, a compromise is called for here. Too much granularity creates excessive cost and too much data to be captured and processed. It is suggested that pallet-level monitoring is going to be the appropriate compromise for a wide range of foods. This level of monitoring can provide a reasonable uniformity level on inputs 1 and 2 together, coupled with acceptable operational cost/simplicity. Furthermore, logistical decisions made based on temperature history are likely to be applied to an entire pallet of goods. Finally, assuming that the temperature-monitoring RFID tags are ultimately readable by the same readers that would read passive (EPC-only) RFID tags on pallets, the pallet-level tagging can fit within existing automated data capture infrastructure without adding a repetitive manual data capture process.

11.6 Challenges

The discussion above suggests that a kind of step change could occur, moving from temperature-limit-based truckload monitoring using data loggers, transponder networks, or chart recorders to shelf life monitoring and FEFO logistics at the pallet level using temperature-monitoring RFID tags. As with any step change, initiation can be a big problem: getting from here to there without an unacceptable situation in between. The following discussion is focused on challenges associated with this change.

11.6.1 *Shelf life accuracy*

In the preceding section, two inputs to a shelf life calculation were identified: (1) time–temperature history and (2) temperature sensitivity rules. The discussion also suggested that moving from truckload-level monitoring to pallet-level monitoring would be advisable on the grounds of accuracy (or validity) of these inputs and their resulting outputs. This is a way to set the stage for accurate shelf life monitoring, but would not ensure it. Accuracy of input 1 would require accurate data loggers (RFID tags) judiciously placed in close association with the goods monitored. Accuracy of input 2 would require a continuation of the learning process already well under way, which serves to define the most realistic set of rules for shelf life calculation on each product monitored. Overall, what is needed is a dedication to learning what is really needed, then how to achieve it efficiently. It is to be expected that the quality of shelf life estimations will improve alongside the efficacy of the logistical processes that make use of them.

11.6.2 *Costs*

RFID tags may currently offer a significant reduction in cost relative to low-end data loggers. However, large-scale pallet-level monitoring requires at least a several-fold reduction in cost. For the short term, RFID tags will likely be used in much the same way that data loggers are used today, perhaps just on a larger number of products. Presumably, a smooth transition can occur along the price–volume curve toward pallet-level monitoring, beginning with the highest-valued or most temperature sensitive products. This gradual adoption scenario works best if there is no significant barrier to the initial adoption of the RFID tags as a drop-in substitute for data loggers and chart recorders.

11.6.3 *Forced infrastructure*

In fact, there may well be a significant barrier to adoption of RFID tags as a drop-in substitute for data loggers and chart recorders, in the form of forced infrastructure. As mentioned in the overviews above, chart recorders and data loggers are manually initiated by push buttons, pull tabs, etc. No computing infrastructure is needed, assuming the devices are appropriately preconfigured before use. Furthermore, chart recorders and data loggers, as enforcers of temperature limit rules, can be read by a manual process centered on visual examination (check a display or scan a strip chart). An RFID tag could be designed with buttons and displays, but that becomes a cost addition issue that may ultimately work against the system design simplifications and cost reductions that the RFID communication interface can offer. While data loggers with push buttons and displays can be modified to include an RFID interface, this will probably not set the stage for pallet-level monitoring due to the persistently high logger cost.

It seems likely that large-scale adoption of RFID tags for temperature monitoring will carry some trade-off of lower tag cost against higher infrastructure cost for tag initiation and data capture. The key will be in keeping the infrastructure simple and inexpensive, to make this trade-off an acceptable one at all stages in the evolution.

11.6.4 Tag/reader interoperability

The vision for pallet-level monitoring, over the long term, assumes the use of fixed RFID reader infrastructure already in place for EPC-based RFID track-and-trace applications. Recall the second driver cited at the outset. This can only happen when a single reader system can handle a variety of tags and vice versa — a situation known as interoperability. While current RFID air interface standards are sufficient to ensure that data can flow between tags and readers, this does not necessarily ensure a successful information exchange.

It is useful to consider the use of USB devices with computers. The fact that a USB device plugs in to the computer via a standardized connector interface means that, in principle, the device will operate on that computer. However, the wide variety of devices possible under the USB standard has led to a need for specific drivers, or software programs, designed to mediate the operation of a particular USB device with a particular operating system. Some of these drivers are preloaded into an operating system and some must be installed by the user. In either case, without drivers, the standardized USB plug interface delivers no successful applications. A similar situation can arise with RFID readers if they follow the air interface standards but are not able to direct the complex and customized information exchanges that will take place with data-logging tags and other higher-functionality RFID devices.

There is much work to be done on standards to help ensure interoperability, and these discussions are getting under way, as mentioned previously. Before this process can yield standards for interoperability, effective tag/reader solutions must be established in practice. Technically, the means for a reader to dynamically switch between processing of sensing and nonsensing (EPC-only) tags must be realized without a significant performance penalty to either process in isolation. This can be a real engineering challenge and a basis for differentiation in the future among RFID reader manufacturers.

It is important to realize that, over the short term, interoperability will not become a "do or die" imperative for the use of RFID tags in a temperature-monitoring system. As mentioned previously, all systems in use today rely on proprietary readers and software.

11.6.5 *Software integration*

The final challenge comes just downstream from the capture of data (including perhaps a shelf life estimate) by an RFID reader system. This challenge is the integration of shelf life data into a FEFO logistics process, or any other company-wide business process. This may require a major effort to update the business rules that drive logistics, mediated by enterprise software and middleware solutions. Until a high enough percentage of products offer shelf life values upon arrival, there would be little value in reconfiguring site-wide business rules and software systems for FEFO logistics. This could represent a chicken-and-egg problem. In fact, however, one need not necessarily have time–temperature data in order to start down this path. Any perishable product will have a standard expiration date. Therefore, the development of FEFO logistics can begin with capture of standard expiration dates from the producer (mediated by RFID tagging perhaps), then move seamlessly and gradually toward the use of recalculated expiration dates provided, upon receipt of the goods, by the temperature-monitoring tag/reader system.

11.7 *Conclusion*

The potential for RFID-enabled temperature-monitoring tags across the produce supply chain is significant, and new tag/reader technologies are rapidly developing in this direction. New types of RFID tags with sensors can give rise to pallet-level monitoring, shelf-life-based logistics, and tag reader interoperability standards that will harness existing RFID data capture infrastructure. This step change on the horizon will inevitably lead to a reduction in losses due to spoilage. While several challenges confront a transition to temperature-monitoring RFID tags and the associated applications, the expectation is that stepwise adoption will occur, with real value delivered to innovative end users at each step of the way.

chapter twelve

Selecting authentication and tracking technologies for packaging

David Phillips

Contents

12.1 Authentication

Authentication may be separated into three specific categories: overt, covert, and forensic level.

Overt security features are apparent and visible and do not require additional readers or instruments to detect them. The general public or untrained personnel can often verify them.

Covert features are concealed; they are not immediately apparent and may require a relatively simple reader or verifier (such as a UV lamp or magnifier) to locate and identify them. Covert tagging is particularly useful to customs inspectors and company investigators so long as they are kept fully informed of what they should be looking for and are trained to use the readers and testing kits.

Forensic-level features are extremely covert and are often present on a need-to-know basis only. They may comprise the addition of unique taggant material or imperceptible changes to a substrate or print that requires a very specialized reader to detect them. The quantity of taggant is so small that it cannot be detected using available analytical techniques and requires a specific test method to determine its presence. One of the main benefits of forensic-level tagging is that it can provide unequivocal evidence that a seized product is or is not genuine. This information can be very valuable in the prosecution of counterfeiting cases where the counterfeit products and packaging are very similar to the genuine article. Authentication features may be combined with a wide range of substrates, carriers, and application processes. These can include direct application into packaging films, metals, or glass as an integral part of the material or may involve addition to a material such as foil, adhesive, or ink that is subsequently applied to the substrate.

Examples of overt authentication features include:

- Optically variable coatings that change color when the viewing angle is changed
- Print and coatings directly applied to tablets and capsules
- Holographic foils attached to labels and tamper-evident seals
- Tear tapes containing printed, colored, or optically variable effects
- Thermochromic inks and coatings that decolorize on warming
- Perforations, embossings, or watermarks

Covert authentication features are particularly useful to company inspectors and enforcement agencies. They represent the second line of defense against counterfeiters and are often ignored by criminals, who may not realize they are present or find them too difficult to reproduce.

Examples of covert authentication features include:

- Microscopic particles of specific colors or colored layers
- Tiny planchettes or narrow threads containing microtext
- Labels printed with color combinations or line structures that will not resolve on a normal scanner or color copier
- Holograms containing microtext that is only readable under magnification
- Inclusions or print containing materials with characteristic spectroscopic properties that are either activated and authenticated visually or detected using a dedicated verifier

Forensic-level authentication may involve the chemical identification of a particular component within the product, such as an active ingredient or a particular binder or excipient. Alternatively, a taggant may be added during the manufacturing of the product or packaging.

Examples of forensic-level authentication features include:

- Paper and packaging containing ppm levels of a taggant that is undetectable by conventional analysis but may be extracted and identified using a dedicated test procedure
- Identifying the isotopic composition of naturally occurring materials providing information about the authenticity and source of the material
- Infrared (IR) analysis compared against a library of spectra held on a database
- Elemental analysis using x-ray fluorescence
- Additions of DNA fragments to products and packaging

12.2 *Coding*

Coding is defined as the ability to apply numbers, letters, or indicia in a structured format that can be deciphered to produce variable information. The bar coding of products has been a very successful way of identifying and tracking products. There are a number of agreed upon standards that allow universal use of bar codes across the manufacturing and retail spectrum. They have three major limitations:

- Numeric or linear bar codes can hold a relatively small amount of data. This has been partially addressed by the introduction of two-dimensional bar codes. However, these require different readers, so the conversion cost can be somewhat prohibitive.
- The information is fixed, and as a result, no further data can be added.
- The coding systems normally comprise black ink structures. Thus, they are easily reproduced by counterfeiters or alternatively located and removed by diverters.

The development of low-cost laser systems that can operate at production line speeds and write variable data has resulted in the permanent encoding of containers and packaging with lot numbers and sell-by dates. These laser systems may now be installed alongside ink jet printers and provide a way of combating criminals who remove and replace ink-jet-printed information or labels.

Identity or recognition information may be embedded directly into a design during the printing process. For example, when an image is scanned or viewed using a digital camera, the image is resolved into its components. A number of digital watermarking companies have devised a technique whereby these components are changed in a controlled manner. The applied alteration has a negligible effect on the final image but allows subsequent scanning (using the correct software) to identify the change and associate this with a particular product or batch. Because the verification process is digital, it is possible to combine it with additional actions, such as the automatic activation of a specific website or the opening of a computer file or spreadsheet.

12.3 Tamper evidence

Food products make use of a variety of tamper evidence technologies to resist the fraudulent alteration of product codes and sell-by dates. This chapter will only describe systems of tamper evidence that complement the areas of authentication and tracking.

Examples of tamper-evident security include:

- Caps and closures that fit over a bottle or container that contain tear strips that must be broken the first time the product is opened.
- Seals that fit inside the closure of a bottle or container that are broken the first time the product is used. These seals can contain a full range of overt, covert, and forensic-level security features.
- Labels that are applied over a bottle closure (e.g., Banderol) or over the closure of a carton so that they must be broken when the package is opened.
- Tear tapes containing authentication features that are used to open outer packaging and are distorted or irreparably damaged when the packaging is removed.
- Blister packs containing a variety of authentication features that cannot be reproduced when the seals are broken.

12.4 Tracking

Tracking is defined as the addition of a feature to the product or packaging that provides information about the origin of the product, its manufacture, or its authorized destination. It is difficult to differentiate between the security value of a device that allows authentication of a food product or packaging and one that provides information about its source or destination. Control of the supply chain is of paramount importance to deter unauthorized reintroduction of food products.

12.5 Combinations of authentication and tracking features

Tracking is most commonly carried out by the application of variable data, such as ink jet codes or bar codes directly to the packaging. In order to identify the removal of data and reprinting of codes, the inks may contain security taggants such as spectroscopic features or chemical characteristics that can be verified using specialist equipment or techniques.

Spectroscopic taggants can comprise inks that may be UV absorbers, emitting in the visible spectrum, or upconvertors that are irradiated by IR and emit in either the near-IR or visible spectrum. More complex spectroscopic taggants make use of particular properties of the emitting substance, such as the spectral decay rate, which is measured using bespoke detectors.

Spectroscopic taggants may be incorporated into particles, fibers, or planchettes, or security threads are embedded directly into paper or packaging. Batch tagging may also be achieved by the addition of specific biological, chemical, or particulate features to both the product itself and the packaging. Health and safety and FDA approval must be considered where appropriate.

Biological taggants may include strands of specific DNA or the addition of chemicals that use biological techniques in their verification. For example, one company has developed a technique of producing monoclonal antibodies of particular molecules. A very low concentration (parts per million) of the taggant is dispersed throughout the product or packaging. In order to verify its presence, a small sample is taken and the taggant extracted, a few drops of which are subsequently placed onto a lateral flow device. The liquid flows up the slide and comes into contact with the monoclonal antibody. If the taggant is present, then a visual indication appears on the lateral flow device.

Chemical taggants may involve indicators that are pH sensitive or are detected using precise analytical techniques such as IR spectroscopy or x-ray fluorescence. Here an indication of the concentration of the taggant may be made if the product has been tampered with or diluted.

Physical taggants were originally developed to identify explosives after detonation. One example comprises microscopic plastic particles that are only visible under magnification and that contain colored layers or colored sections. The colors represent a numeric code allowing rapid authentication without complex equipment.

Print design may also provide a vehicle whereby security can be embedded directly into packaging. Digital watermarking is a term used to describe a technology directly associated with print design. There are a number of companies that employ differing technologies but offer the end user a similar result.

Simplistically, if an image can be converted into digital format, then it is possible to manipulate the components of the digital data. This allows a characteristic to be embedded into the design that does not affect the visual appearance of the printed image but can subsequently be recognized and read using digital equipment such as a scanner. This means that a normal litho, flexo, or gravure image can have a "watermark" embedded into it without adding any additional feature or process during manufacture. This could have particular advantages for companies selling highly branded products and who do not want security features interfering with the visual appearance of the packaging.

In this format the embedded data represent a simple tag. However, if the printer was able to put down a variable image using a digital printer or as a result of multiple plate changes, then a series of tags could be embedded into identical images and the system used to batch-track the product.

12.6 Electronic tracking

Conventional bar codes require the printed code to be visible to the reader and readers are able to only distinguish one code at a time.

Radio frequency identification (RFID) was initially introduced as an asset recognition system and has been developed into smart labels that data can be electronically written to and read from using a remote transponder. These smart labels contain an integrated circuit (chip) and may be either passive (no power source) or active and include an integral battery. Reading protocols allow very rapid sequential reading of tags, in effect providing simultaneous reading of multiple tags.

The high cost of RFID tags has resulted in the development of a number of lower-cost chipless electronic tags that provide short-range remote-sensing capability and batch- or low-level unique coding. Examples of chipless tags include magnetic and electromagnetic materials in the form of fibers, threads, or patches that may be embedded directly into packaging or closures during manufacture. Most of these tags are designed to be read either at short distances (less than 1 mm), although some will still operate up to 1 m from the reader.

The Auto-ID Center at MIT (the precursor to EPC Global) has proposed a specification for a unique electronic coding system called ePC (electronic product code) that involves an electronic tag with a minimum data capacity of 64 bits that can be read remotely containing a code that is allocated when the associated product is manufactured and stays with the product throughout its life cycle. MIT has not specified a particular supplier of tags, stating that the system is open to any company whose tag meets the technical specification.

12.7 Security feature synergies

Some food products are of high value, requiring a high level of consumer confidence and integrity. Variable pricing across geographies and markets creates an opportunity for criminals to divert product from its authorized destination back to higher-cost markets and replace the diverted products with counterfeits.

There is an increasing need to maintain complete visibility throughout the manufacturing process and supply chain until the food product reaches the consumer. All of this must be achieved without alarming the consumer or undermining the integrity of the food companies or their products.

12.8 Additional information

More information can be obtained from the Product Surety Working Group Initiative. Contact www.productsurety.org.

chapter thirteen

Perceptions of consumer needs for active packaging

Cris Tina Spillett

Contents

The premise behind this chapter is: to develop a product that will be purchased, you need to find a consumer need and fill it. Your business will probably be more profitable if you first make sure that you are filling a really important need. This chapter will discuss the process for extracting consumer needs and then translating them into a product.

But first pretend you are a consumer in the market for an apple to eat. What do you want when you buy an apple (list your wants)? What is the order of your wants? When you do select an apple, how do you tell if each want (need) is met? How do you judge how well each want (need) is met? Chances are you have never thought about your apple selection process at this level of detail before, or at least not for a very long time. If you are like most consumers, you have a system for selecting apples (or anything else), but it is probably done without you even having to think about it. Your system may be different from another consumer's system, and that could be based on your having different needs or wants, or it could be due to what you know or have been trained to do, or both. So who is the consumer who will determine your list of important needs, their rank order, and how they will be judged? Figuring out the most important needs for developing a product, while it seems like it should be simple, often can be more difficult than expected. This chapter will give you a high-level view of how to figure out the following:

1. Who is your consumer really?
2. What are your consumer's needs?
3. What is the order (hierarchy) of those needs?
4. How are the needs judged?
5. How well are the needs met?

I will also list some current trends for consumers.

13.1 Who is your consumer really?

For the active packaging producer your customer (the purchaser of the active packaging) may or may not be your real consumer. Certainly take your customer's product demands to develop a product, but do not stop there. Figure out who your real consumer is and include those needs to create a truly superior product. Start with your customers and ask them who buys their product with your active packaging. Then go to their customers and ask what they do with the product and active packaging; follow the chain until you find out who is the real consumer of the product and the active packaging. Chances are if you please the end user, everyone is happy. Here is a (highly) hypothetical scenario. You are the developer of active packaging for bananas and your customer is a packing company in Ecuador. The packing company sells to a banana distributor, who then sells to several food distributors. One food distributor sells to the largest grocery chain in the nation and another high-end grocery store. The end consumers at the large grocery chain are looking for the best deal they can get. The end consumers at the high-end grocery store are also looking for the best deal they can get. However, pleasing these two groups of end consumers is different. The large grocery chain shoppers' needs are nutrition with a secondary need for convenience. They judge nutrition by taste and they select based on looks. They buy large bunches of bananas at a time so they do not have to go shopping more than once a week. The exclusive grocery shoppers' needs are nutrition with a secondary need for fitting in with their healthy lifestyle. They judge nutrition by taste and they select based on looks. They buy one or two bananas at a time since they shop every day or every other day to ensure they have the freshest produce on hand at all times. We will discuss the product development implications for this case study in the banana example below.

13.2 What are your consumers' needs?

Once you have found the end consumer group or consumer segment through interviewing your customers, their customers, and so on, you will want to set up interviews with 8 to 15 end consumers and ask them questions while they are using the product with the active packaging. In some cases, the active packaging has already served its purpose and the consumer is unaware that it was even used; in that case, you will just discuss the product.

Simplistically, interview your consumers (8 to 15 people) with probing questions. Basically, ask a lot of why, what, or how questions. Do not ask leading questions.

For example, in the hypothetical banana case:

> *Large-chain customer at the banana stand. Observation woman is selecting a large bunch of green bananas.*
>
> Q: Why did you select that bunch of bananas?
>
> A: I need a lot to last a week.
>
> Q: Why did you pick that particular bunch instead of another?
>
> A: Well those other bunches have brown spots already. I want these to ripen without brown spots.
>
> Q: Why is that?
>
> A: Well if they have brown spots or bruises when they're green, they don't taste good when they ripen.
>
> Q: Why don't you buy bananas that are already ripened?
>
> A: I guess I want them to ripen perfectly and not too quickly so I can put them in the fridge when they are the perfect ripeness.
>
> *And so on.*

You would probably also want to interview some of these people in their kitchen as they are unloading their bananas and some when they are about to use their bananas. You are looking to get as many needs as possible rather than reinforce that everyone wants the same thing, that is, you are going for breadth rather than depth at this stage.

In more rigorous studies you would video or at least record the voices and transcribe them and then gather the needs (wants) in consumer language. It is important to keep the language in the form that the consumer would use, not the more technical terms that you might use. However, you will need to understand what they mean so you can translate, if necessary, for your product developers later. In the ideal case, the product developers would do this research so they understand for themselves the nuances the consumers are looking for. Ensure that the needs are all in the positive format for the next phase. For the above example, here are some of the extracted needs: large bunch, brown spot-free, bruise-free, taste good, perfectly ripened.

13.3 What is the order (hierarchy) of those needs?

Once you have interviewed your group of consumers and extracted the needs, the next phase is to arrange them into a decision tree. In the simplest scenario you could get the same group of consumers back together to group the needs you have extracted from their interviews. However, any consumers of the same type should prioritize similarly for you. This is done as a group exercise. You would not use the high-end grocery consumers to prioritize the large-chain consumer needs since there is the possibility that they may prioritize them differently. However, you probably could lump the needs extracted from either group together for a prioritization exercise. In this exercise you would have each individual need on a card and then tell the consumers to group the needs that belong together into piles and either create a header that describes the pile or use one of the needs as the descriptor.

In a simple case, you would then have the consumers rank the needs at the header level from most important to least with weighting (dividing 100 points among them is one good way). Then have them rank within each of the header groups. For more rigorous studies you would do a large quantitative survey among each of the consumer segments, but these studies can run into the hundreds of thousands of dollars.

At the end of this phase you should have a prioritized list of consumer needs.

13.4 How are the needs judged?

Make sure in your interviews that you understand how your consumers judge their needs. Chances are they are judged by one of the five senses. In the case of food products, sight and smell are used most often, but for certain items, touch or even sound may be weighted more heavily. The ultimate test, of course, is going to be the experience they have when they taste the product, but they usually do not get the chance to experience that until after they have already purchased it. (Again, I am assuming that taste is the ultimate number one need, but this may not always be the case.)

It is your job to understand what tests they are using to judge how a product will taste or whatever their number one need is. These tests may or may not correlate. Your understanding of this will guide product development. For instance, in our hypothetical banana case study, say you could prevent all brown spots from occurring in bananas with your active packaging. However, the brown spots have nothing to do with taste in blind taste tests. You might still want to include a brown spot preventative if in consumers' minds brown spot-free bananas are the biggest indicator of great taste. This can give you and your customers a competitive advantage. On the other hand, if you have something that prevents brown spots and in consumers' minds that has nothing to do with taste or any top needs, then why include something that will cost money if it is for free or a cost savings you may reconsider.

Once you thoroughly understand how the consumer needs are judged, go through your list of consumer needs and list all the technologies that will deliver against each need or order of ability to deliver against that need. You probably will want to stop with a short list of consumer needs or the list can become unwieldy. You may find that some technologies will meet multiple needs. Now prioritize the list of technologies that will best impact the top consumer needs.

There may be trade-offs that are required between technologies. This is where technical and consumer judgment comes into play. You probably will need to check in with your consumers to verify your assumptions. For example, if the number one need is flavor and the number two need is prolonging freshness and you have a technology that will prolong freshness by five times but give a slight off-flavor, is that acceptable? Eventually you can set targets for meeting your consumers' needs and acceptable limits and understand how your technologies correlate with that.

13.5 How well are the needs met?

It is also desired to understand how well consumer needs are met with your current product offering and any competitors' offerings as well. If your product is already meeting the top two needs, you may want to focus on making an improvement on the third need (without sacrificing anything on the first two needs). Make sure you are making a consumer-noticeable difference with any product improvement; otherwise, your efforts are wasted. For instance, in our hypothetical case, if you were able to extend the freshness of bananas as indicated by yellow color (and that was the number one need) and the consumer noticed that they stayed fresher longer, then you would have a winner.

13.6 Consumer trends and facts

The following is a list of facts and consumer trends for food shoppers that may also influence your product development:

- ~75% are female.
- ~45% know what they are looking for.
- ~55% care about what they serve their family.
- Due to an upward trend in bulk/club shopping, consumers store more food at home than a few years ago.
- 66% of dinners are prepared at home.

- Consumers have more educated palates and desire great taste more than ever.
- Consumers are more willing to pay for convenience now more than they were a decade ago. Convenience = avoiding frustration. What are major frustrations?
    - Throwing out food
        - This makes consumers feel guilty, wasteful, and angry.
        - ~90% of households discard produce over a 2-week period.
        - ~25% of all stored food is thrown out.
    - Lost time spent in food management
    - Worry over uncertainty of food safety

Do you have intelligent packaging solutions that address the trends for better taste and convenience? Guidance with these trends and your dive into understanding your consumers will help you develop on-target products without overengineering them.

chapter fourteen

Needs for active packaging in developing countries

Elhadi M. Yahia

Contents

14.1 Introduction

There is no clear definition of the term *developing countries* (DCs), which encompasses a wide range of countries with diverse challenges. The food industry in DCs is faced with tremendous growth, but with minimum or no regulatory action. The application of refrigeration, packing and packaging, transportation, modified (MA) and controlled (CA) atmospheres has improved significantly over the last two to three decades.[36] Rising income and widespread urbanization have been the most important determinants behind the increase in food, especially fresh fruits and vegetables consumption, and have also increased the importance of postharvest handling of fresh commodities.[26,31,32,37] However, the establishment of an adequate cold chain remains the major obstacle facing the handling of foods in general and fresh commodities in particular.[4,36,37] More improvements in other handling techniques are also still needed.[38] Food packaging is improving, but remains one of the most important causes of losses of perishable foods. MA and CA are used in different applications, such as for marine transport of several commodities and for storage of apples and pears. Modified atmosphere packaging (MAP) for fruit and vegetable salads is used in very few DCs, but no intelligent packaging, involving the use of time–temperature indicators, gas modification, controlled seal leakage, and altered package O_2 or CO_2 permeability by sensing and responding to changes in temperature, is currently used. The challenges facing the application of safe MAP and CAP for perishable foods in DCs are enormous. A major concern in most DCs is the ultimate safety of packed products, and thus the consumers.[16,27,34] There is no regulatory activity yet related to the use of MA and CA. The application of MAP and CAP (controlled atmosphere packaging) in DCs is feasible and needed; however, the challenge of implementing them will depend on the improvement of the cold chain, the handling of food in general, and the hygienic status, and the understanding of the multidisciplinary nature of the technology.[37]

Intelligent packaging is an emerging technology that uses the communication function of the package to facilitate decision making to achieve the benefits of enhanced food quality and safety.[6,10,21,40] The latest advances in smart package devices, including bar code labels, radio frequency identification tags, time–temperature indicators, gas concentration indicators, and biosensors, have contributed very significantly to developing adequate packaging systems. As it is the trend in developed countries, consumers' interest in functional foods has been increasing in some DCs and will continue to increase.[20] However, the situations in some other DCs are different, as food resources are limited, and thus to help solve such complex problems, not only new technologies but also conventional technologies have to be mobilized.[23,38]

14.2 Developing countries

There is no clear definition of the term *developing countries* (DCs). This term has been used with some other terms, such as *underdeveloped countries, less developed countries,* and *third world countries,* and encompasses a wide range of countries with diverse challenges. This term has been used for many years, without very much knowledge of its significance and without clear agreement between the people and the organizations that use it. In fact, the definition/classification has always been controversial. The criteria used by the United Nations to define and classify development in member countries include:

1. Less developed/developing countries (LDCs). Indicates a country showing (1) a poverty level of income, (2) a high rate of population increase, (3) a substantial portion of its workers employed in agriculture, (4) a low proportion of adult literacy, (5) high unemployment, and (6) a significant reliance on a few items for export. According to this definition, diverse countries such as India and Pakistan, Egypt and Yemen, Mexico and the Dominican Republic, Chile and Jamaica, Indonesia and the Philippines, are all classified in the same category.
2. Least developed countries. Some 36 of the world's poorest countries are considered by the UN to be the least developed of the less developed countries. Most of them are small in terms of area and population, and some are landlocked or small island countries. They are generally characterized by low per capita income, literacy levels, and medical standards, subsistence agriculture, and a lack of exploitable minerals and competitive industries. Many suffer from aridity, floods, hurricanes, and excessive animal and plant pests, and most are situated in the zone 10 and 30° north latitude. These countries have little prospect of rapid economic development in the foreseeable future and are likely to remain heavily dependent on official development assistance for many years. Most are in Africa, but a few, such as Bangladesh, Afghanistan, Laos, and Nepal, are in Asia. Haiti is the only country in the Western hemisphere classified by the UN as least developed.
3. Developed countries. A country that is technologically advanced, highly urbanized and wealthy, and has generally evolved through both economic and demographic transitions, although the complete list of these countries is not fully known.
4. Organization for Economic Cooperation and Development (OECD) member countries. Twenty countries signed the constitution of the Organization for Economic Cooperation and Development on December 14, 1960. Since then, 10 more countries have become members of the organization.

The World Bank, in its World Development Report (2004), classified the world into three groups:

1. Low-income countries, with gross national per capita income of up to U.S.$735.
2. Intermediate-income countries, with more than U.S.$735 and up to U.S.$9076. These are further classified into intermediate–low income, between U.S.$735 and 2935, and intermediate–high income, between U.S.$2936 and 9076.

3. High-income countries, with higher than U.S.$9076.

This classification system does not relate to development, and therefore it is difficult, for example, to define the developmental status of a country such as the United Arab Emirates, with a very high per capita income, according to this system.

Another class of countries that has emerged recently is "countries with economics in transition," including countries that show dynamic economic growth and play an important role in the manufacturing and financial sectors of the world.

The International Association of Lawyers established a list of 118 countries considered to be developing countries.

Therefore, different classification systems are used, and international organizations do not agree on a single uniform classification.

In these days there are many types of DCs, and therefore the problems in these countries are very diverse and different. The postharvest status and problems in these countries are also very diverse.[36]

In the absence of an adequate definition/classification of DCs and due to the overwhelming use of the term, this chapter will follow suit, hoping that serious attempts should start redefining this term.

The agricultural sector in DCs is undergoing rapid changes as a consequence of both technological progress and economic forces that call for an increased market focus, competitiveness, and higher productivity.[3,11,15]

Land in crop production in the DCs, excluding China, may expand from the 760 million ha in 1988–1990 to 850 million ha in 2010, an increase of 90 million ha, or about 5% of the 1.8 billion ha of the world's still uncultivated land with rain-fed crop potential.[3,33] The growth rate of world agricultural production was 3% a year in the 1960s, 2.3% in the 1970s, 2% in 1980–1992, now is 1.8% and will continue to drop in the period to the year 2010.[33]

In early 2003, the world population reached approximately 6.3 billion inhabitants; 5.1 billion inhabitants, i.e., 81% of the global population, live in DCs. It was estimated that 35% of the population of DCs live in cities, where food supply problems are becoming increasingly acute and where the development of refrigeration is essential. In 2020, the world population will most likely reach 7.6 billion, an increase of 31% over the mid-1996 population of 5.8 billion. Approximately 98% of the projected population growth over this period will take place in DCs. It has also been estimated that between the years 1995 and 2020 the developing world's urban population will double. There is also an increase in the middle-class sector in DCs (for example, there are 290 million in China, 91 million in India, and 58 million in Brazil), which contributes very significantly to more expenditure on more expensive food items such as fruits and vegetables.[28] This overall increase in population, the urban population in particular, poses great challenges to food systems. Intensification of agriculture and animal husbandry, more efficient food handling, processing, and distribution systems, and introduction of newer technologies, including packaging, will have to be exploited to increase food availability to meet the needs of growing populations. Some of these practices and technologies may also pose potential problems of food safety and nutritional quality and call for special attention in order to ensure consumer protection.

Rapid urbanization has led urban services to be stretched beyond their limits. This scenario further stresses food distribution systems as greatly increased quantities of food must be transported from rural to urban locations in an environment that is not conducive to hygiene and sanitation. Population and income growth are raising demand for food in DCs, much of it from import.[37]

14.3 The food industry in DCs

The food industry in DCs, especially that of the fresh horticultural crops industry, is faced with tremendous growth and market demand, both for the local markets and for export, but with minimal or no regulatory action for enforcing minimum quality standards and implementation of food safety assurance practices.[25,27,36,37] This situation, especially for the local markets, has dictated that the whole responsibility lies on the food manufacturers and handlers to deliver safe and wholesome food to the consumers.

Rising income in several DCs, and the impact on levels of food consumption, has been one of the most important determinants behind the increase of fresh fruit and vegetable consumption in DCs. The rate of income in the so-called low-income countries has increased 221% between 1960 and 1998, which surpassed that of higher-income countries.[30] With the exception of roots and tubers, food supplies substantially increased in middle-income countries. Per capita availability of fresh fruits and vegetables in low-income countries remains far below that of middle- and high-income countries.

Widespread growth in urbanization has also increased food consumption in general and fresh fruits and vegetables in particular, and has increased the importance of postharvest handling of fresh commodities in DCs.

The organic food market is growing in most developed countries and also in some DCs. Estimates indicate that annual sales growth rates will range between 5 and 40% over the next few years, depending on the market. Consumers, including those in DCs, are increasingly aware of health and environmental issues. DCs produce a wide range of organic products and many are exporting them successfully. World markets for organic foods will offer DCs good export opportunities; however, the success of DCs in providing these types of foods to developed, as well as to internal, markets will depend upon overcoming constraints, including certification, market intelligence, and technical know-how, especially packaging technology.

DCs must ensure food quality and safety, starting from agricultural production through storage, processing, and distribution using good manufacturing and proper food handling practices. Governments, the food industry, and consumers have to play their roles effectively and, in a concerted manner, ensure that the quality and safety of food supplies are not compromised and losses in the food system are minimized.

In all countries the food industry bears the responsibility of meeting food quality and safety regulatory requirements. The food industry encompasses the activities of small-scale farmers through medium- to large-scale producers, food storage, processing, and wholesale and retail marketing. Food preservation, processing, and packaging systems can be minimal or highly sophisticated, but ensuring food quality and safety in all situations should be a constant. Industry must play its role in ensuring food quality and safety through the application of quality assurance and risk-based food safety systems utilizing current scientific knowledge. The implementation of such controls throughout production, handling, processing, and marketing leads to improved food quality and safety, increased competitiveness, and reduction in cost of production and wastage. Through national food control systems, governments should provide a supporting infrastructure and assume an advisory and regulatory role.

Some developing countries have adopted and implemented comprehensive national food quality and safety standards based on the international recommended Codex Alimentarius Commission standards guidelines and codes of practice. These countries have benefited from higher levels of investment in the food sector, better acceptance by consumers of higher-quality and safer domestically produced raw and processed foods, and greatly improved access to foreign markets for their food exports. Meeting these Codex-based standards has also increased efficiency in food production, processing, and distribution, promoted a lower-cost domestic supply of good-quality and safe foods, reduced food loss problems, and greatly increased export earnings. A training package for strengthening national food safety systems through enhanced participation in the Codex process is being prepared to help developing countries better understand the process, thus enhancing the quality and usefulness of their participation. It provides information that can be used by countries to develop training programs to suit their specific needs, and thus enhance their capability to participate in the work of the Codex Alimentarius Commission. The package has already been field-tested in a range of countries/regions, including Fiji, Suriname, and Tanzania (with participation of trainees from Kenya and Uganda). The production of the training package is supported by the governments of Canada and Switzerland.

Most DCs have a warm (equatorial, tropical, or Mediterranean) climate, which means that refrigeration is very important. The application of refrigeration, packing and packaging innovations, transportation technology (especially marine transport), and modified and controlled atmospheres has improved significantly in some DCs over the last two to three decades, and therefore their high-value fresh horticultural crops are exported to different markets.

However, food systems in DCs are not always as well organized and developed as they are in the industrialized world. Moreover, problems of growing population, urbanization, lack of resources to deal with pre- and postharvest losses in food, and environmental and food hygiene mean that food systems in DCs continue to be stressed, adversely affecting quality and safety of food supplies. People in some regions of some DCs are therefore exposed to a wide range of potential food quality and safety risks.

Access by DCs to food export markets in general, in the industrialized world in particular, is increasing, and its growth will depend on their capacity to meet the regulatory requirements of importing countries. For most DCs, agriculture lies at the center of their economies and food exports are a major source of foreign exchange and income generation for rural and urban workers in agriculture and agroindustrial sectors. The long-term solution for DCs to sustain a demand for their products in world markets lies in building up the trust and confidence of importers in the quality and safety of their food supply systems. This requires improvement within national food control systems and within industry food quality and safety programs. Such efforts will greatly help in increasing the relatively small share of DCs in the international food trade.

Food systems are complex. In the case of DCs, they are highly fragmented and predominated by small-scale producers and handlers. This has its own socioeconomic advantages. However, as large quantities of food pass through a multitude of food handlers and middlemen, extending the food production, processing, storage, and distribution chain, control is more difficult and there is a greater risk of exposing food to contamination or adulteration. Lack of resources and infrastructure for postharvest handling, processing, and storage can lead to losses of food quality and quantity, and greater potential for microbial contamination.

On the whole, DCs focus more on increasing agricultural production than on preserving these agricultural products. This explains why postproduction losses are so high, whether in the field or during transport, storage, and distribution, as shown in Table 14.1.[18,36] In several DCs, there is still a debate over food preservation. Some think that traditional methods (salting, curing, storage in the ground, etc.) are most appropriate, as they are inexpensive. Others think that populations in DCs have the right to the food preservation technology that has been tried and tested in developed countries, particularly refrigeration technology.

Table 14.1 Estimated Postharvest Losses of Fresh Horticultural Food Crops in Developing Countries

Location	Range (%)	Mean (%)
From production to retail sites	5–50	22
At retail, food service, and consumer sites	2–20	10
Cumulative total	7–70	32

Source: Kader, A.A., *Acta Hort.*, 682, 2169–2175, 2004.

There are many export opportunities for food products that are of a better quality and in great demand in developed countries. The requirements for exported products include compliance with standards and regulations, traceability, hygiene, etc. The setting up of appropriate cold chains for exported foodstuffs promotes the adoption of similar systems for local trade.

14.4 Some of the characteristics of DCs regarding food and food consumption

There are some major distinctions between DCs and developed countries regarding food handling and consumption, which are important to understand, especially when technologies are suggested for implementation in DCs. Some of these distinctions include:

1. People in low-income (DCs) countries spend on average much more (47%) of their total income on food than high-income (developed) countries (13%).
2. Consumers shop for food much more frequently (often daily) in DCs than in developed countries (often weekly).
3. Refrigerated food and awareness for the need for refrigeration, although they have improved significantly, are still not very common in DCs. In fact, refrigerated foods in some DCs may indicate lack of freshness, and thus can be considered of low quality for some consumers in some DCs.

14.5 Postharvest food handling in DCs

Estimates of postharvest (PH) losses in DCs are two to three times those of the developed countries.[18,36] Both quantitative and qualitative losses take place in horticultural crops between harvest and consumption. The reduction of postharvest losses involves: (1) understanding the biological and environmental factors involved in deterioration and (2) use of postharvest technology procedures that can control senescence and maintain the best possible quality.[18] Qualitative losses are much more difficult to assess than quantitative losses, especially because standards for quality and consumers' purchasing power in DCs are different from those in developed countries. For example, elimination of defects for a given commodity before marketing is much less rigorous in DCs than in developed countries. However, this is not necessarily bad, because appearance quality is somewhat overemphasized in developed countries. A fruit or vegetable that is misshapen or has some blemishes may be as tasty and nutritious as one that looks perfect. Minimizing postharvest losses of horticultural perishables is a very effective way of reducing the area needed for production or increasing food availability.

Postharvest technologies refer to the stabilization and storage of unprocessed or minimally processed foods from the time of harvest until final preparation for human consumption.[5] There is a special emphasis on seasonal crops and simple, labor-intensive, capital-sparing technologies suitable for DCs where food spoilage rates are high and malnutrition is prevalent. In fresh horticultural crops the main spoilage vectors are bruising, rotting, senescence, and wilting. Bruising is avoided by careful handling and use of shock-resistant packaging. Rotting is controlled by good housekeeping, gentle handling to avoid breaking the skin, cool storage, and use of preservatives. Senescence is retarded by refrigerated handling or modified/controlled atmospheres. Wilting is controlled by high humidity and cold storage.

There has been significant attention to postharvest (PH) handling and preservation of food in DCs, especially in the last few years, although there are indications[29] of less emphasis on PH, at least in some parts of the developed world. The attention to PH in DCs is due to the persistent problem of losses of food in quantity and in quality suffered in these countries, increasing demand for better and healthy products in the local markets, need for export in almost all DCs, demanding importing markets and consumers, demand for quality and healthy products, etc. In addition, there are three major global trends currently affecting all sorts of life, including PH, which also influence the attention given to PH in DCs:

1. Globalization. Global and regional integration and trade liberalization facilitated food flow to and from DCs, and therefore increased the importance of PH.
2. Urbanization, where an increasing percentage of the population is concentrated in urban centers, with food habits shifting to convenience and more diversified diets and more emphasis on consumption of fruits and vegetables, and where a great amount of food for these urban centers in the developed and developing world is coming from DCs. In the future, urbanization will primarily affect DCs. In 1960, developed countries accounted for about one third of the global urban population, but by 1988 they accounted for only about one fourth of the global urban population of about 3 billion. The urban population in DCs had doubled to nearly 4 billion by 2002, and thus the effect of urbanization on future food consumption changes will be most evident among DCs.
3. Food safety. Although food safety concern in DCs is still very limited compared to that in developed countries, the effect is spelling to these countries due to different reasons, particularly the concern of consumers in imported countries. Demand for food in DCs is changing, and the interest in the consumption of fresh fruits and vegetables is growing, although not as much as in several developed countries. Consumer demand for improved food quality is improving in several DCs and has led private and a few public sectors to develop and implement mandatory and voluntary quality control, management, and assurance schemes. Quality assurance schemes and standards for production, processing, and handling of food need to be

developed and improved in most DCs. Safety from risks of disease-causing organisms still needs major attention in DCs. Detection technology and mitigation methodology are still rare in most DCs. Food safety standards still need to be developed in most DCs and applied adequately in almost all other countries.

Postharvest technology and handling of food in DCs has improved significantly during the last few decades, especially due to increasing interest in export, increasing competition from external markets, and more demand and awareness from local consumers. However, PH losses in quality and quantity are still staggering.[18]

Temperature control and fluctuation remain the major problems facing the handling of foods in general and fresh horticultural crops in particular in DCs. The cold chain is very weak in almost all of DCs. Other technical problems facing the handling of fresh horticultural crops in DCs are numerous and include packaging, transport, markets and marketing, quality control and assurance, hygiene, and food safety.[36]

14.6 The fresh-cut (minimally processed) industry in DCs

In DCs sales of fresh-cut produce (minimally processed), such as cut lettuce and cut carrots, have increased in the past few years.[2] A large percentage of these products are packaged such that the levels of relative humidity (RH), O_2, and CO_2 are significantly modified relative to their levels in ambient air. Although MAP can potentially extend shelf life, it cannot be expected to overcome the negative effects of temperature abuse. However, due to variation in respiration and permeation, limitations imposed by available polymer films, and the chance of exposure to elevated temperatures during handling and distribution, there is no assurance that atmospheres attained will be consistent among packages. Due to lack of adequate control, O_2 concentrations could fall below the safe level in the current MAP. Risks include not only the loss of product quality through fermentative metabolism, but also the growth of potential human pathogens that can thrive under anaerobic conditions. The gas levels generated in MAP are a function of film permeability chosen and the respiratory behavior and gas exchange characteristics of the enclosed living, breathing plant product. The traditional objective of MAP has been to select films with permeabilities that allow gas levels favorable for shelf life extension. Commercially, claims of MAP systems are plentiful, although the specific nature of the modified atmosphere is not generally mentioned. Several commercial MAP designs are now available, although in practice, gases may not achieve expected levels and there are few public data available for verification.

The accumulation of ethanol indicates that the package designs are not capable of consistently maintaining aerobic conditions. The presence of fermented odors and flavors is easily detected in these packages. This suggests that the use of MAP may be limited to those products that can tolerate a degree of fermentation without significant loss of quality. Products such as cut broccoli or cauliflower that produce offensive off-odors under low O_2 levels would presumably not be suitable if there was a risk of anaerobic conditions. In addition, there remains the question of whether these low O_2 levels increase health risks.

The fresh-cut industry in DCs has seen some development, although not as much as in developed markets. This development is especially clear for those horticultural products that are intended for export to developed markets.[11]

14.7 *Food packaging in DCs with emphasis on horticultural crops*

As in developed countries, changes in the way foods are produced, distributed, stored, and retailed, reflecting the continuing increase in consumer demands in many DCs for improved quality and extended shelf life for packaged food, are placing ever-greater demands on the performance of food packaging. Many consumers in DCs, especially in urban centers, want to be assured that the packaging is fulfilling its function of protecting the integrity, quality, freshness, and safety of foods. To provide this assurance and help improve the performance of the packaging, innovative active and intelligent packaging concepts are also important for consumers. However, in almost all DCs, the development and application of active and intelligent packaging systems have been very limited thus far. The main reasons are lack of knowledge of the consumers, and even in the industry, on application and efficacy of such systems and the economic and environmental impact they may have.

Global packaged food sales reached over U.S. $1254 billion in 2003, up by 4.1% in current terms from 2002. This increase, especially in North American and Western European markets, was driven by increased demand for both healthy and easy-to-prepare packaged food, and premium and indulgent brands in other food sectors, such as dairy and bakery products. Value sales growth was also evident in regions such as Asia-Pacific and Latin America (Figure 14.1). The key trends driving growth in developed packaged food markets are convenience, functionality, and indulgence. The most successful new product developments reflect increasing consumer demand for convenient, portable, easy-to-prepare meal solutions, which ease the effects of faster-paced, urban-based living and the consequent constraints on food preparation and shopping time that affect consumers. In spite of the prevailing trend toward healthier products, the consumption of premium indulgent packaged food has experienced a significant rise over the last years. Underpinning the trend toward luxury packaged food has been rising, disposable incomes in developed as well as in some developing markets, with consumer tastes growing in sophistication due to wider international travel, dining

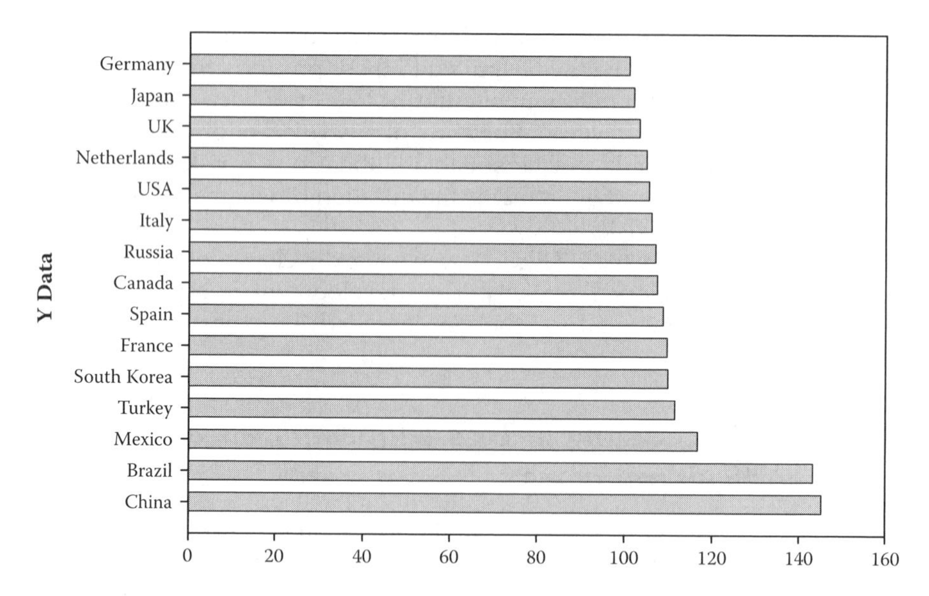

Figure 14.1 Total packaged food value sales growth index in some developed and developing markets (1998/2003) in percent constant local currency growth.

out, and lifestyle media programming. The increased availability of packaged food in developing markets derives from market liberalization, improved distribution networks, and global manufacturer consolidation. Urbanization in many developing countries is driven by the creation of new jobs in manufacturing and services in large metropolitan centers. In such markets, the younger urban generation prefers to shop in comfort in modern large-surface retail outlets for meat, fish, fruit, and vegetables, instead of at the traditional wet markets.

Multinational food manufacturers have increasingly sought strategic alliances with local producers and distributors, and therefore the consolidation among national and international packaged food producers contributed significantly to growth in emerging markets such as China, Russia, and Brazil, and further consolidation of multinational producers will result in greater penetration of these developing markets.

In most developing markets the focus will also be on the continued development of manufacturing and retail networks, increasing the necessity for packaged food to vast regional populations. As it has happened in developed markets, packaging will certainly play an important role in stimulating growth of the food sector, increasing brand recognition and technological innovations. However, the development of new food packaging technologies in DCs is still quite a challenge. Generally, there is still no clear distinguishing (even among many people in the industry) among (1) passive packaging, (2) active and intelligent packaging, and (3) new techniques for the design and presentation of packages. Active packaging reacts to the internal atmosphere by initiating a change that is beneficial to the product shelf life. A

Figure 14.2 Use of modified atmosphere for banana in the Dominican Republic.

well-known example is the use of oxygen scavengers. Intelligent packaging incorporates a sensor that reacts to the external atmosphere, thus monitoring the circumstances under which the product was stored. In other words, the respective labels inform the user about the freshness of the product.

14.8 The use of modified and controlled atmospheres in DCs

Modified atmospheres (MAs) and controlled atmospheres (CAs) are used in different forms in DCs (Figure 14.2, Figure 14.3, and Figure 14.4).[35,38] MAs and CAs are used in several countries for marine transport of several commodities, such as cantaloupe, banana, mango, avocado, pome, and stone

Figure 14.3 Application of modified atmosphere for marine transport of some fruits in Mexico.

fruits.[35,36,38] CA is used for the storage of apples, pears, and kiwifruit in some DCs such as Mexico, Chile, Argentina, Uruguay, Brazil, and Jordan.[36] CA storage is used fairly adequately in most DCs, but MAs and CAs for transport are still facing some difficulties, mainly originating from the lack of a system of standardization by service providers, not employing adequately trained technicians, and the lack of research to establish optimum requirements for several of the crops grown and exported from DCs.

14.9 The use of modified (MAP) and controlled (CAP) atmosphere packaging in DCs

Some MAP for fruits and vegetables, especially for vegetable and fruit salads, is used in a few DCs, such as Mexico. The limited use of MAP for foods in DCs so far has been promoted by major supermarket chains (Figure 14.5, Figure 14.6). Supermarkets are increasing very significantly in several DCs; however, there are several DCs, such as India, where supermarkets have not been established in a large scale.

Figure 14.4 Controlled atmosphere storage of apples in Mexico.

The challenges facing the application of safe MAP and CAP for fresh perishables in DCs are enormous. The use of MAP in DCs can be limited for several reasons:

1. The unavailability of appropriate films that can provide safe atmospheres, especially under abusive temperature conditions.[19] Packages that provide safe atmospheres at one temperature may result in anaerobic conditions at higher temperatures.[9]
2. The expense of using MAP technology, on the basis of both film cost and the modification of packing line systems.[19]
3. Problems associated with maintaining package integrity during handling (transport, storage, marketing, etc.). Package materials must be flexible and easy to use but strong enough to survive handling abuse, especially common in DCs.

Figure 14.5 Experimental use of modified atmosphere packaging for vegetables export from the highlands of Guatemala.

Figure 14.6 Some films used for modified atmosphere packaging of different foods in Colombia.

Films currently used in MAP in DCs are mostly polyethylene and polyvinyl chloride (PVC). Among the most common plastics, PVC creates and emits chlorinated organic chemicals. PVC could be replaced by chlorine-free plastics, which is commercially available, although prices may increase. PVC is the most hazardous packaging material produced in Mexico.[1] Emission of vinyl chloride, the key carcinogen released in PVC manufacturing, is high in Mexico. Some films developed for the use of MAP are sourced from some foreign companies. Very little or practically no work has been done on the characterization of the properties of the films used, and the data are taken from the description of the sourcing companies.[12–14,39] The selection and use of plastic films for packaging of fresh products in DCs is commonly done on the basis of their prices. The most commonly used material is low-density polyethylene (LDPE), because of its low cost. Technological developments in this field in DCs are very few.[39] Gravure-printed color indicators on common nylon/polyethylene (Ny/PE) films were newly developed for kimchi (fermented vegetable products in Korea) packaging to examine the degree of fermentation of such products nondestructively during distribution and sale.[17] A printing ink consisting of Ca (OH)2 as a CO_2 absorbent, bromocresol purple (BP) or methyl red (MR) as a chemical dye, and a mixture of polyurethane and polyester dissolved in organic solvents as a binding medium was applied on Ny films by the gravure process and then laminated with PE films to form the printed indicator. During storage at 0, 10, and 20°C, kimchi fermentation was assessed for titratable acidity (TA) and the color of the indicators was determined as Hunter values and expressed as total color difference (TCD). Although the TCD value of the BP type indicator was slightly less than that of the MR type, color changes in both types of indicators correlated well to TA values of kimchi. The results suggest that gravure-printed color indicators have great potential as an intelligent packaging system for monitoring the degree of fermentation of commercial kimchi products.

It has long been recognized that increases in temperature during shipping, handling, or retailing MA packages could cause a decrease in package O_2 levels because respiration tends to increase more than permeation through polymers. Using trucks that are not precooled, improper handling at transfer points, and retail display under nonrefrigerated conditions contribute to this problem.

For MA systems designed to maintain optimum O_2, some of the most permeable films have desired properties for produce with low (apple, carrot, celery, cabbage, and green pepper) or medium (cauliflower) respiration rates. Only combinations of polymeric and perforated films could potentially provide fluxes and selectivity for commodities with high respiration rates. When exposed to temperatures above optimum, MA packaging could cause depletion of O_2 and accumulation of CO_2, resulting from a greater increase in respiration than in permeability of membranes. This could result in a transient anoxic atmosphere that could damage the fruit or vegetable.

The use of MAP and CAP for fresh perishable foods in DCs should put consumer safety first and freshness and shelf life prolongation second.

The factors that are needed to ensure the successful use of MAP and CAP of perishable foods in DCs include (1) efficient distribution system, (2) good quality of raw ingredients, (3) low initial microbial load, (4) proper hygiene conditions during processing and handling, and (5) proper temperature control throughout the whole food chain.

Researchers in DCs have attempted to develop MAP for horticultural crops, especially for tropical fruits, but results have been variable.[35] Some of the reported benefits may be due to maintaining a humid atmosphere around the commodity rather than to modification of oxygen and carbon dioxide concentrations. However, effects of gas modification in addition to humidity control resulted in a significant increase in the postharvest life of several crops, such as cactus stems.[14,35] Variable results of the research on MAP in DCs (as is the case in some developed countries) are due to little experimental control, especially regarding the different types of polymeric films used without appropriate characterization.

Most consumers in DCs have not yet shown interest in the environmental credentials of the products that they consume. However, export markets (especially in Europe) are insistent regarding the environmental impact of the product coming from DCs, and this will certainly open the eyes of consumers in these countries.

Polymeric films currently used in the industry for MAP have been selected primarily on the basis of applied research and practical considerations. Only a limited number of films are available with adequate O_2 permeability to prevent anaerobic conditions. This situation is in contrast to the number of high-barrier films that are used for MAP of meats and other products that require a totally anaerobic environment. Most films used have been polyvinyl chloride and perforated, thin LDPE for bagged products. Perforations can greatly enhance permeability but are very unresponsive to temperature increases. Actually, most films are proprietary in nature, and it is difficult to ascertain the composition and obtain accurate determinations of permeability. In addition to high O_2 permeation, films suitable for MA packages in DCs should have a predictable permeability (e.g., not overly affected by the presence of water) and be consistent, inexpensive, easy to use, workable, pliable for moving through packaging machines, sealable (heat sealable is preferred), resistant to tears and punctures, and transparent for a view of the package interior. Ultimately, it will be desirable to use recyclable materials, which could eventually preclude the use of colaminates or other nonrecyclable polymers.

Currently, films with adequate O_2 permeability, adequate response to increases in temperature, or both are few. However, even if they were available, variation in respiration of the packaged produce can cause significant variability in the O_2 levels attained in MA packages.

Controlled packaging has been used to describe the addition of O_2 scrubbers, CO_2 scrubbers and emitters, and ethanol emitters, and some of these formulas have been used on a limited scale in some DCs. Various possibilities exist, although precise control of O_2 levels in MA and CA packages is still not possible. Fewer applications have been introduced for precut produce, although many new possibilities are being tested. However, there appears to be marked resistance from the industry to add to the cost of packaging materials.

The use of polymeric packages that are not readily biodegradable or recycled, which are the most commonly used, are considered in few DCs as an environmental problem. Plastic films used in packaging of fresh fruits and vegetables constitute a very small portion of the plastic used in DCs.

14.10 Strategies to deal with the problem of the deficient cold chain in DCs with regard to its effect on MAP and CAP

The effect of temperature is much more significant than the effect of low O_2 or high CO_2 in preserving the quality of fresh horticultural crops.

The deficiency, abuse, or fluctuation of the cold chain (temperature and RH) in DCs not only increases the deterioration of the product, but also can trigger safety problems. High temperature or fluctuating temperature can increase RH and may cause water fluctuation in the package, which would greatly increase the proliferation of spoilage microorganisms. A possible technique to reduce water condensation without affecting the RH in the package is through the use of a silicon-based coating on the inner surface of the packaging material. This coating does not prevent the condensation but reduces the size of the droplets and disperses them over a large surface area. The permeability of most polymers is usually an order of magnitude higher for water than for respiratory gases. For example, the permeability of LDPE to H_2O is 3.5×10^{-7} m^2/h, whereas to O_2 it is 4.6×10^{-9} m^2/h. Polymers with high rates of vapor transmission have some potential in DCs to reduce water condensation in situations of deficient cold chain (high or fluctuating temperature and RH). As the temperature rises around the package of a fresh horticultural product, respiration increases, thus decreasing the O_2 concentration and increasing the CO_2 concentration. Normally, as the temperature increases, the respiration rate of the product increases more rapidly than the rate of permeation.[7,8] Therefore, packages that are usually optimized at low temperature may become anaerobic at higher temperatures. The ideal situation to maintain adequate O_2 and CO_2 in the package is to ensure a control of temperature. However, in most DCs where temperature control is not commonly easy, a valid alternative is to develop and use packages that function at the highest temperatures typically encountered. As the temperature of the packaged product increases, the influence of low temperature is reduced and the influence of the modified atmosphere is increased.

Side-chain-crystallizable acrylic polymers (Landec Co., Menlo Park, CA) form highly permeable films of up to 100 times more than LDPE and are formulated to achieve a rapid increase in permeability as the temperature increases. The polymer is designed to undergo a phase transition at physiological temperatures, and thus the polymer molecules shift from a somewhat ordered, more crystalline state (less permeable) to a more amorphous or disordered state (more permeable) as the temperature increases.[9] These polymers can also be designed so that their permeability increases faster than the product respiration. These films are used on a very limited scale as packaging materials, because their physical properties are not suitable, and they are only used as patches applied to packages.[9,22]

Another suggestion to deal with the problem of a fluctuating or deficient cold chain in DCs is the developing and using of a "sense and respond" or smart package system that senses either the environment or the physiological status of the enclosed product and responds by modulating its own permeability to gases. Cameron et al.[8] proposed that some components of the package would sense a signal, such as an increase in temperature or an increase in headspace ethanol, and initiate a marked increase in permeability to compensate for the change. A system for increasing permeability in response to increasing temperature was suggested. The system is based on opening holes originally blocked by solid hydrocarbons with melting points between 10 to 30°C. Once the melting point is exceeded, permeability increases dramatically as the hydrocarbons melt and are wicked from the holes. The objective is to provide adequate O_2 levels to prevent excessive fermentative induction during inadvertent exposure to elevated temperatures. Although irreversible in its current form, this system could be useful for certain MA applications. Work also has been reported on the development of a system that could sense and respond to fermentation of the packaged product. During periods of exposure to low O_2 levels, plant products will produce ethanol. In many cases, it is only after extended periods that off-flavors become significant, and in fact, small amounts of ethanol may be desirable. An active component of a package could be designed to sense ethanol and respond by increasing film permeability, perhaps in a manner analogous to the system described above for temperature. Ethanol appears in the headspace after the lower O_2 limit is exceeded. A simple biosensor system that changes color in response to an increase in headspace ethanol can be used. Development of other types of sense-and-respond systems is an interesting possibility. For instance, it may eventually be realistic to identify the presence of specific human pathogens in packages if appropriate; inexpensive biosensors can be constructed to detect specific volatiles. A patent was developed and granted in 1991[24] for the creation of a patch that responds to temperature by opening a pore, and thus increasing the capacity of O_2 influx. This system depends on the sealing of a porous patch or a hole with a compound that melts just below the abusive temperature, although it is reversible when the temperature goes down again.

An alternative approach to providing relatively high gas exchange needed for fresh horticultural crops is film with holes or pores. FreshHold, a microporous film technology, P-Plus microperforated technology, and proprietary microperforation technologies have been used for fresh-cut fruit products. Microporous and microperforated films allow sufficient gas exchange, more than would normally be possible through plastic films, but films with pores or small holes have some physical limitations. Carbon dioxide diffuses through plastic films two to six times faster than oxygen (O_2), and therefore, CO_2 exits a package much faster than O_2 enters, resulting in an equilibrium atmosphere of low O_2 and relatively low CO_2. A range of CO_2/O_2 permeability ratios among plastic films can provide a range of CO_2/O_2 proportions inside packages. Films with holes or pores admit O_2 and CO_2 at similar rates, and therefore the ratios of gases that can result inside such packages are limited. Landec Corporation (California) has developed side-chain polymer technology that allows the film OTR to increase rapidly as temperature increases, thereby avoiding anaerobic conditions subsequent to loss of temperature control. In addition, these polymers can provide very high OTRs, an adjustable CO_2/O_2 permeability ratio, and a range of moisture vapor transmission rates.

14.11 *Future feasibility of MAP and CAP in DCs*

The challenge of implementing or increasing the use of MAP and CAP, and thus active and intelligent packaging, in DCs will depend on the cost–benefit ratio and improvement of the cold chain, as well as the improvement of the handling of food in general, especially the hygienic status.

Active packaging has been investigated for more than 40 years, ever since passive packaging embracing oxygen and water vapor barriers became important to the protection of food during distribution. There has been use in some DCs of technological innovations for control of specific gases within a package involving the use of chemical scavengers to absorb a gas, or alternatively other chemicals that may release a specific gas as required. This includes the use of oxygen scavenging for several types of packaged foods, such as cereals and nuts, water vapor removal in some dried foods, and ethylene removal in the packages of some horticultural products. These applications usually involve the inclusion in the package of a small sachet that contains an appropriate scavenger. The sachet material is highly permeable, and diffusion through the sachet is not a serious limitation. The reacting chemical commonly used for ethylene removal is potassium permanganate. Carbon dioxide or sulfur dioxide generating materials can be used for table grapes. However, temperature control is very important because when the temperature of the packed grapes rises due to inadequate control, the slow-release system fails, releasing high levels of sulfur dioxide and carbon dioxide, which can lead to fruit injury and problems; in addition, it can lead to illegal residues of sulfur dioxide. Some work has been done to develop systems that gradually release sulfur dioxide and are less sensitive to high temperature and moisture than those commonly used. Other systems of active packaging that can be used in DCs include sachets containing iron powder and calcium hydroxide that can scavenge both oxygen and carbon dioxide, film containing microbial inhibitors, including metal ions and salts of propionic acid, and specially fabricated films to absorb flavors and odors or, conversely, to release them into the package.

14.12 Specific needs

- Improved cold chain.
- Regulatory actions related to the use of MAP and CAP in foods in general, but particularly in minimally processed foods.
- Consumer awareness and education. There is practically a total lack of serious consumer advocate organizations in most DCs. In addition, regulations to help the consumer evaluate the status of the food and to maintain it with minimum safety problems are needed. Regulations such as for labeling (i.e., "keep refrigerated," "sell by," "use by," etc.) are of particular need.
- Familiarity with the changes of the gas composition inside the package due to the use of the new technology, especially during the distribution, and retail system conditions where temperature variation is common will help estimate the organoleptic and microbial quality.
- Familiarity with the microbiological flora of the packaged products, and how this flora can be affected by the use of the new technology.
- Establishment of quality control systems. Well-designed quality control systems are in great need in DCs. These are still difficult to establish and even more difficult to maintain in most DCs.
- The lack (or difficulty) of temperature control during the postharvest handling chain in DCs puts much more emphasis on the development of packages and film materials that constantly interact with the food as well as with the outside environmental conditions (especially temperature and relative humidity). Some of the desirable characteristics include:
    1. Ability to control moisture vapor transmission rate in order to prevent vapor accumulation and condensation problems
    2. Ability to reduce or eliminate barrier properties in case of a temperature increase
    3. O_2/CO_2 permeability ratios that are drastically different than 1:4
    4. Variation in O_2/CO_2 permeability ratio (adjustable preferential gas permeability) as a function of temperature
    5. Developing of consumer warning signs to indicate when the product quality is reduced and, more important, when abuse conditions have occurred and might result in safety problems
- Key logistic changes to the packaging industry will need to be implemented in DCs to increase supply chain management efficiency and safety of products.

- Adequate collaboration between companies providing the different services (packaging materials, packaging equipment, gases, analyzers and other instruments, etc.), which are mostly from developed countries, and the clients in DCs (MAP and CAP users) is essential. Information regarding the characteristics of the material or system, mechanisms of action, limitations, possible disadvantages, and possible problems need to be clearly delivered to the clients.

14.13 Conclusions

The application of MAP and CAP in many DCs is feasible and needed. In fact, several applications of the technology are already in use in some DCs for several types of food. The complexity of MAP and CAP dictates that a team approach should be implemented in order for the technology to be used adequately. Experts from different areas (namely, postharvest physiology and technology, food science and technology, packaging, microbiology, and food safety) need to collaborate in a multidisciplinary manner. Unfortunately, team/multidisciplinary work is not very common in many DCs, although it is essential for the successful application of technologies such as this. The application of food safety principles and the appropriate maintenance of the cold chain are strict requirements for the successful application of MAP and CAP in DCs. It will take major efforts to improve processing, safety, and temperature control, as well as the appropriate knowledge in choosing the fitting packaging materials.

Produce handlers in DCs need to use active packaging if it improves their ability to sustain or expand their markets and if the benefits are greater than the costs of such packaging. This is more likely to be the case for high-value intact or fresh-cut fruits, vegetables, herbs, and flowers that are exported from DCs to markets in developed countries.

No packaging will replace the need for proper management of product temperature throughout the postharvest handling system. Thus, produce handlers in DCs should focus first on the basics (such as protecting the product from physical damage, quality uniformity within packages, cooling and other temperature management procedures, etc.) before evaluating the cost/benefit ratio of active packaging as a supplemental procedure. If they are already doing the basics of maintaining quality and safety, then their need for active packaging is not different from the need in developed countries.

Acknowledgment

Thanks to Dr. Adel A. Kader for reading the text.

References

1. Ackerman, F. Environmental impacts of packaging in the US and Mexico. *Phil. Tech.*, 2, 1–7, 1997.
2. Anon. Fresh-cut Sales of Retail Produce Approaching 4 Billion a Year. *Fresh Cut*, Columbia Pub. and Design, Yakima, WA, November 2003.
3. Baumgartner, B. and Belevi, H. *A Systematic Overview of Urban Agriculture in Developing Countries*. EAWAG–Swiss Federal Institute for Environmental Science & Technology SANDEC–Department of Water and Sanitation in Developing Countries, 2001.
4. Billiard, F. New Developments in the Cold Chain: Specific Issues in Warm Countries. *Ecolibrium*, July 10–14, 2003.
5. Bourne, M.C. Selection and use of postharvest technologies as a component of the food chain. *J. Food Sci.*, 69, R43–R46, 2004.
6. Brody, A.L. Active and intelligent packaging: the saga continues. *Food Technol.*, 56, 65–66, 2002.
7. Cameron, A.C., Beaudry, R.M., Banks, N.H., and Yelanich, M.V. Modified atmosphere packaging of blueberry fruit: modeling respiration and package oxygen partial pressures as a function of temperature. *J. Am. Soc. Hort. Sci.*, 119, 534–539, 1994.
8. Cameron, A.C., Patterson, B.D., Tlasila, P.C., and Joles, D.W. 1993. Modeling the risk in modified-atmosphere packaging: a case for sense-and-respond packaging, in *Proceedings of the Sixth International Controlled Atmosphere Research Conference*, Cornell University, Ithaca, NY, 1993, pp. 95–102.
9. Clarke, R. and De Moore, C.P. The future of film technology: a turnable packaging system for fresh produce, in *CA'97 Proceedings*, Vol. 5, *Fresh Cut Fruits and Vegetables and MAP*, Postharvest Horticulture Series 19. University of California, 1997, pp. 68–75.
10. de Kruijf, N., van Beest, M., Rijk, R., Sipilainen-Malm, T., Losada, P.P., and De Meulenaer, B. Active and intelligent packaging: applications and regulatory aspects. *Food Add. Contam.*, 19, 144–162, 2002.
11. Dolan, C. and Humphrey, J. Governance and trade in fresh vegetables: the impact of UK supermarkets on the African horticulture industry. *J. Dev. Stud.*, 37, 147–176, 2000.
12. Guevara, J., Yahia, E.M., Cedeño, L., and Tijskens, L.M.M. Modeling the effects of temperature and relative humidity on gas exchange of prickly pear cactus (*Opuntia* spp.) stems. *Lebensm. Wiss. Technol.*, 39, 796–805, 2006.
13. Guevara, J.C., Yahia, E.M., and Brito de la Fuente, E. Modified atmosphere packaging of prickly pear cactus stems (*Opuntia* spp.). *Lebensm. Wiss. Technol.*, 34, 445–451, 2001.
14. Guevara, J.C., Yahia, E.M., Brito, E., and Biserka, S.P. Effects of elevated concentrations of CO_2 in modified atmosphere packaging on the quality of prickly pear cactus stems (*Opuntia* spp.). *Postharvest Biol. Technol.*, 29, 167–176, 2003.
15. Hagen, J.M. Agri-Food Innovation in Developing Countries: The Role of Retailers. IAMA, Department of Applied Economics and Management, Cornell University, Ithaca, NY, 2003.
16. Holton, W.C. Innovations. Fresh ideas for food safety. *Environ. Health Perspect.*, 108, 516–519, 2000.

17. Hong, S.I. Gravure-printed color indicators for monitoring kimchi fermentation as a novel intelligent packaging. *Packaging Technol. Sci.*, 15, 155–160, 2002.
18. Kader, A.A. Increasing food availability by reducing postharvest losses of fresh produce. *Acta Hort.*, 682, 2169–2175, 2004.
19. Kader, A.A. and Watkins, C. Modified atmosphere packaging: toward 2000 and beyond. *HortTechnology*, 10, 483–486, 2000.
20. Kwon, T.W., Hong, J.H., Moon, G.S., Song, Y.S., Kim, J.I., Kim, J.C., and Kim, M.J. Food technology: challenge for health promotion. *Biofactors*, 22, 279–287, 2004.
21. Labuza, T.P. and Breene, W.M. Applications of "active packaging" for improvement of shelf-life and nutritional quality of fresh and extended shelf-life foods. *J. Food Proc. Preserv.*, 13, 1–69, 1989.
22. Lange, D.L. New film technologies for horticultural products. *HortTechnology*, 10, 487–490, 2000.
23. Parthasarthy, R.P. and Hammond, T. Technical Innovation: Options for Developing Countries. *The Hindu*, Business Section, May 24, 2001. Available at http://www.the-hindu.com),http://www.indiaserver.com/thehindu/2001/05/24/stories/0624000h.htm.
24. Patterson, B.D. and Cameron, A.C. Modified-Atmosphere Packaging. Australian Provisional Patent Application PK6567, filed June 1991. Commonwealth Scientific and Industrial Research Organization (CSIRO), Patterson, B., and Michigan State University, Cameron, A.C., International Patent Application PCT/AU92/00267, filed October 6, 1992.
25. Rae, A. and Josling, T. Processed Food Trade and Developing Countries: Protection and Trade Reform, NZ Trade Consortium Working Paper 15. 2001.
26. Remi, Ed. *Changing Structure of Global Food Consumption and Trade*, Agriculture and Trade Report WRS-01-1. USDA, Economic Research Service, May 2001.
27. Schill, P. and Quenther, D. Developing country requirements for the improvement of food quality and safety. In *Challenges to Organic Farming and Sustainable Land Use in the Tropics and Subtropics*. Deutscher Tropentag, Witzenhausen, October 9–11, 2002.
28. Senaur, B. and Goetz, L. The growing middle class in developing countries and the market for high-value food products. In *Proceedings of the Workshop on Global Markets for High-Value Food*, USDA, Economic Research Service, Washington, DC, February 14, 2003.
29. Shewfelt, R.L. and Henderson, J.D. The future of quality. *Acta Hort.*, 604, 49–59, 2003.
30. The World Bank. *World Development Indicators*, 2001.
31. United Nations Food and Agriculture Organization (FAO). *Compendium of Food Consumption Statistics from Household Surveys in Developing Countries*, Vol. 1, Asia. FAO Economic and Social Development Paper 116/1, 1993.
32. United Nations Food and Agriculture Organization (FAO). *Compendium of Food Consumption Statistics from Household Surveys in Developing Countries*, Vol. 2, *Africa, Latin America and Oceania*. FAO Economic and Social Development Paper 116/2, 1994.
33. United Nations Food and Agriculture Organization (FAO). *FAOSTAT Database*. Rome, 2003.
34. World Health Organization. Food safety in developing countries building capacity. *Releve Epidemiol. Hebdomadaire*, 18, 175–180, 2004.

35. Yahia, E.M. Modified and controlled atmospheres for tropical fruits. *Hort. Rev.*, 22, 123–183, 1989.
36. Yahia, E.M. Postharvest research and reality in developing countries. *Acta Hort.*, 682, 1655–1666, 2004.
37. Yahia, E.M. Postharvest technology of food crops in the Near East and North Africa (NENA) region. In *Crops: Quality, Growth and Biotechnology*, Dris, R., Ed. WFL Publisher, Helsinki, Finland, 2005, pp. 643–664.
38. Yahia, E.M., Barry-Ryan, C., and Dris, R. Treatments and techniques to minimize the postharvest losses of perishable food crops, in *Production Practices and Quality Assessment of Food Crops*, Vol. 4, *Postharvest Treatment and Technology*, Dris, R. and Jain, S.M., Eds. Kluwer Academic Publishers, Dordrecht, The Netherlands, 2004, pp. 95–133.
39. Yahia, E.M., Guevara, J., Tijskens, L.M.M., and Cedeño, L. The effect of relative humidity on modified atmosphere packaging gas exchange. *Acta Hort.* 674, 97–104, 2005.
40. Yam, K.L., Takhistov, P.T., and Miltz, J. Intelligent packaging: concepts and applications. *J. Food Sci.*, 70, R1–R10, 2005.

chapter fifteen

Regulation of new forms of food packaging produced using nanotechnology

*Michael F. Cole and Lynn L. Bergeson**

Contents

* Lynn L. Bergeson is managing director of Bergeson & Campbell, P.C., a Washington, D.C. law firm focusing on conventional and engineered nanoscale chemical, pesticide, and other specialty chemical product approval and regulation, environmental health and safety law, chemical product litigation, and associated business issues. Michael F. Cole is of counsel to the firm.

15.1 Introduction

This chapter outlines the challenges facing the U.S. Food and Drug Administration (FDA) as it braces to regulate the use of nanotechnology in advanced forms of food packaging. Food packaging is a key focus of companies seeking to promote the commercialization of nanotechnology. One respected industry analyst has reported that there are already 250 packaging products on the market incorporating substances manufactured using nanotechnology, often referred to as nanopackaging. These products generated over $860 million in sales worldwide last year, and the same analyst projects that within 10 years, nanopackaging will be a $30 billion market.*

The Center for Food Safety and Applied Nutrition (CFSAN) is the arm of the FDA responsible for the regulation of packaging materials for food products. CFSAN has substantial experience in regulating materials at the molecular level, since one of its principal concerns is the possible migration of substances from packaging material to the food with which they may come into contact. The migration occurs in very small amounts. For example, one of the criteria for exemption from regulation is whether the use of the material results in a dietary concentration of 0.5 parts per billion (ppb). Consequently, the guidance documents and other materials issued to assist industry in preparing submissions are written to address materials at that level. However, the issue will be whether the present system is adequate to cope with the novel considerations raised by nanotechnology. Those considerations revolve, in part, around whether the nanoscale version of known substances or the nanomanufactured version of new substances that do not occur naturally present different issues relating to toxicity, route of exposure, and transmigration potential than those presented by the macrosized versions of the substances. The second and equally important issue is whether there are differences in how the nanoscale materials migrate into food, and if so, whether such differences are significant.**

In addressing these issues, the chapter will first briefly discuss the regulatory scheme for the regulation of food packaging. Then, the food packaging uses being investigated and promoted will be examined, followed by

* "Nanotechnology Makes Packaging Intelligent, Smart and Safe" (May 26, 2005), available at www.nanotechwire.com/news.asp?nid=1961&ntid=116&pg=9. The Helmut Kaiser Consultancy study referenced in this article, "Nanotechnology in Food and Food Processing Industry 2003-2006-2010-2015," is available for purchase at www.hkc22.com/Nanofood.html.

** The FDA has adopted, for the present, the definition of nanotechnology developed by the National Nanotechnology Initiative:

1. Research and technology development at the atomic, molecular or macromolecular levels, in the length scale of approximately 1–100 nanometer range.
2. Creating and using structures, devices and systems that have novel properties and functions because of their small and/or intermediate size.
3. Ability to control or manipulate on the atomic scale.

See "FDA and Nanotechnology Products: Frequently Asked Questions," available at www.fda.gov/nanotechnology/faqs.html.

an assessment of the adequacy of the system in place to address the novel issues embodied in the use of nanotechnology.

15.2 Food packaging regulation

Substances incorporated in food packaging are regulated by the FDA as food additives or food contact substances. Section 321(s) of the Federal Food, Drug and Cosmetic Act (Act)* defines a food additive, in relevant part, as:

> Any substance the intended use of which results or may reasonably be expected to result, directly or indirectly, in its becoming a component or otherwise affecting the characteristics of any food (including any substance intended for use in producing, manufacturing, packing, processing, preparing, treating, packaging, transporting, or holding food … if such substance is not generally recognized, among experts qualified by scientific training and experience to evaluate its safety … to be safe under the conditions of its intended use; except that such term does not include —
>
> …
>
> Any substance used in accordance with a sanction or approval granted prior to September 6, 1958, pursuant to this chapter.**

The implementing regulations, 21 CFR Section 170.3(e), further define the term:

> Food additives include all substances not exempted by section 201(s) of the act, the intended use of which results or may reasonably be expected to result, directly or indirectly, either in their becoming a component of food or otherwise affecting the characteristics of food. A material used in the production of containers and packages is subject to the definition if it may reasonably be expected to become a component, or to affect the characteristics, directly or indirectly, of food packed in the container. … If there is no migration of a packaging component from the package to the food, it does not become a component of the food and thus is not a food additive.
>
> Uses of food additives not requiring a listing regulation. Use of a substance in a food contact article (e.g., food-packaging or food-processing equipment) whereby the substance migrates, or may reasonably be expected to migrate, into food at such levels

* 21 USC §§ 301–397.
** 21 USC § 321(s).

> that the use has been exempted from regulation as a food additive under Sec. 170.39, and food contact substances used in accordance with a notification submitted under section 409(h) of the act that is effective.

Chemical substances incorporated in packaging are indirect food additives, a term coined by the FDA to indicate that the additives are not intended to have a technical or functional effect on food. Any effect on food itself only occurs if the substance migrates from the packaging material.*

15.3 Safety of food additives

All food additives, including indirect additives, must be established to be safe before they can be used in packaging materials.** The definition of safe or safety is:

> That there is a reasonable certainty in the minds of competent scientists that the substance is not harmful under the intended conditions of use. It is impossible in the present state of scientific knowledge to establish with complete certainty the absolute harmlessness of the use of any substance. Safety may be determined by scientific procedures or by general recognition of safety.***

The primary safety concern with indirect additives is whether the substances incorporated will migrate from the packaging or storage container to the food. The issues become the toxicity of the materials, the amount that might become a part of the food, and the effect that it might have on the health of the consuming public.

A manufacturer must demonstrate to the FDA that an additive is safe, unless the manufacturer is seeking to distribute an additive that meets one of the exceptions listed in the statute and regulations. Even the exceptions, in effect, require proof of safety because a manufacturer has to demonstrate that the material does not migrate to any appreciable extent, or the manufacturer has to prove that the safety information is already available or a decision on safety has already been made.

* FDA, "Guidance for Industry: Preparation of Food Contact Notifications: Administrative, FINAL GUIDANCE" (May 2002), available at www.cfsan.fda.gov/~dms/opa2pmna.html#II-B.
** 21 USC § 348(c)(3)(A).
*** 21 CFR. § 170(i).

15.4 *Proof of safety: the necessary filings*

The manufacturer must submit a filing to CFSAN to demonstrate safety, or be able to document that the substance used for its intended purpose meets an exception to filing. There are two possible applications that a manufacturer might file, depending on the circumstances: a Food Additive Petition (FAP) or a Food Contact Notification (FCN). The mechanisms for doing so are outlined below.

15.4.1 *Food additive petition**

Until 2000, the FAP was the procedure to use for seeking approval of all indirect food additives that required a filing. After Congress enacted legislation creating the FCN,** CFSAN promulgated regulations requiring FAPs in only two cases.***

When a FAP is appropriate, the manufacturer files a petition for approval of the indirect food additive. If the FDA approves the substance for the intended use, a formal rule-making proceeding is commenced. When a final rule is promulgated, it is published in the *Federal Register* and then in the Code of Federal Regulations (CFR), which lists the substance, the intended use, and any limitations and specifications that are applicable. Any manufacturer can market the substance based on the rule.

The FAP is a notice and comment rule-making process that is unusually time consuming. For that reason, in the late 1990s Congress provided for a new process, the FCN.

15.4.2 *Food contact notification*****

The FCN is the most common form of application to secure permission to market an indirect food additive in use today. A manufacturer or supplier submits an FCN, and CFSAN then reviews it. If it has no objection, CFSAN

* 21 CFR. §§ 170–171.

** See Food and Drug Administration Modernization Act of 1997 (FDAMA).

*** (1) When the use of the food contact substance will increase the cumulative estimated daily intake (CEDI) of the food contact substance from both food and food contact uses to a level equal to or greater than 1 part per million (ppm) (i.e., 3 mg/person/day) for a substance that is not a biocide or, in the case of a biocide (e.g., it is intended to exempt microbial toxicity), to a level equal to or greater than 200 parts per billion (ppb) (i.e., 0.6 mg/person/day), or (2) When existing data include one or more bioassays on the food contact substance that the Agency has not reviewed already and such studies are not clearly negative for carcinogenicity.

See Guidance for Industry: Preparation of Food Contact Notifications: Administrative, FINAL GUIDANCE, available at www.cfsan.fda.gov/~dms/opa2pmna.html#II-C.
21 USC § 348(h); see also 21 CFR § 170.100.

**** See FDA, Inventory of Effective Food Contact Substances, available at www.cfsan.fda.gov/~dms/opa-fcn.html.

declares the notification to be effective and adds the substance to the inventory of effective notifications on its web page.* The inventory identifies the substance, the manufacturer, the intended use, all limitations and specifications applicable to the substance, and the effective date.

The food contact substance is effective for the intended use. It is not referred to as approved, although having an effective FCN is equivalent to having an additive listed in the CFR. Only the submitter and any distributors it lists can market the substance based on an effective FCN. Anyone else seeking to market the substance must file a separate FCN. Congress conferred this limited period of exclusivity as a means to encourage the filing of FCNs, including FCNs for materials that manufacturers may have marketed previously under one of the exemptions.**

15.4.3 Threshold of regulation exemption

CFSAN has indicated that it will exercise its enforcement discretion and not regulate substances where there is competent evidence to prove that the substance will migrate into the food, if at all, in amounts below a stated threshold of regulation (TOR).***

To be eligible for this discretionary exemption, the food contact substance cannot be a carcinogen according to the FDA test set forth in 21 CFR Section 170.39(b); raise any other health or safety concerns; or be present in food at any level above 0.5 ppb. A manufacturer, called a requestor, files a request for exemption. If CFSAN grants the request, the requestor receives a letter from CFSAN stating that the food contact substance is exempt. A list of exempt substances is maintained at the CFSAN Division of Dockets Management.

15.4.4 Generally recognized as safe

A manufacturer does not have to file a FAP or an FCN if the substance is generally recognized as safe (GRAS) for the intended purpose, if it is the subject of a prior sanction, or if it will not migrate into food. A substance that is GRAS is not considered a food additive, and a manufacturer does not have to comply with the rules applicable to food additives.

The safety standard is the same as that for an FCN or a FAP. The difference is that a manufacturer is claiming that the publicly available scientific literature shows the substance to be safe, and that qualified and experienced experts concur. The FDA does not need to concur officially with this conclusion, however, as the manufacturer makes this determination. The manufacturer can file, seeking FDA concurrence, if it so chooses.****

* See FDA, Guidance for Industry: Preparation of Food Contact Notifications: Administrative, at paragraph II.B.2, available at www.cfsan.fda.gov/~dms/opa2pmna.html#II-E-1.

** 21 CFR §§ 170.39, 174.6.

*** See generally FDA, GRAS Notification Program, available at www.cfsan.fda.gov/~dms/opa-noti.html.

**** See FDA, Summary of All Gras Notices, available at www.cfsan.fda.gov/~rdb/opa-gras.html.

A manufacturer making a GRAS self-determination is subject to challenge at any time by CFSAN, or the manufacturer can seek CFSAN's agreement that the substance is GRAS.

A person that files a GRAS for a particular substance is considered the notifier. The notice includes a GRAS exemption claim and supporting information. If CFSAN concurs, it issues a letter stating that it has no questions and includes the name of the substance and a copy of its letter to the notifier on its web page.*

15.4.5 Ingredient subject to a prior sanction

A prior sanctioned ingredient is exempt from the definition of food additive.** This grandfather provision excludes ingredients explicitly approved by the FDA or Department of Agriculture for a specific use prior to September 6, 1958, when the food additive legislation was enacted.*** Since the original approval for the ingredient occurred a half century ago, the regulations provide that a prior sanction can be amended to impose whatever limitations are needed for the safe use of the ingredient. CFSAN can also ban use of the ingredient, if warranted.****

15.4.6 No-migration determination

According to 21 CFR Section 170.3(e)(1), "if there is no migration of a packaging component from the package to the food, it does not become a component of the food and thus is not a food additive." Judicial interpretation has also expanded the definition to exempt ingredients where there is migration, but only in insignificant amounts.*****

Informally, FDA has characterized an insignificant amount of migration as less than 50 ppb.****** A manufacturer makes the determination regarding insignificant migration, preferably based on a properly conducted migration study or calculation. CFSAN can challenge that determination at any time.

* 21 USC § 321(s).

** 21 CFR §§ 181.1, 181.5.

*** 21 CFR § 181.1(b).

**** *Monsanto v. Kennedy*, 613 F.2d 947, 955 (D.C. Cir. 1979).

***** The Food Additive Problem of Plastics Used in Food Packaging, speech by L.L. Ramsey, Assistant Director for Regulatory Programs, Bureau of Science, Food and Drug Administration, for presentation at the National Technical Conference of the Society of Plastics Engineers (November 4–6, 1969).

****** ElAmin, A. Nanotechnology Targets New Food Packaging Products (October 12, 2005), available at www.foodnavigator.com/news/ng.asp?id=63147; Research Pushes Possibility of Intelligent Packaging (June 4, 2005), available at www.foodproductiondaily.com/news/ng.asp?n=59220&m=1FPD406&c=; Nanotechnology and Food Packaging, published in *Advantage Magazine* (February 2004), available at www.azonano.com/details.asp?ArticleID=857; Nanologue.net, Nanologue Background Paper, Section 2.3, Food Packaging, at 20 (September 2005), available at www.nanologue.net/custom/user/Downloads/Nanologue BackgroundPaper.pdf.

The above section describes the various mechanisms that will permit the marketing of a substance used in food packaging. To determine which of the mechanisms will prove useful in regulating substances produced using nanotechnology, it is necessary to examine the several applications where nanotechnology is a relevant factor, in terms of both products in commerce and products being studied for possible commercialization.

15.5 Nanopackaging

Nanoscale materials, generally but not always consisting of nanoparticles and structures of one form or another, are believed to improve the mechanical and heat resistance properties of packaging to prolong shelf life, and to increase the barrier properties of packaging by affecting gas and water vapor permeability.* Active packaging is being developed to facilitate antimicrobial and antifungal surfaces to decontaminate packaging and protect the food contents. Smart packaging is also under commercial consideration. Sensors would be incorporated into packaging materials to signal microbiological and biochemical changes. Other smart packages would be used to track and trace the location of food products in commerce.**

The first products in the commercial market were a variety of organo clay fillers used primarily in a nylon matrix resin. Foils or membranes are also available that offer adjustable gas permeability. Materials exhibiting antimicrobial properties imparted by nanoparticles of silver, zinc oxide, and magnesium oxide are in distribution, and dirt-repellent coatings are being developed to prevent the invasion of microorganisms.***

Nanotechnology in food packaging involves the manipulation of chemical substances and particles at the molecular level to engineer materials with novel, unique properties that address vexing packaging problems. For example, polymers are not inherently impermeable to gases such as oxygen, carbon dioxide, and water vapor. Oftentimes, multilayer films composed of different plastic materials have to be developed to provide both oxygen and water impermeability, while providing barrier properties. Nanoclays and other nanoparticles engineered at the molecular level have been developed that will greatly reduce both oxygen and water vapor release, while at the same time providing barrier protection. The new

* Nanotechnology Targets New Food Packaging Products, *supra* note 23; Nanotechnology and Food Packaging, *supra* note 23; Nanologue Background Paper, *supra* note 23.

** Nanotechnology Targets New Food Packaging Products, *supra* note 23.

*** Choudhury, N.R. and Dutta, N. Abstract: Development of Multilayered Packaging Materials with Controlled Barrier Properties, available at www.unisa.edu.au/iwri/futurestudents/honsprojects/developmentofmultilayered.asp; Selke, S.E., Nanotechnology and Packaging (PowerPoint presentation), available at www.msu.edu/~ifas/downloads/Nanotechnology_and_Packaging.pdf; Downing-Perrault, A. (Mohan, A.M., Ed.), Nanotechnology Offers Big Benefits for Packaging, published in *Packaging Digest* (October 2005), available at www.packagingdigest.com/articles/200510/50.php.

material will be only one layer thick, while still meeting the needed performance specifications.*

Manipulation of materials, chemical substances, and structures at the molecular level can cause physical and chemical changes compared to the substance at the macrolevel. The reduction in the size of the material means that the resulting small mass makes gravitational forces negligible. Instead, electromagnetic forces are dominant in determining the behavior of atoms and molecules. Second, materials at the nanoscale express quantum mechanical phenomena, rather than classical behaviors, which are more predictable. Third, the size of the nanoparticle further creates a large surface area-to-volume ratio, which makes nanoscale materials "staggeringly reactive" in comparison to their macroscale counterparts.** The bandgap — or distance between electron energy levels in an atom — also morphs at the nanoscale, changing the electrical resistance and chemical reactivity of a nanoparticle (e.g., the nanoparticle could become a conductor, an insulator, or a semiconductor).*** Lastly, nanoscale materials are affected far more than their macroscale counterparts by random molecular motion. These changes combine to produce at the nanoscale properties that can differ fundamentally from those same materials at the macroscale.**** These changes include color and related interactions with light, electrical conductivity, magnetization and polarity, melting points, hardness, resistance, and strength.***** Any or all of the foregoing properties can differ fundamentally from the same properties at the macroscale.

A number of the nanomaterials being developed are atomically modified materials that do not exist in nature. Many are familiar compounds that have never been marketed on the nanoscale. As stated above, the present regulatory framework contains lists of substances permitted for various uses in food packaging. Whether those approvals apply without qualification to the nanotechnology-produced versions of the substances is a key open issue, as is whether all the new substances should be treated as new, requiring some form of clearance, and what data are relevant to that clearance.

The answers to these questions hinge on whether there is reason to believe that the nanomaterials are fundamentally different from their mac-

* Nanoscience: Working Small, Thinking Big, published by Australian Academy of Science in *Nova Science in the News* (September 2003), available at www.science.org.au/nova/077/077key.htm.

** Niemann, J., Electrical Measurements on Nanoscale Materials, published by Keithley Instruments (July 2005), available at www.keithley.com/support/data?asset=50358.

*** Lesson 3: Unique Properties at the Nanoscale, published by nanosense.org (May–June 2005), available at nanosense.sri.com/activities/sizematters/properties/SM_Lesson3Teacher.pdf.

**** Nanoscience: Working Small, Thinking Big, *supra* note 27.

***** For more information, see the following: How Nanotechnology Will Revolutionize Plastic, available at www.phillyspe.plastics.com/nano2.pdf; Nanotechnology Targets New Food Packaging Products, available at www.foodproductiondaily.com/news/ng.asp?n=63147-nanotechnology-food-packaging-research-and-development; Packaging on the Molecular Level, available at www.pmtdirect.com/website/article.asp?id=1242; Securely Wrapped, available at www.research.bayer.com/medien/pages/2999/polyamides.pdf; *Nanotech Briefs* (March 2005), available at www.nanotechbriefs.com.

roscale counterparts in ways that are biologically relevant. Do they migrate in the same fashion, or do they accumulate in the body, as some have posited? There are suggestions that the materials will be able to cross the blood–brain barrier, which presents a host of possible issues. If such is the case, then the materials should likely be considered new substances, since the nanocharacteristics of the material have a different effect on the risk–benefit analysis that led to the permission to market the macroparticle form of the substance.

Therefore, the main question is: Is the present regulatory scheme administered by CFSAN capable of making these determinations, or must changes be made in how food contact substances produced using nanotechnology are regulated?

15.6 The sufficiency of the present regulatory framework

Even the staunchest advocates of the benefits of nanotechnology will admit that there is a paucity of hard scientific data on the toxicity of nanoscale materials and their biological potency. Research is not widely available to demonstrate whether the pattern of migration of nanoparticles is similar to that of their macroscale counterparts. There is some discussion to the effect that nanoscale materials may accumulate to a greater degree after migration, but this is unconfirmed, as is the possibility of crossing the blood–brain barrier. Some of the relevant considerations are:

- Is there sufficient information available to establish the long- and short-term stability of nanoscale materials and structures in a polymer matrix?
- Are there robust and reliable detection and measurement technologies to determine and monitor potential migration of nanoscale materials?
- Are there sufficient data on how the toxicology of nanoscale materials work, and are the current toxicological screens sufficient?

It is not entirely clear how CFSAN will go about addressing these questions. One point that is important to remember is that CFSAN, like the other FDA centers, does not regulate in a vacuum. It must decide policy in the course of reviewing specific applications for specific uses of particular substances or materials. It takes a considerable amount of time to build up precedent that can have general applicability in these circumstances. This is apparent when one considers the different forms of nanomaterials that may be used, and the purposes for which they are employed.

15.7 High-performance barrier materials/films*

Uses for nanoclays, nanotubes, and metal oxides include:

- Increased tensile strength
- More efficacious gas/water barrier
- Decreased weight
- Increased durability
- Higher temperature/UV resistance
- Maintain freshness for increased shelf life
- Preservatives for flavor and color
- Self-cleaning surfaces (some with antimicrobial functions)
- Improved transparency
- Improved secondary processing characteristics
- Capable of handling greater weights
- Aesthetic improvements of surface materials (e.g., glossy finishes)
- Radiation resistance
- Self-cleaning/dirt-repellent packaging
- Increased flexibility
- Easier to use and transport

15.8 Programmable barrier properties**

Uses include:

- Control of internal moisture
- Control of atmospheric environment
- Self-assembling nanocoatings (for pulp/paper)

15.9 Intelligent packaging***

Uses include:

- Alert to food expiration
- Tamper-proofing
- Nutritional content alerts

* See Big Opportunities With Tiny Technology, available at www.paperloop.com/db_area/archive/p_p_mag/2004/0005/comment.html.

** See Nanotechnology Makes Packaging Intelligent, Smart, and Safe, available at www.chemie.de/news/e/46235/?pw=a&defop=and&wild=yes&sdate=01/01/1995&edate=05/25/2005; Research Pushes Possibility of Intelligent Packaging, available at www.foodproductiondaily.com/news/news-ng.asp?n=59220-research-pushes-possibility.

*** See Color-Coded Pathogens Offer Safer Food Formulation, available at www.foodnavigator-usa.com/news/news-ng.asp?n=60665-color-coded-pathogens.

15.10 Nanobarcodes (e.g., biosensors) for pathogen identification*

Uses include:

- Biosensors can detect pathogens in food/packaging.
- Nano-silver, nano-zinc oxide, and nano-magnesium oxide.
- Antimicrobial nanoparticles for food packaging.

15.11 Solar cells**

Nanotechnology applications in solar cells can help to power intelligent packaging. Uses include:

- High conductivity
- Increased power output
- Economical value
- Increased sensitivity
- The projects generally are intended to:
    - Enhance the durability of packaging to provide longer shelf life
    - Improve the barrier properties of packaging by affecting gas and water vapor permeability
    - Provide antimicrobial and antifungal activity to decontaminate packaging or food
    - Implement smart packaging by incorporating sensors in packaging to detect and identify contamination or spoilage

A wide variety of materials, all with differing properties to achieve different results, will be used in these applications and will raise discrete, sometimes unique, issues. CFSAN, and indeed any regulator, does not appear to have sufficient information to form overarching policies for the consideration of issues involved in the nanoscale materials used in these product applications. It will have to review a host of submissions before making a judgment as to whether the use of nanotechnology changes the risk–benefit analysis in general terms, instead of on a product-by-product basis. This is one of the reasons why CFSAN personnel have indicated, unsurprisingly, that they will take a "wait-and-see" attitude toward the need for additional regulation or guidance. That seems to be an appropriate response, if CFSAN acts to ensure that it will be apprised of the fact that nanotechnology has been employed to produce the substance in question so that it can begin to develop a database. CFSAN personnel and others in the FDA have remarked that they will not know if nanotechnology has been

* See Nanotechnology May Give Solar Cells a Boost, available at www.spacedaily.com/news/nanotech-04zzzd.html.

** 64 *Fed. Reg.* 27666 (May 21, 1999).

used unless the submitter notifies them, and it has indicated that such notification does not always, or perhaps even routinely, happen. That should be an early requirement imposed by CFSAN going forward, so that it will be able to take advantage of every opportunity to accumulate information that will aid in developing a nanotechnology policy.

One of the first questions that will have to be resolved in considering new applications is whether the nanoscale version of a substance is the same material as the macroscale version. This is a crucial determination since so many substances are already approved by CFSAN for a wide variety of uses, and manufacturers will be anxious to take advantage of such rulings to market without the need for submissions. That is as it should be, if the original basis for the permission to market still applies with the same force to the nanoparticle version.

There is little for CFSAN to go on by way of FDA precedent on this issue. At present, the only relevant decision by the FDA on how to approach the question of the status of a nanoparticle is the final ruling by the Center for Drug Evaluation and Research (CDER) for sunscreen products for over-the-counter human use. The rationale for the ruling should be of value to CFSAN and the manufacturer in the food packaging context. The issue is whether the substance is safe. It is not presumed to be safe unless it has been approved or declared effective, or unless it is GRAS, meets the TOR criteria, or does not migrate into food. So the manufacturer is attempting to demonstrate that the substance produced using nanotechnology is safe because it is the "same" chemical as one previously considered in one of the above ways. That was the issue with the sunscreen determination. Specifically, the issue was whether the micronized titanium dioxide proposed for use was the same titanium dioxide that had previously been found safe for over-the-counter drug use. CDER determined that it was, albeit a specific grade of the chemical, because it was a refinement of particle size distribution that raised no safety concerns.* The reasoning put forth by CDER is instructive:

> Another comment asserted for the following reasons that micronized titanium dioxide is a new ingredient with several unresolved safety and efficacy issues: (1) [i]t does not meet the definition of a sunscreen opaque sunblock, (2) there is no control of particles to agglomerate, which is critical to effectiveness, (3) no standards exist to ensure integrity of coatings, (4) there are no performance-based standards of identity; micronized titanium dioxide is not included in the USP, (5) its photocatalyst potential, and (6) the potential for the smaller particle size to accumulate under the skin.

* *Id.* at 27671.

> The agency finds the data with the comments supportive of monograph status for micronized titanium dioxide. Acute animal toxicity, irritation, sensitization, photoirritation, photosensitization, and human repeat insult patch and skin penetration studies revealed no deleterious effects. SPF values for four product formulations containing from 4.4 to 10 percent micronized titanium dioxide were from 9 to 24 and support effectiveness as a sunscreen ingredient.
>
> The agency is aware that sunscreen manufacturers are using micronized titanium dioxide to create high SPF products that are transparent and esthetically pleasing on the skin. The agency does not consider micronized titanium dioxide to be a new ingredient but considers it a specific grade of the titanium dioxide originally reviewed by the Panel. Fairhurst and Mitchnick … note that "fines" have been part of commercially used titanium dioxide powders for decades, and that a micronized product simply refers to a refinement of particle size distribution. Based on data and information presented at the September 19 and 20, 1996, public meeting on the photobiology and photochemistry of sunscreens … the agency is not aware of any evidence at this time that demonstrates a safety concern from the use of micronized titanium dioxide in sunscreen products. While micronized titanium dioxide does not meet the proposed definition of a sunscreen opaque sunblock, the agency has not included the use of this term in the final monograph.… The potential for titanium dioxide particles to agglomerate in formulation, which could result in lower SPF values, is addressed by the final product SPF test.*

The key to the ruling is that the advocate for considering the versions to be the same was able to present test data of the various types mentioned to show that the nanoparticle did not present different safety issues from the form previously used. The micronization did not alter the risk–benefit analysis. That will be the way these issues are eventually resolved, and the question will be: Does CFSAN leave it to the submitter to design and run the tests to show the lack of additional harm, or does CFSAN dictate the tests to be done? Even though the sunscreen ruling was made 8 years ago, it is still a topic of current discussion, and how the matter is finally resolved may bear on the development of CFSAN policy.

The sunscreen ruling is particularly relevant because six activist groups filed a Citizen's Petition with the FDA on May 16, 2006, requesting specific

* The petition can be viewed on the International Center for Technology Assessment (ICTA) website at www.icta.org/doc/Nano%20FDA%20petition%20final.pdf. A report on cosmetics and sunscreens prepared by another petitioner, Friends of the Earth, is available at www.foe.org/camps/comm/nanotech/nanocosmetics.pdf.

action with regard to sunscreen drug products composed of engineered nanoparticles.* The specific actions sought are to remove all sunscreen products containing nanomaterials from commercial distribution. The sunscreen products are over-the-counter drugs containing ingredients listed as a part of the Over-the-Counter Drug Review. Petitioners want the administrative record of the Final Over-the-Counter Sunscreen Drug Product Monograph reopened to receive information on engineered nanoparticles of zinc oxide and titanium dioxide used in the sunscreen products, but they really want more. Petitioners also request that the Final Monograph be amended to declare that sunscreens containing engineered nanoparticles are not covered, making them new drugs, requiring new drug applications. They also want currently marketed sunscreen products containing engineered nanoparticles of zinc oxide or titanium dioxide declared an imminent hazard to public health and immediately recalled from the market until the manufacturers secure New Drug Application approvals and comply with FDA product testing regulations yet to be promulgated.

The 80-page petition documents what the petitioners characterize as the "scientific evidence" of nanomaterial risks stemming from the unpredictable toxicity and seemingly unlimited mobility of the nanoparticles. For example, petitioners allege that a 2004 study showed rapid brain damage in fish exposed to a type of manufactured nanoparticle (fullerenes or buckyballs) used in some cosmetics. Petitioners also state that other studies suggest that nanoparticles can trigger unpredictable inflammatory and immune responses, and have found that nanoparticles can penetrate cells and move within the body freely, even crossing the blood–brain barrier. They allege that the preeminent U.K. Royal Society and the Royal Academy of Engineering concluded in 2004 that nanoparticles' unique hazards warranted a moratorium on their release into the environment.** Most recently, petitioners cited an incident where an aerosol spray bathroom cleaner marketed as a nanoproduct was recalled by German authorities after approximately 77 people reported severe respiratory problems and six were admitted to the hospital with fluid in their lungs.***

The petition focuses on the FDA's regulation of nano-sunscreens, sunscreens composed of engineered nanoparticles that, due to the nanoparticles' fundamentally different properties, appear transparent or "cosmetically

* In fact, the report, entitled "Nanoscience and Nanotechnologies: Opportunities and Uncertainties," recommends "that the use of free (that is, not fixed in a matrix) manufactured nanoparticles in environmental applications such as remediation be prohibited until appropriate research has been undertaken and it can be demonstrated that the potential benefits outweigh the potential risks" (Section 5.4, paragraph 44), available at www.nanotec.org.uk/finalReport.htm.

** The *Washington Post* reported that the product "Magic Nano" may not have even contained nanoparticles; see Nanotech Product Recalled in Germany, available at www.washingtonpost.com/wp-dyn/content/article/2006/04/05/AR2006040502149. html.

*** ICTA, Consumer, Health, and Environmental Groups Launch First-Ever Legal Challenge on Risks of Nanotechnology (May 16, 2004), available at www.icta.org/press/release.cfm?news_id=19.

clear." Petitioners claim that the engineered nanoparticles of zinc oxide and titanium dioxide used in nano-sunscreens raise red flags for scientists who have found that they can induce free radicals and cause DNA damage, and are uncertain how easily the tiny particles can penetrate and circulate throughout the body.* The FDA has not acted on the petition as of this writing.** The claims made by the petitioners echo the concerns raised by many with the effects of nanotechnology in other areas. For example, commentators in the environmental area have urged the same sort of caution. They urge the regulator to be proactive and require submissions, in no small part because of confusion as to how the requirements of statutes such as the Toxic Substances Control Act (TSCA) apply to nanoscale materials.***

A final consideration is the vehicle that CFSAN will utilize to review products using nanotechnology. Most submissions should be in the form of the aforementioned FCN. The procedure seems suited for the task. It is a very flexible process, allowing CFSAN to request submission of almost any kind of data or information where appropriate. CFSAN has prepared a series of guidance documents to aid the submitter in gathering the requisite information. Those guidances are not rules, but rather nonbinding recommendations to manufacturers on what to include in submissions. CFSAN can amend them at will, without the need to go through the protracted notice and comment rule-making procedures. Having the ability to do that, CFSAN can target issues involving nanotechnology that might arise in one application and provide for the submission of data on the issue in subsequent applications. Manufacturers can object if they believe that the recommendations in the guidance documents are not supportable, but that does not ordinarily happen. By that process, CFSAN will indirectly build a precedent file, implementing the rules dealing with nanotechnology. New data requirements can be imposed dealing with the toxicity, stability, and migration study questions posed earlier in the chapter, and once any objections are sorted out, CFSAN will have in place a mechanism to gather the data it believes it needs.

It does not appear that the GRAS process will be of much use in bringing products to market. At this stage in the evolution of nanotechnology, there is scant agreement among experts as to the safety of any particular nanoscale material, so in the short-term GRAS will not be helpful. The TOR process may be useful once issues are resolved regarding the significance of migration data. Until then, manufacturers may find it more productive to use the FCN and obtain the short-term exclusivity that comes with the declaration that a FCN is effective.

* See letter from Richard A. Denison, Ph.D., senior scientist, and Karen Florini, senior attorney, Environmental Defense to the Honorable Susan B. Hazen, acting assistant administrator, Office of Prevention, Pesticides and Toxic Substances, EPA (September 2, 2004), available at www.environmentaldefense.org/documents/4457_ NanotechLetterToEPA.pdf.

** See also Lynn L. Bergeson & Joseph E. Plamondon, TSCA and Engineered Nanoscale Substances, 4.1 *Nanotechnology L. Bus.*, forthcoming, 2007.

*** *Monsanto v. Kennedy*, 613 F.2d 947, 955 (D.C. Cir. 1979).

That leaves the no-migration determination, possibly the biggest regulatory headache for CFSAN going forward. Many estimate that a large number of food contact substances are marketed on the basis of these self-determinations that the material is not a food contact substance because it cannot reasonably be expected to become a part of the food due to the lack of any migration from the packaging material. The process is a throwback to the days of the protracted FAP and has operated as a safety valve to avoid inordinate delays in the processing of applications for materials that are likely to be no safety threat. The risk–benefit analysis regarding nanoscale materials, however, is not yet a matter of consensus, and so the degree and form of migration generally acceptable for macroscale materials may not apply to nanoscale materials. This is an important issue. The no-migration determination is made when there is insignificant migration, as well as when there is no migration at all. The court case mentioned above established the principle that FDA had to show that migration that took place was significant and raised some issues. It is entirely unclear if a lesser amount of migration than the currently allowed 50 ppb would be considered significant because the migrating particles accumulate differently, for example. Similarly, it is unclear whether it will be considered significant at a lesser amount if it is shown that the materials are more toxic or pass across the blood–brain barrier.

It can be postulated that CFSAN does not presently know the answers to these questions, but the self-determinations continue to be made, at least some presumably for substances manufactured in whole or in part by nanotechnology. Companies considering a filing are likely to continue applying the sunscreen reasoning in deciding that a substance is the same as the macroscale version. Companies appear to be applying the same criteria for what is significant migration.

There have been no reports of calamities as a result of the use of the no-migration determination over the years. Most, if not all, companies can be expected to document their decisions not to file, although some companies will likely be more thorough than others. The companies involved, however, are not singing from the same page of the hymn book. There is no standard methodology to employ, so companies and their consultants are on their own. Some may miss the boat altogether, but others can be counted on to be innovative and develop methods that will further the knowledge about the effect of using nanoscale particles. The latter companies, however, are operating in a void. As the process works now, CFSAN will not be the beneficiary of this accumulated experience, information that could aid in assessing the central question regarding the effect on the classic risk–benefit analysis. Whether CFSAN should act to compel submission of this information and, if so, how are unclear. Any such move would lead to almost certain litigation, with FDA's own words and the *Monsanto* case used to justify maintaining the status quo. CFSAN could ask companies voluntarily to report the results of studies they had done, but that would put CFSAN in possession of at least general knowledge of the products for which companies are making self-determinations, and that would lead to confusion and a possible reticence to provide information to CFSAN. Such activism by CFSAN is unlikely and, without much more information, probably unwise. A useful program could be compromised on the basis of what are at present vague suspicions.

One way the matter could likely resolve itself is if all the testing done by companies coalesced to demonstrate that nanoscale materials do not present a safety risk. Companies would make sure this information is known, to allay any consumer fears about the material, to advance the attraction of nanotechnology to capital markets, and to ensure that the no-migration determination continued as a viable path to market. If the testing shows that there are legitimate concerns, as have been bandied about almost from the beginning, the product would likely be abandoned, so there would be no compulsion to notify CFSAN of the results. Companies would be reticent about adverse publicity because of the possibility of cooling off the market for capital, so the information might be slow getting to CFSAN. In the months ahead, it will be interesting to see if CFSAN tries to obtain the information in the manufacturers' files, and how it will deal with the no-migration determination in the course of doing so.

15.12 Conclusion

CFSAN has a good clearance mechanism for dealing with issues that may arise regarding the use of nanotechnology to produce food contact materials. The FCN is a flexible tool that allows CFSAN to ask for almost any kind of data and information to aid in its review. CFSAN has a series of guidance documents that can be amended without protracted delay to address subjects that might arise in its review of products resulting from applications of nanotechnology. This process will take time, however, since CFSAN must first address the specifics presented in each notification or similar submission. The body of precedent will accumulate slowly. The process might be accelerated if the data from studies of the no-migration determinations were made available to CFSAN. It remains to be seen, however, how CFSAN will address this issue.

chapter sixteen

Active and intelligent packaging: a European anomaly

Jerome Heckman

Contents

In an interesting paper published in the October 2005 edition of *Food Additives and Contaminants*,[1] A.R. DeJong and a group of coauthors presented information about new systems being developed to make it possible to use packaging materials to actively preserve foods or to indicate when foods are spoiled in some way. A proposition set forth in this paper, which is worthy of serious legal discussion, is the authors' assertion that "most active and intelligent concepts that are on the market in the USA and Australia cannot be introduced in Europe yet, due to more stringent EU legislation."[2]

It is the purpose of this chapter to question this assertion and the need for the special provisions now included in the latest Framework Regulation for Food Contact Materials of the European Community.[3]

16.1 Background

As this observer has said previously in a European forum, in the year 2000,[4] it is difficult, if not impossible, to understand why any special legislation is needed to clear active or intelligent packaging from the legal perspective, unless one assumes that the law demands some showing of functionality for some such systems.[5] In the U.S., when Congress was considering the 1958 Food Additives Amendment,[6] there was a point in 1957 when the FDA's legislative proposal, H. R. 4014, included a provision that would have required proof of functionality for obtaining a clearance. This proposal was opposed by industry and the then secretary of commerce, Sinclair Weeks, advised the relevant subcommittee of the House Committee on Interstate and Foreign Commerce that:

> The Department considers highly undesirable any blanket requirement such as is contained in H. R. 4014 that a chemical additive be shown to serve a purpose which will be useful to the consuming public before its use is permitted. Such a restriction would be undesirable and it might impose a great burden upon industry without any countervailing benefit to the consuming public.[7]

This point was discussed even more thoroughly during the hearings on the legislation, with industry's position being that the marketplace could be relied upon to weed out useless additives and that government intrusion into the usefulness arena was completely unnecessary.[8]

The idea of requiring showings of functionality was dropped and has never been revived. It was, of course, ultimately agreed that no additive should be used that would give rise to consumer deception or adulteration of food, so there is a provision in the Food Additives Amendment that specifically precludes clearance of any additive that would bring about such deception or fraud.[9]

In the same vein, we are not aware of any provision in the European Union's various statutes or framework regulations relating to packaging materials that speak to the issue of functionality. Thus, as to packaging materials, there is no difference between Europe and the U.S. where functionality is concerned.

16.2 *What is the real issue?*

If this is correct, it is submitted that the thesis of the DeJong paper is in error to the extent that it declares it impossible to bring active and intelligent packaging to market in Europe due to legal proscriptions. It is, however, possible that some legislative modification might be necessary to permit active packaging where a system results in additions of substances so significantly that they are, in fact, direct food additives.

Stated another way, where a substance will be added to food from packaging in concentrations that exceed the general migration limitation of 60 mg/kg of food or 10 mg/dm^2, it is conceivable that it could not be justified as technologically required, and this might require some sort of special dispensation since the direct additive provisions in Europe do require showings of technological requirement.

It is here submitted that if a substance is designed to be added to foods in some amount that exceeds the general migration limitation, albeit via a package, it is a direct additive and should be dealt with as such. (In the U.S., this was the case with butylated hydroxytoluene (BHT), which was once incorporated into cereal packages, the intent being for it to migrate into cereals and thereby extend shelf life.) If it is simply designed to help protect the food from deterioration, contamination, or other adulteration, and does not add anything to food, it is a classic food contact substance or indirect additive, already required to be regulated by existing law.

Indeed, the authors of the DeJong paper state in various places that substances used to effect active and intelligent packaging do have to be cleared by appropriate food contact or direct additive legal requirements in accordance with existing law. Why then the notion of another type of special clearance?

Some proponents of the special treatment have been heard to point to the fact that the EU Framework Regulation, newly revised in 2004,[10] calls for packaging materials to be inert and theorize that active and intelligent packaging is not intended to be inert. In this connection, the Framework Directive actually provides as follows:

> (3) The principle underlying this Regulation is that any material or article intended to come into contact directly or indirectly with food must be sufficiently inert to preclude substances from being transferred to food in quantities large enough to endanger human health or to bring about an unacceptable change in the composition of the food or a deterioration in its organoleptic properties.

Thus, the term *inert* is not any more absolute in Europe than in the U.S. Its use simply requires that there be no transfer of any quantity of a substance that exceeds the 60 μg/kg of food criterion, or that will endanger human health or change the character of the food in some significant way. This definition aptly covers all packaging materials where there is no intent to add anything to food, i.e., to have a technical effect in food. Since essentially all packaging is also intended to protect food from contamination and to keep it wholesome, it is hard to see how so-called active packaging is distinguishable from other packaging. Indeed, the drafters of the Framework Directive seem to have encountered this same problem because it took them the following several paragraphs to try to draw a distinction:

> (4) New types of materials and articles designed to actively maintain or improve the condition of the food ("active food contact materials and articles") are not inert by their design, unlike traditional materials and articles intended to come into contact with food. Other types of new materials and articles are designed to

monitor the condition of the food ("intelligent food contact materials and articles"). Both these types of materials and articles may be brought into contact with food. It is therefore necessary, for reasons of clarity and legal certainty, for active and intelligent food contact materials and articles to be included in the scope of this Regulation and the main requirements for their use to be established. Further requirements should be stated in specific measures, to include positive lists of authorized substances and/or materials and articles, which should be adopted as soon as possible.

(5) Active food contact materials and articles are designed to deliberately incorporate "active" components intended to be released into the food or to absorb substances from the food. They should be distinguished from materials and articles which are traditionally used to release their natural ingredients into specific types of food during the process of their manufacture, such as wooden barrels.

(6) Active food contact materials and articles may change the composition or the organoleptic properties of the food only if the changes comply with the Community provisions applicable to food, such as the provisions of Directive 89/107/EEC on food additives. In particular, substances such as food additives deliberately incorporated into certain active food contact materials and articles for release into packaged foods or the environment surrounding such foods, should be authorised under the relevant Community provisions applicable to food and also be subject to other rules which will be established in a specific measure. In addition, adequate labeling or information should support users in the safe and correct use of active materials and articles in compliance with the food legislation, including the provisions on food labeling.

(7) Active and intelligent food contact materials and articles should not change the composition or the organoleptic properties of food or give information about the condition of the food that could mislead consumers. For example, active food contact materials and articles should not release or absorb substances such as aldehydes or amines in order to mask an incipient spoilage of the food. Such changes which could manipulate signs of spoilage could mislead the consumer and they should therefore not be allowed. Similarly, active food contact materials and articles which produce color changes to the food that give the wrong information concerning the condition of the food could mislead the consumer and therefore should not be allowed either.

In our view, this tortuous attempt to draw a distinction between ordinary packaging and active packaging is inapt. The fact is that the only thing that distinguishes one type of packaging as active from another type that simply protects the food from spoilage or contamination is an apparent desire to make a marketing claim that the substances "added" will somehow enhance the characteristics of the food. As long as a substance used in a package does not result in untoward migration into a foodstuff, does not give rise to taste or odor problems, and is not used to mislead the consumer in any way, it is simply a food contact substance and should be treated as such.[11]

A prime purpose of essentially all packaging is to protect food from contamination or undue exposure, so all of it is active. The only possible distinction here between packaging generally and so-called active packaging is that, in the latter case, some substances may be intended to be added to food in amounts that exceed the overall migration limits. Where this is so, the substance is really a direct additive and should be cleared as such.

16.3 Intelligent packaging

What is termed intelligent packaging could present a different case. It has not been the subject of such strained attempts to define the nature of the concept. It is merely presented as a class of materials "designed to monitor the condition of the food."

Thus, presumably, such materials would serve no ordinarily protective purpose in the food package (i.e., would not help protect against contamination or adulteration of the food) other than to provide notice if the food had spoiled or was otherwise adversely affected.

This does, perhaps, present a practical distinction, but not because of the inertness criterion since it would still not be tolerated if it added anything to foods that would make them unhealthy. Here, the distinction is that something would be permitted that, at least in theory, is not required to make the package a better package per se, and that will not be a part of the food either.

To the extent that proponents of legislative change have some reason for wanting such materials specially cleared or classified, they could be considered unique and worthy of some special regulatory form. Perhaps one reason for such a classification could be that regulators might otherwise be bothered by the fact that the substance cannot be considered reasonably necessary or functional as to the integrity of the package.

In our view, a substance can be incorporated in a package if it is safe, and the government has no statutory basis for examining or passing on its functionality, or the value thereof, it being very reasonably presumed that no one will want to add to the cost of a packaging material by using something that is unnecessary.

16.4 Conclusion

In sum, it would appear that those who seek to establish new classifications for certain types of packaging materials are doing so for commercial reasons and not to fill a regulatory gap.[12] They can accomplish this purpose in the normal way by the use of marketing techniques, such as special labeling, public education, or advertising. It is suggested here that trying to do so by adding regulatory complications where there is no public interest requirement is of questionable value and should be avoided.

When statutes are changed, one must be careful of what he wishes for lest he set up new complications, the ultimate thrust of which might be to grant governments the right to impose added responsibilities in areas where none are necessary.

References

1. DeJong, Boumans, Slaghek, Van Veen, Rijk, and Van Zandvoort, Active and intelligent packaging for food: is it the future? *Food Additives and Contaminants*, 22, 975–979, 2005.
2. Ibid., p. 979.
3. 1935/2004/EC, published on October 27, 2004.
4. See Heckman, Why Active and Intelligent Packaging Systems Present No Special Regulatory Concern in the United States, paper presented at the TNO Nutrition and Food Research Institute and European Union Sponsored Actipak Conference on Active and Intelligent Packagings, Zeist, The Netherlands, April 18, 2000.
5. Indeed, a careful review of the 79-page "Final Report on Task 5 of the 'Actipak' Project, Recommendations for Legislative Amendments" indicates to this author that this presentation, the basis for the new EU proposals, does not indicate the need for changes to the fundamental EU packaging law provisions, although it does indicate ways in which some sort of legislative imprimatur might enhance the marketing of such systems. FAIR-Project PL 98-4170.
6. Food Additives Amendment of 1958, PL 85-929, 72 Stat. 1784, 21 U.S.C. 301–395.
7. Hearings before a subcommittee of the Committee on Interstate and Foreign Commerce of the House of Representatives, 85th Congress, 1958, p. 26.
8. See, e.g., the testimony of George Faunce Jr. on behalf of the American Bakers Association and his exchange with Congressman John Dingle on this point. Ibid., pp. 248–259.
9. Food Additives Amendment of 1958, PL 85-929, 72 Stat. 1784, 21 U.S.C. 301–395. Section 409(c)(3)(B) provides that the FDA shall deny the clearance of any additive if the evidence provided for it "shows that the additive would promote deception of the consumer … or would otherwise result in adulteration or misbranding of food…."
10. Regulation (EC) 1935/2004, the "Framework Directive," was published in the *Official Journal of the European Union* on November 13, 2004, and became effective December 3, 2004.

11. Of course, if the use of the additive were to result in overall migration exceeding the prescribed limit of 10 mg/dm^2 (EC Directive 2002/72, Article 2), it should be classified as a direct additive requiring conventional food additive clearance.
12. In fact, the document urging the EU to adopt special language for active and intelligent packaging, "Final Report on Task 5 of the 'Actipak' Project, Recommendations for Legislative Amendments," cited above in note 5, contains an Appendix II that clearly shows no member of the EU has heretofore found it necessary to have such special language and several of the countries have indicated they see no need for it. See p. 39 *et. seq.* of the referenced report.

chapter seventeen

ACTIPAK in Europe

William D. van Dongen and Nico de Kruijf

Contents

17.1 Introduction

Active packaging may be defined as packaging that changes the condition of the packaged food to extend shelf-life or to improve safety or sensory properties, while maintaining the quality of the food. In the literature other definitions are also given.[1,2] Active packaging systems are sometimes compared with modified atmosphere packaging (MAP). The difference between MAP and active packaging is that MAP is a passive system, whereas active packaging plays an active role during storage and transport. Besides, active packaging systems employ a wide range of technologies, each selected to deal with specific problems, while MAP is based on one technology only.

This chapter will be focused on the current position and future perspectives of active packaging in Europe. Active packaging is not only restricted to packaging materials used to wrap the food, but also any form, shape, or size of active materials and articles are considered.

First an overview of the available active packaging technologies will be given, followed by a short overview of the relevant European regulations and requirements of these regulations (legislative aspects) and their role in the introduction and acceptance of active packaging. Finally, future perspectives of active packaging in Europe will be discussed.

17.2 Overview of existing active packaging systems

Active packaging systems can be distinguished into active scavenging systems (absorbers), active releasing systems (emitters), and others (antimicrobial films, self-venting packaging, and susceptor laminates for microwave application and self-heating/-cooling systems). Scavenging systems remove undesired components, such as oxygen, excessive water/moisture, ethylene, carbon dioxide, other specific food constituents, and taints. Releasing systems actively add compounds to the packaged food, such as carbon dioxide, water, antioxidants, or preservatives. Both scavenging and releasing systems are aimed at extending shelf-life or improving food quality.[3] Active packaging systems can be used for numerous applications. Table 17.1 gives an overview of some current applications of active packaging systems.

17.2.1 Active scavenging systems

17.2.1.1 Oxygen scavengers

Oxygen can exert considerable detrimental effects on foodstuffs. Oxygen is a reactant in several chemical reactions such as lipid oxidation, resulting in the formation of off-flavors in fatty foods, enzyme-induced oxidation in sliced vegetables and fruits, discoloration caused by oxidation of pigments, and oxidation of components like tocopherol (vitamin E) and ascorbic acid (vitamin C), which will result in nutritional losses. Growth of many spoilage organisms like molds and yeast relies on the availability of oxygen. Reducing the oxygen concentration in the headspace of a package will generally inhibit the growth of spoilage organisms and reduce several shelf-life-limiting processes. Oxygen also affects the physiological condition of respiring foods like fruits and vegetables as well as infestation. Reduction of oxygen decreases respiration rate and ethylene production and inhibits the growth of insects, for example. A possible disadvantage of the use of oxygen scavengers is a higher risk for growth of pathogenic microorganisms. Most

Table 17.1 Overview of Active Packaging Systems, Their Objectives, and Possible Food Applications

Active Packaging System	Objective	Food Applications
Oxygen scavengers	Prevention of microbial infections (aerobic bacteria, yeast, and molds), oxidation (e.g., lipid oxidation, rancidity), discoloration, and infestation Slows ripening of climacteric fruits and vegetables	Coffee, tea, meat, potato chips, fish, poultry, prepared dishes, cheese, bakery products, pizza, pasta, nuts, fried food, milk powder, dried products, cake, cookies, herbs, spices, beans, cereals, beverages, snack food products
Moisture absorbers	Prevention of microbial growth Removal of dripping water Prevention of fogging	Fish, meat, poultry, snack food products, cereals, dried foods, sandwiches, bakery products, fruits and vegetables
Microwave susceptors	Can provide microwave cooking benefits such as crisping and browning Delivers cooked-product textures similar to those of conventional cooking	Ready-made meals, bread and bakery items, french fries in microwave
Ethylene scavengers	Delays ripening of climacteric fruits and vegetables Prevention of postharvest disorders	Fruits and vegetables and other horticultural produce
CO_2 scavengers	Removal of carbon dioxide after packing	Coffee, cheese
Flavor/odor absorbers	Absorption of off-flavors	Fish, fatty foods, fruit juices, biscuits, cereals
Antimicrobial packaging	Prevention of growth of microorganisms (bacteria, molds, yeast)	Bakery products, bread, fruit, vegetables, cheese, meat, fish, snack food products, prepared dishes, wine, flour, grain, beans
Antioxidant-releasing systems	Prevention of oxidation (e.g., lipid oxidation)	Breakfast cereals, fatty foods
Flavor-releasing systems	Release of flavor to improve the sensory quality and avoid flavor scalping	Ice cream, orange juice
Self-heating/-cooling	For heating food, hot drinks in the field where no oven or microwave is available	Coffee, tea, ready-made meals

pathogenic microorganisms only grow under anaerobic conditions. With the use of oxygen scavengers, an anaerobic atmosphere is created in the headspace of a package.

Existing oxygen-scavenging technologies utilize one or more of the following mechanisms: iron powder oxidation, ascorbic acid oxidation, photosensitive dye oxidation, enzymatic oxidation, and unsaturated fatty acids or immobilized yeast on solid material. Different forms of oxygen scavengers exist. The best-known scavengers are small sachets containing iron powder. They can easily be inserted into food packages. The main advantage of these sachets is that they are able to reduce the level of oxygen to less than 0.01% in a relatively short time. Their disadvantage is that there is always a possibility, despite the label "Do not eat," that consumers accidentally ingest the contents of the sachet. An alternative to sachets is the integration of the active scavenging system into packaging materials such as films or closures. In general, the speed and capacity of oxygen scavengers incorporated in films or closures are lower than with the iron-based oxygen-scavenging sachets.[4,5]

Oxygen scavengers can be applied to various foods, such as bakery products, mayonnaise, powdered milk, fruit juices, beer, fresh meat, and fish.[5–9]

17.2.1.2 *Moisture Scavengers*

Several foodstuffs require a strict control of water. For example, a high water content causes softening of dry, crispy products. On the other hand, very low levels of water can decrease the shelf-life of fatty foods due to an increase of lipid oxidation. According to Rooney,[10] there are two distinct manners to regulate the moisture content of packed foods: liquid water control and humidity buffering. Excess water can be controlled by applying drip-absorbent sheets, usually composed of a superabsorbent polymer between two microporous layers. Polyacrylate salts and copolymers of starch are suitable absorbing agents. The sheets can be used as pads, for example, under meat and fish.

Another way to control excess moisture in packed food is to regulate the relative humidity of the packed food (humidity buffering). The humectants (e.g., propylene glycol) can be placed between two plastic films. For a wide range of dry foods, desiccants such as silica gel, calcium oxide, and active clay are successfully used.[11,12]

17.2.1.3 *Ethylene scavengers*

Ethylene is a growth-stimulating hormone that accelerates ripening and senescence by increasing the respiration rate of climacteric fruits and vegetables, thereby reducing shelf-life. Ethylene also accelerates the rate of chlorophyll degradation in leafy vegetables and fruits.[13] Therefore, to prolong shelf-life and maintain an acceptable visual and sensory quality, accumulation of ethylene in the package should be avoided.[14] Different ethylene-absorbing packaging systems currently exist, most of which are supplied as sachets or incorporated in films. The active ingredient is often potassium permanganate on an inert mineral substrate. The most well known, inexpensive, and used ethylene-absorbing system uses silica as substrate. The silica absorbs etylene, and potassium permanganate oxidizes it to ethylene glycol. Silica is kept in a sachet highly permeable to ethylene, or it can be incorporated in packaging film. Potassium permanganate, however, is not allowed to be used in food contact materials due to its toxicity.[1] Minerals such as zeolite and clays, silica gel, and active carbon powder incorporated in packaging materials such as films and bags are also applied to absorb ethylene.[3,14] When zeolites impregnated with potassium permanganate carrier are coated with a quaternary ammonium cation, this active packaging system is also capable of absorbing other organic compounds, such as benzene, toluene, and xylene from the packed foodstuffs.[15]

17.2.1.4 Other scavengers

Other active absorbing packaging systems are developed to remove, for example, undesirable flavor constituents or carbon dioxide. Carbon dioxide is sometimes formed due to deterioration or respiration reactions. An active packaging system that scavenges both oxygen and carbon dioxide has been used to package fresh roasted coffee to avoid deterioration or package destruction. Generally, flavor scalping by the package is detrimental to food quality, but sometimes it is useful to selectively absorb unwanted odors or flavors such as aldehydes and amines. Amines are formed from protein breakdown in fish muscle, and aldehydes are reaction products of auto-oxidation of fats and oils. For both components, packaging systems have been developed that are able to absorb these. Other commercial systems absorb odors due to the formation of mercaptans and hydrogen sulfide (H_2S).[16]

It is important that the aforementioned technologies are not misused to mask the development of microbial off-odors.[17]

17.2.2 Active releasing systems

In active releasing systems components migrate from the packaging material to the packaged food with the aim to extend shelf-life or improve the quality of the packed food.

17.2.2.1 Antimicrobial packaging

The demand for minimally processed foods and easily prepared and ready-to-eat fresh food products, globalization of food trade, and distribution from centralized processing pose major challenges for food safety and quality. Microbiological growth is one of the major modes of deterioration of fresh foods. Besides the removal of oxygen or moisture to avoid growth of microorganisms, the release of specific antimicrobial agents is a potential application of active packaging. When spoilage of microorganisms is strictly due to surface growth, a film or sachet that emits an antimicrobial agent can be of value. The principal action of antimicrobial films or sachets is based on the release of antimicrobial entities. The major potential food applications include meat, fish, poultry, bread, cheese, fruits, and vegetables.[11]

Antimicrobial packaging systems with different applications and active agents are on the market. In Japan, Ag-substituted zeolite is the most common antimicrobial agent incorporated in plastics. Ag ions inhibit a range of metabolic enzymes.[18] Other active substances that can be used include ethanol and other alcohols, carbon dioxide, sorbate, benzoate, propionate, bacteriocins, fungicides, and enzymes.[19,20] Each antimicrobial agent has its own target microorganism and hence its own application. For example, ethanol is effective against surface growth of molds and can be used, for example, for bakery products.[5,21]

There has been little commercial activity in the field of antimicrobial packaging in North America or Europe. Due to its very broad range of fresh foods, Japan has historically been a leader in antimicrobial use. Allylisothiocyanate, used in foods such as sushi and horseradish, are used as antimicrobials in commercial packaging in Japan but are limited outside of Asia by their unpleasant odor. Chlorine dioxide is used commercially in sachets, but commercial products are difficult to find. Antimicrobials such as silver, nisin, and herbal extracts (rosemary) are being investigated, but commercial application is limited by technical and regulatory barriers.

A significant problem is limited effectiveness, since antimicrobials act only at the surface, while microbial activity is often not limited to the surface. The effectiveness of an antimicrobial is influenced by several factors, including use temperature, moisture levels, the nature of the antimicrobial agent, and how it is released.

The use of antimicrobial packaging might lead to less hygienic working conditions in plants. Another disadvantage could be that microorganisms may build up resistance against the antimicrobial compound by excessive use.

17.2.2.2 Antioxidant-releasing systems

Antioxidants can be incorporated in plastic films for polymer stabilization in order to protect film from degradation. However, when the antioxidant migrates into the food, it may have an additional positive effect on the shelf-life of the packed food. In the U.S., release of BHA and BHT into breakfast cereals and snack products has been applied. Vitamins E and C have been suggested for integration in polymer films by virtue of their antioxidant effect. Vitamin E is a safe and effective antioxidant for cereal and snack products where the development of rancid odors and flavors is often the shelf-life-limiting factor.[22]

17.2.2.3 Flavor-releasing systems

Flavor-releasing systems can be used to mask off-odors coming from the food or the packaging. Other applications of flavor-enriched packaging materials include the possibility to improve the sensory quality of the product by emitting desirable flavors into the food and to encapsulate pleasant aromas that are released upon opening. It should be mentioned that flavor-releasing systems should not be misused to mask microbial off-odors, thus introducing risks to consumers.[16,17]

17.2.3 Miscellaneous active packaging systems

17.2.3.1 Microwave susceptors

Susceptor materials include paperboard with embedded aluminum patterns as well as paperboard laminated to metallized polyester film. The susceptor absorbs microwave energy from the oven, and that energy is transferred to the food as heat. The special metallized inner layer of the laminate can have different patterns. The pattern used depends on the products since they determine the temperature. Some need a higher temperature than the 200°C (maximum 250°C) of the standard safety susceptor. These allow the microwave to heat the product to 200°C, which enables the products to crisp and brown.

The use of microwave susceptors also delivers cooked-product textures similar to those of conventional cooking. For example, bread and bakery items microwaved in susceptor materials avoid problems common to microwaving in nonsusceptor packaging, such as coming out chewy/tough or soggy/mushy. Susceptors also make it possible to microwave food products that include an uncooked protein component, such as fish.

17.2.3.2 Other active packaging systems

Other active packaging systems are temperature control packaging systems and packaging materials with foaming properties. Temperature-controlling materials include the use of innovative insulating materials and self-heating and self-cooling packaging systems.

17.3 Legislation of active packaging in Europe

In the U.S., Japan, and Australia, active packaging is already being successfully applied to extend shelf-life while maintaining nutritional quality and ensuring microbiological safety. In Europe, however, only a few of these systems are applied and the global market share is relatively small (Figure 17.1).

The main reason for this is that EU legislation is more stringent than in other parts of the world. Active releasing materials were not allowed before 2004 since the regulations at that time set an overall maximum migration limit of 60 mg/kg of food from the packaging into food for all packaging material, including active packaging. This limit was not appropriate for active releasing materials, since it is their aim to usually release substances above this limit. The active systems that were not limited by the legislation

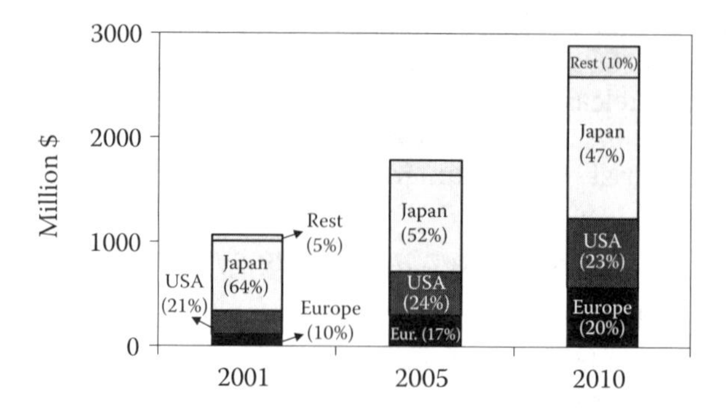

Figure 17.1 Global markets for active packaging in 2001, 2005, and 2010. (From Climpson, J., *The Future of Active Packaging in Food and Drink Markets*, Pira International Leatherhead, Surrey, UK, 2005.)

at that time were absorbing materials (oxygen scavengers and moisture absorbers) and susceptor laminates (microwave heating), since they did not interfere with the legislation at that time as long as the toxicological properties and quantities of migrants of the active absorber/susceptor materials were acceptable.

A European Commission-funded project (FAIR Project CT-98-4170)[23] known by the acronym ACTIPAK came up with recommendations that were taken up in the drafting of amendments to the EU Framework Directive for food contact materials (89/109/EEC).[24] This resulted in the adoption of a new framework regulation (1935/2004/EC)[25] in which the use of active and intelligent packaging systems are now authorized, provided the packaging can be shown to enhance the safety, quality, or shelf-life of the packaged foods. The Framework Regulation for Food Contact Materials (1935/2004/EC; published on October 27, 2004)[25] is a regulation instead of a directive, which means that it immediately became valid in all EU member states. All active and intelligent packaging systems are also subject to preexisting food contact material (FCM) regulations, e.g., plastics, ceramics, etc.

In addition, some of the systems are subject to regulations on food additives, biocides, labeling, environment/waste, modified atmosphere, food hygiene, safety, and weight and volume control (a gas absorber or releaser can theoretically influence the volume of a packed food product). All new active and intelligent packaging systems initially need to be evaluated by the European Food Safety Authority (EFSA). EFSA guidelines are under preparation but will most likely contain the following issues:

- Much attention will be given to the efficacy of active and intelligent materials. The idea is that if there is no effect, there is no benefit, and it should not be used as it approaches the area of misleading the consumer even if there are no foreseeable health problems. It could be imagined that EFSA will ask for the capacity of an oxygen absorber in relation to the headspace of the packaged food. It makes sense that in practice, the right capacity should be applied, but in the end the commission has to authorize a system, and that means they have to know what are the possible failures.

- For releasers, the status as food additive is crucial. In case there is no status as food additive, there will be authorization of the active component.
- Migration may not be so easy to determine, and dedicated migration methods may be required to demonstrate that migration of active or intelligent components does not endanger human health.
- The authorization will require proper labeling of the material or article on its functionality, amount of released substances, etc.
- If toxicological data are needed, then the SCF guidelines will be applied. It would be a good opportunity to apply exposure principles for this new area of materials. Whether this will be accepted is questionable.

Based on the outcome of that evaluation, the commission (DG SANCO) will grant a petitioner authorization for the submitted active ingredients/systems, which will be entered in the regulation. The general requirements for food contact materials (1935/2004/EC)[25] also apply to active and intelligent packaging systems, and consequently they shall not endanger human health. However, releasing systems are allowed to change the composition of the food, provided the released substance is an authorized food additive (94/34/EC).[26] Labeling should comply with the food additive directive. The release or absorption of substances should not mislead the consumer. Antimicrobial components incorporated in FCMs and articles intended to be released into the food (directly or via the headspace) must be approved food additives and will need evidence of their effectiveness. Any claim made for the active system should be proven. Preservatives may also be added to FCMs to obtain an antimicrobial surface. These applications are not considered active materials. These systems should not release the preservative into the food, and thus have no effect on the microbial conditions of the food. If the material is a plastic, it must comply with the current Framework Regulation (1935/2004/EC)[25] and the directives on plastics (2002/72/EC).[27]

A specific regulation (EMB/973 Rev. 5B)[28] for active and intelligent packaging is approaching its final stage. The final regulation most likely will not differ significantly from the document now available. The new regulation will authorize the use of active packaging, provided the packaging can be shown to enhance the safety, quality, or shelf-life of the packaged foods. Active releasing compounds will be allowed provided the final food complies with the food additive regulation. The final draft is expected to be published sometime in 2007. Although the main hurdle for the use of active and intelligent packaging in the EU, i.e., migration of active components into food, has been taken away, Europe's active packaging future will still move relatively slowly (Figure 17.1). It will take another 18 months for the EFSA to draw up a list of approved substances for first approvals of active releasing materials, keeping the industry insecure about what will be allowed. In Table 17.2 an overview of the current European legislation is given.

17.4 Future trends

17.4.1 Global trends

The trend globally of active packaging will be growth. The total market in 2001 was approximately $1 billion, and a little less than $2 billion in 2005 — an annual growth of 14%. It is expected that the growth will be at least 10% per year until 2010, reaching a market of $2.5 to 3.0 billion (Figure 17.1).[29]

The main drivers will be shelf-life extension and convenience foods. This becomes apparent from the fact that oxygen scavengers (shelf-life extension) currently have by far the largest market share (approximately 40%), followed by moisture scavengers (shelf-life extension), and this will be the case at least until 2010. Number three, at least until 2010, is susceptor laminates for microwave use (convenience). In Japan, active packaging is already generally accepted (52% of the global market); the reason for this is the very broad spectrum of fresh and nonprocessed foods and the Japanese affinity for high-tech gadgets. This is illustrated by the fact that in Japan the market share for oxygen scavengers, moisture absorbers, and microwave packaging is only 19%, leaving 81% for many different active technologies and applications. In the U.S., active packaging is becoming more and more accepted, and Europe is still behind.

Table 17.2 Overview of Current Regulations and Requirements Relevant for Active and Intelligent Packaging

Regulation	Issues related to active packaging
Framework Regulation (EC) 1935/2004 *Abstract* State that all FCM may: Not endanger human health Not effect an unacceptable change in composition Not effect deterioration of the organoleptic characteristics Not mislead the consumer	*General* Authorization of active materials subject to EFSA evaluation; these are in accordance with "general food law" (Regulation (EC) 178/2002) Compliance must be demonstrated to relevant authorities Active materials change the composition or organoleptic properties of food; to account for this, specific provisions are included in Article 4 *Article 4* Allows changes in food composition/organoleptic properties, provided they comply with 89/107/EEC on food additives Misleading of consumer (intelligent) and masking of spoilage not allowed Labeling of presence of active materials/released substances/nonedible parts

Table 17.2 Overview of Current Regulations and Requirements Relevant for Active and Intelligent Packaging (Continued)

Regulation	Issues related to active packaging
Draft regulation on active materials (EMB/973 Rev. 5B) *Abstract* Final regulation most likely will not differ significantly from Regulation (EC) 1935/2004; deals with authorization of active components; can be considered a starting point on requirements on active and intelligent packaging	*Authorization* Draft regulation requires individual authorization of active and intelligent components *Listing*: Active and intelligent components are inserted on a list of authorized components *Suitability*: Demands that active and intelligent packaging "are suitable and effective for the intended purpose" and are authorized *Certification*: Active and intelligent materials must demonstrate to be safe as FCM under specified conditions; documentation that proves the validity of the certificated must be available *Migration* *Carriers*: Should comply with safety requirements of Framework Regulation (EC) 1935/2004 and implemented EU/national measures *Active releasing*: Substances should *not* be included in overall migration, special protocols are needed; specific migration limits may be exceeded provided that the final food complies with processed foods rules and restrictions
EFSA Guidelines (authorization requirements) *Abstract* General guidelines for food contact materials (EFSA Note for Guidance, 2001) and particular plastics; guidelines not generally applicable to active packaging; EFSA guidance available at the moment the regulation on active packaging is ready for implementation	Opinions of EFSA are based on a risk assessment; EFSA will require all data needed to make a proper safety assessment *Releasing materials*: Authorization as a food additive is required; released component must show effectiveness to comply with food additive requirements; carrier of releasing material should not migrate at unacceptable quantities *Absorbing materials*: Focus will be on the toxicological properties and quantities of migrants; efficiency/capacity of absorber will be considered
Relevant measures to be considered	*General food law (Regulation 178/2002)*: Sets general requirements in respect to food and feed safety *Food additives (89/107/EEC + 94/34/EC)*: Releasing systems are subject to this directive *Biocides (98/8/EC)*: Notes that biocides are not food additives *Labeling (2000/13/EC)*: Released substances should be labeled *Hygiene (Regulation 852/2004)*: Active materials to ensure food integrity *General product safety (2001/95/EC)* *Weight and volume control (1976/211/EEC)*

17.4.2 Europe

Differences in legislation across the world did and will continue to have a major impact on the sales and market penetration of active packaging. The EU is the only region in the world that has legislation specific for active packaging substances. However, delays in drawing up the legislation have hampered the introduction of active packaging (especially for releasing systems) in Western European food markets. As explained in the previous section, the reason Europe is behind is that only after formulation of EU Framework Regulation 1935/2004[25] the use of active releasing materials was allowed. The applied active systems until 2004 were absorbing materials (oxygen scavengers and moisture absorbers) and susceptor laminates (microwave heating), since they did not interfere with the legislation at that time. This is also illustrated in Figure 17.1: the European market share in 2001 was only 10%, growing to 17% in 2005 and expected to slowly grow to 20% in 2010.[29]

The next hurdle to take is the conservative consumer attitude in Europe towards active packaging. European consumers are not very willing to pay for it. In contrast to Japan, consumers in Europe have no affinity with this type of technology, and the perception is that although active packaging is increasing the shelf-life, it decreases the quality and freshness.

Therefore, it is anticipated that the trend will be, especially in Europe, to obtain approvals for existing active packaging concepts, and development efforts will be mainly in cost reduction and in identifying the benefits that will justify the additional packaging costs. Also, instead of sachets containing the oxygen and moisture scavengers, more consumer-acceptable formulations, such as labels, films, and cards, will be developed and marketed.[15]

Susceptor laminates for microwave packaging will also grow relatively fast because of the increasing demands for convenience products in Europe. With respect to active releasing packaging, there is a small growth anticipated in the future (Table 17.3).

An outcome of the ACTIPAK project[23] did demonstrate that acceptance is not gained purely on the extension of shelf-life, but that the nature of the

Table 17.3 European Market for Active Packaging in 2005 and 2010

	2005	2010
European market (in million $)	300	570
Oxygen scavengers (beer, soft drinks, meat)	39%	43%
Moisture scavengers (meat)	37%	26%
Microwave (self-venting and susceptor laminates)	17%	20%
Antimicrobial	2%	3%
Rest	5%	8%

Source: Climpson, J., *The Future of Active Packaging in Food and Drink Markets*, Pira International Leatherhead, Surrey, UK, 2005.

applied technology and releasing chemicals is very essential for consumer acceptance. Releasing compounds that are, for instance, of nonbiological origin or have gained in other applications a negative image are less acceptable for the consumer than, for instance, the use of vitamin C as releasing antioxidant.

On the counterbalance, the industry's need to effectively and safely pack foods for transport and storage while maintaining the quality, along with increasing demands from consumers for fresher, more convenient, minimally processed, and safer foods with a prolonged shelf-life, presents a bright future for active packaging in the mid- and long term.[29] Interest in active packaging and the number of food-related applications are expected to increase significantly in the years to come.

References

1. Rooney, M.L. Overview of active food packaging. In *Active Food Packaging*, Rooney, M.L., Ed. Blackie Academic & Professional, London, 1995, pp. 1–37.
2. Ahvenainen, R., Hurme, E. Active and smart packaging for meeting consumer demands for quality and safety. *Food Additives and Contaminants*, 14, 753–763, 1997.
3. Labuza, T.P. An introduction to active packaging for foods. *Food Technology*, 50, 68–71, 1996.
4. Day, B.P.F. Underlying principles of active packaging technology. *Food, Cosmetics and Drug Packaging*, 23, 134–139, 2000.
5. Floros, J.D., Dock, L.L., Han, J.H. Active packaging technologies and applications. *Food, Cosmetics and Drug Packaging*, 20, 10–17, 1997.
6. Smith, J.P., Ooraikul, B., Koersen, W.J., Jackson, E.D., Lawrence, R.A. Novel approach to oxygen control in modified atmosphere packaging of bakery products. *Food Microbiology*, 3, 315–320, 1986.
7. Gill, C.O., McGinnes, J.C. The use of oxygen scavengers to prevent the transient discoloration of ground beef packaged under controlled oxygen-depleted atmosphere. *Meat Science*, 41, 19–27, 1995.
8. Schozen, K., Ohshima, T., Ushio, H., Takiguchi, A., Koizumi, C. Effects of antioxidants and packing on cholesterol oxidation in processed anchovy during storage. *Food Science and Technology*, 30, 2–8, 1997.
9. Berenzon, S., Saguy, I.S. Oxygen absorbers for extension of crackers shelf-life. *Food Science and Technology*, 31, 1–5, 1998.
10. Rooney, M.L. Active packaging in polymer films. In *Active Food Packaging*, Rooney, M.L., Ed. Blackie Academie & Professional, London, 1995, pp. 74–110.
11. Labuza, T.P., Breene, W.M. Applications of active packaging for improvement of shelf life and nutritional quality of fresh and extended shelf-life foods. *Journal of Food Processing and Preservation*, 13, 1–69, 1989.
12. Day, B.P.F. Active packaging of foods. *CCFRA New Technologies Bulletin*, 17, 23, 1998.
13. Knee, M. Ethylene effects in controlled atmosphere storages of horticultural crops. In *Food Preservation by Modified Atmospheres*, Caldoron, M., Barkai-Golan, R., Eds. CRC Press, Boca Raton, FL, 1990, pp. 225–235.

14. Zagory, D. Ethylene-removing packaging. In *Active Food Packaging*, Rooney, M.L., Ed. Blackie Academie & Professional, London, 1995, pp. 38–54.
15. Ozdemir, M., Floros, J.D. Active food packaging technologies. *Crit Rev Food Sci Nutr*, 44, 185–193, 2004.
16. Nielsen, T. *Active Packaging: A Literature Review*, SIK Report 361. 1997, p. 20.
17. Vermeiren, L., Devlieghere, F., Beest van, M.D., Kruijf de, N., Debevere, J. Developments in the active packaging of foods. *Trends in Food Science and Technology*, 10, 77–86, 1999.
18. Hotchkiss, J.H. Safety considerations in active packaging. In *Active Food Packaging*, Rooney, M.L., Ed. Blackie Academie & Professional, London, 1995, pp. 238–255.
19. Han, J.H., Floros, J.D. Casting antimicrobial packaging films and measuring their physical properties and antimicrobial activity. *Journal of Plastic Film and Sheeting*, 13, 287–298, 1997.
20. Padgett, T., Han, I.Y., Dawson, P.L. Incorporation on food-grade antimicrobial compounds into biodegradable packaging films. *Journal of Food Protection*, 61, 1130–1335, 1998.
21. Smith, J.P., Hoshino, J., Abe, Y. Interactive packaging involving sachet technology. In *Active Food Packaging*, Rooney, M.L., Ed. Blackie Academie & Professional, London, 1995, pp. 143–173.
22. Wessling, C., Nielsen, T., Leufven, A., Jägerstad, M. Mobility of α-tocopherol and BHT in LDPE in contact with fatty food simulants. *Food Additives and Contaminants*, 15, 709–715, 1998.
23. FAIR Project PL 98-4170. "Actipak": Evaluating safety, Effectiveness, Economic-Environmental Impact and Consumer Acceptance of Active and Intelligent Packaging. Duration, 1998–2001. Final report, 2003.
24. Council Directive 89/109/EEC of December 21, 1988, on the approximation of the laws of the member states relating to materials and articles intended to come into contact with foodstuffs. *Official Journal of the European Union*, L 040, 11/2/1989, 0038–0044.
25. Regulation (EC) 1935/2004 of the European Parliament and the Council of October 27, 2004, on materials and articles intended to come into contact with food and repealing Directives 80/590/EEC and 89/129/EEC. *Official Journal of the European Union*, L 338, 11/13/2004, 4–17.
26. Council Directive (and European Parliament) 94/34/EC of June 30, 1994, amending Directive 89/107/EEC on the approximation of the laws of member states concerning food additives authorized for use in foodstuffs intended for human consumption. *Official Journal of the European Union*, L 237, 10/9/1994, 0001–0002.
27. Corrigendum to Commission Directive 2002/72/EC of August 6, 2002, relating to plastic materials and articles intended to come into contact with foodstuffs. *Official Journal of the European Union*, L 220, 8/15/2002, 0018–0058; L 0392, 2/13/2003, 01–42.
28. EMB/973 Rev. 5B, Working Document on a draft regulation on active and intelligent materials and articles intended to come into contact with foodstuffs. Version updated May 16, 2005.
29. Climpson, J. *The Future of Active Packaging in Food and Drink Markets*. Pira International Leatherhead, Surrey, UK, 2005.

Index

Numbers

A

B

C

D

E

F

G

H

I

N

O

P

R

S

T

U

V

W

X